Statistical Physics
and the
Atomic Theory of Matter

From Boyle and Newton
to Landau and Onsager

Princeton Series in Physics

edited by Arthur S. Wightman
and Philip W. Anderson

Quantum Mechanics for Hamiltonians Defined as Quadratic Forms
by Barry Simon

Lectures on Current Algebra and Its Applications
by Sam B. Treiman, Roman Jackiw, and David J. Gross

Physical Cosmology *by P.J.E. Peebles*

The Many-Worlds Interpretation of Quantum Mechanics
edited by B. S. DeWitt and N. Graham

The $P(\Phi)_2$ Euclidean (Quantum) Field Theory *by Barry Simon*

Homogeneous Relativistic Cosmologies *by Michael P. Ryan, Jr., and
Lawrence C. Shepley*

Studies in Mathematical Physics: Essays in Honor of Valentine Bargmann
edited by Elliott H. Lieb, B. Simon, and A. S. Wightman

Convexity in the Theory of Lattice Gases *by Robert B. Israel*

Surprises in Theoretical Physics *by Rudolf Peierls*

The Large-Scale Structure of the Universe
by P.J.E. Peebles

Quantum Theory and Measurement *edited
by John Archibald Wheeler and Wojciech Hubert Zurek*

Statistical Physics and the Atomic Theory of Matter,
from Boyle and Newton to Landau and Onsager *by Stephen G. Brush*

Statistical Physics
and the
Atomic Theory of Matter.
From Boyle and Newton
to Landau and Onsager

by
Stephen G. Brush

Princeton Series in Physics

Princeton University Press
Princeton, New Jersey

ACKNOWLEDGMENTS

This book is based on research sponsored by the National Science Foundation. Permission to reprint previously published material (with substantial revisions in some cases) has been granted by the copyright holders as follows: Chapter Two, *Journal of the History of Ideas*; Sections 5.1–5.6, *Archive for Rational Mechanics and Analysis* (Springer-Verlag); Section 6.4, *Reviews of Modern Physics* (American Institute of Physics); Chapter Seven, *PSA* 1976 (Philosophy of Science Association).

I am grateful to a number of colleagues for suggestions and for pointing out errors in an earlier draft; M. Alexanian; G. A. Baker, Jr.; D. D. Betts; J. R. Dorfman; P. Ehrlich; W.A.B. Evans; M. E. Fisher; V. J. Frenkel; D. ter Haar; J. Kincaid; and G. L. Trigg.

CONTENTS

Statistical Physics
and the
Atomic Theory of Matter

From Boyle and Newton
to Landau and Onsager

I. EARLY DEVELOPMENT
OF THE KINETIC THEORY OF GASES

1.1 INTRODUCTION

The kinetic theory of gases may not be typical of most modern theories in the physical sciences, but those who write and teach about science often use it as an example of what a satisfactory theory should be able to do. The textbook derivation of the ideal gas equation from a simple model of bouncing little billiard balls obviously fulfills a deep need to explain empirical regularities in terms of a visualizable atomic substratum obedient to plausible mechanical laws: if only we could do the same for every other natural phenomenon, we could really say that we understand how the world works!

It can reasonably be argued that the revolution in physical worldview that occurred in the first part of this century, as a result of the establishment of the quantum and relativity theories, has made obsolete the goal of mechanical explanations such as those provided by the kinetic theory. But that is true only in a rather limited sense. We believe that the new theories have to be used when we describe the internal structure of atoms, the behavior of matter at very low temperatures and very high pressures, and phenomena involving very high speeds, strong gravitational fields, electromagnetic radiation, and so forth; and that these theories also determine the range of approximate validity of "classical" (nonquantum, nonrelativistic) theories. Nevertheless there are still an enormous number of situations (including most of the ones encountered in everyday life or technological applications) where quantum and relativity effects are either completely negligible or can be satisfactorily allowed for by appropriate changes of numerical parameters. Even more important is the fact that mechanism has been a congenial mode of thought for scientists and many intelligent laymen for millennia, and seems just as attractive today in spite of recurrent criticisms. A mechanistic theory, no matter how naive, stands a better chance of being regarded as an *explanation* than a theory that does no more than deduce empirical data from a set of mathematical equations. The history of kinetic theory is in part the history of conflicts between these two kinds of theories.

The transition from kinetic theory to statistical mechanics (contrary to what the nonspecialist might guess) represents the subordination of the mechanistic viewpoint to the mathematico-empirical approach. Statistical mechanics is not only completely compatible with quantum theory and relativity, it also helped to give birth to the former. At the same time, it has been extended to cover the entire domain previously governed by kinetic theory in such a way that mechanistic explanation (for example of viscous flow in terms of molecular collisions) is not abolished but enriched. In the 19th century such a shift would have implied the abandonment of atomism; in the 20th century the existence of atoms is no longer in doubt, but they have to be described by an abstract probability calculus which takes account of infinitely many possible states of an unlimited number of particles.

Historians of science have also changed their views on explanation in recent years, but in a direction that may puzzle scientists. For example, the methodological/philosophical issues mentioned in the preceding paragraphs are of little interest to the modern practitioner of statistical mechanics, since today's problems are primarily technical in nature. One knows how to write down a set of equations for any well-defined problem; the only difficulty is finding an accurate solution. The closest thing to a "philosophical" issue in this discipline is the "problem of irreversibility": Is the tendency of natural processes to go in the direction of increasing disorder (entropy) the result of a fundamental law or a statistical artifact? That problem also played an important role in the historical development of the discipline, as will be seen in Chapter Two; but there it will be connected with other problems that are now considered separate, such as the age of the earth and indeterminism. In general, an historian of science likes to deal with many aspects of a subject that may no longer be of interest to the scientist who wants to know only the highlights of the path leading to the present. Conversely the historian often neglects recent discoveries and facts to track down earlier speculations that anticipated them. As a physicist turned historian I have some sympathy for both viewpoints; don't expect to find either one followed exclusively in this book.

1.2 MECHANICAL FOUNDATIONS

This book is limited to the period of "modern" science, which we may consider to have started with Copernicus in the first half of the 16th century. The doctrines of the Greek atomists and of the Roman poet Lucretius[1] were known as part of the body of ancient philosophy being

[1] Birth and death dates, and other biographical data, for persons mentioned in the text are collected in an index at the end of the book.

publicized by Renaissance humanists, but because atomism was associated with atheism it was not regarded as an acceptable foundation for scientific theories. In this case the theological objection seems to have been based on feelings that survived the expulsion of theology from science: it is offensive to our belief in the harmony of the universe to say that all natural phenomena are the result of the random motions and collisions of solid particles moving through empty space. If the world can be explained by nothing more than matter and motion, we can dispense with supernatural agents (as Lucretius argued), but we must also abandon the idea that everything is tied together by a substratum of forces or ethers. Without such a substratum it is hard to understand how there can be any regularity or continuity in the universe.

Early in the 17th century the state of science became more favorable to atomic hypotheses. Some historians credit the writings of Pierre Gassendi with removing the taint of atheism; he argued that atoms do not determine their own motion (as the theory of Lucretius implied) but are set in motion by God. This idea fitted into the doctrine of the "clockmaker God" developed by Robert Boyle and others later in the 17th century: the universe is a clockwork mechanism which, once created and energized, runs forever in a completely determined fashion without any need for further divine intervention. Watching the elaborate clocks and automatic mechanisms that could imitate animal motions, devices that were popular in late medieval Europe, one could convince oneself that a supremely ingenious and omnipotent artisan might construct a machine to imitate almost anything.

To make the clockwork-universe concept the basis of scientific theory one had to have laws and formulae telling how individual parts of the machine would move when pushed or pulled in a particular way. And, especially for a theory of atoms moving freely through space, one needed Galileo Galilei's principle of inertia, later known as Isaac Newton's first law of motion: a body remains at rest or in a state of motion at constant speed in a straight line unless acted on by a force.

Galileo introduced his principle in order to counter the intuitive objection to Copernican astronomy: If the earth is really spinning on its axis and moving around the sun, why don't we notice any effects of such motion? For example, an object thrown into the air or dropped from a tower should get "left behind" as the earth moves under it. Galileo's argument (leaving aside the distinction between truly rectilinear motion and motion along a small part of a large circular path) was that the free-falling object retains the component of motion which it shared with the earth before it was thrown or dropped, and follows along with the earth's motion by inertia until it hits the ground. Thus, to a first approximation, the earth's motion has no direct effect on the observed motion of objects near its surface, and the most serious physical objection to the new heliocentric astronomy was removed.[2]

The new heliocentric astronomy called for a new physics, and even after Galileo was forced by the Catholic Church to stifle his support of Copernicus he continued to develop the physics of matter and motion. His last book, *Two New Sciences* (1638), discusses atomism but, more important, explains how a very simple force—gravity near the earth's surface—acts on simple bodies in the absence of friction or air resistance. First, he showed (by a thought-experiment) that a heavy and a light object dropped simultaneously from the same height will hit the ground at the same time. (The modern interpretation of this result is that the force of gravity F which causes acceleration of the heavy object is greater in the same proportion as the inertial mass m which resists acceleration; so the actual acceleration a, being the quotient F/m by Newton's second law, is the same for both.) Second, Galileo showed that a body moved by a constant force such as gravity will be uniformly accelerated in the sense that it changes its speed by equal amounts in equal time intervals. This principle was illustrated by measuring the total time required for a ball to roll down an inclined plane as a function of the distance it travels; if the acceleration is uniform, the theory indicates that the time should be proportional to the square of the distance. Isaac Newton later generalized these results in his second law of motion (1687).

We cannot jump straight from Galileo to Newton, however, for then we would miss the establishment of the intellectual context in which Newton's discoveries were made and became famous. That context has been characterized by historians as the "mechanical philosophy" (see for example the 1952 essay of Marie Boas) and is associated with the writings of René Descartes and Robert Boyle. In a broad sense it includes the concept of the clockwork universe mentioned above, but it usually signifies a comprehensive program directed toward the explanation of natural phenomena that can be observed and measured in the laboratory rather than a doctrine about God's role in the history of the universe. Moreover, the mechanical philosophers rejected in principle the concept of action at a distance, which they stigmatized by associating it with magical "occult forces." The clockwork universe, on the other hand, enjoyed its greatest triumphs at the price of accepting gravity as an

---

[2] Strictly speaking the object should fall a little ahead of the point on the ground directly below it, since the original horizontal component of motion of an object in *circular* motion at constant angular speed will be greater if it is farther from the center of rotation. That effect was not directly observable in Galileo's time, though it has important geophysical consequences which could later be used to show the effect of the earth's rotation.

attractive force that could extend over huge distances without giving any evidence of being transmitted through an intervening material medium. The success of Newtonian celestial mechanics was an embarrassment for the mechanical philosophy, and it provoked countless futile attempts at a mechanical explanation of gravity; two of the early kinetic theories of gases were indirect results of such attempts (see below, §§1.7, 1.8).

Descartes, in his *Discourse on Method* (1637), proposed that scientific theories should be deduced from self-evident fundamental principles rather than induced from observation. For this he has been widely criticized, though many other scientists, modern as well as ancient, have followed the same procedure without admitting it. For Descartes it was self-evident that space is completely filled with pieces of matter of various sizes and shapes. While disclaiming belief in atoms—for his particles could be broken down into smaller ones without limit—Descartes thus encouraged the growth of atomism by emphasizing particulate explanations.

It was also self-evident to Descartes that the total amount of *motion* in the world must remain constant. The quantity of motion of a particle was defined as its quantity of matter multiplied by its speed; for Descartes speed and motion were *scalar* quantities whose value did not depend on their direction. This principle of conservation of motion ensured that the world-machine would never run down; unfortunately it did not always agree with the results of collision experiments, as Descartes himself must have realized.

Although a single particle in the Cartesian universe would move forever in a straight line at constant speed if it never collided with another particle, this situation could never occur, because each position to which a particle might try to move would already be occupied by another particle that would have to be pushed out of the way. From this Descartes concluded that the only possible kind of motion is the rotation of rings of particles, forming a "vortex." Such vortices, on a very large scale, might provide a basis for explaining the motions of the planets around the sun and around their own axes. On a very small scale, a molecular vortex could, by its rotation, keep other vortices away from a space whose volume increases with the speed of rotation, thereby accounting for the fact that substances expand as they get hotter.

All of these explanations, at least in the original form given by Descartes, break down when examined closely: they are either mathematically defective or in disagreement with quantitative experiments. Hence modern scientists tend to dismiss the Cartesian system as a collection of erroneous ideas that had to be refuted in order for physics to get on the right (that is, Newtonian) track. Even those modern scholars who do study

Descartes tend to concentrate on his philosophical writings, overlooking his immense influence on the history of physics. One indication of this neglect is the fact that, as of this writing, no complete English translation of the *Principia Philosophiae* (1644) has been published—the sections on physics are omitted from the only edition which is generally available. (But see Hall 1970:264–70.)

The reasons for Descartes' importance in the development of physics may be summarized as follows: first, a large amount of the progress made in physics and astronomy between 1650 and 1750 was inspired by attempts to prove or disprove Descartes' theories—this applies to most of Huygens' work, Book II of Newton's *Principia*, and many of the discoveries of Halley, Maclaurin, the Bernoulli family, Euler, the Cassini family, d'Alembert, Clairaut, Leibniz, and Maupertuis. Second, his mechanistic viewpoint survived the defeat of his own special applications of it, and underlies the work of such wildly different scientists as Emanuel Swedenborg and J. J. Thomson. (It is not quite irrelevant to mention that Noam Chomsky calls his theory "Cartesian linguistics.")

The Royal Society of London, chartered in 1662, took as its patron saint Francis Bacon, the proponent of empiricism. Yet one of its first major projects, and indeed its most important achievement as a society, was to establish the laws of collision—in particular the law of conservation of *vector* momentum—thereby revising and completing Descartes' theory of motion. The law, as formulated in 1668 by Christian Huygens, John Wallis, and Christopher Wren, states that the vector sum of the product mass × (vector) velocity has the same value before and after a collision. Note that the law applies to inelastic as well as elastic collisions. In particular it allows the case in which two blobs of clay with momenta $m\vec{v}$ and $-m\vec{v}$ (equal speeds in opposite directions) before the collision stick together and come to rest after the collision: $m\vec{v} - m\vec{v} = 0 + 0$.

Descartes' (incorrect) principle of conservation of motion was thus transformed into the (correct) law of conservation of momentum. But the Cartesians shouldn't have cheered so loudly at this, for their triumph had been obtained only by sacrificing the metaphysical purpose of the original principle. Now the world-machine *could* grind to a halt, losing motion while retaining momentum, through inelastic collisions like the one described above. Only by postulating that all matter is composed of perfectly elastic parts could this fate be avoided. Then one could assume that the loss of motion in inelastic collisions of macroscopic objects is only apparent, and that the motion is really transformed to the invisible particles in those objects. It seems to have been a fairly natural inference that this motion would then be manifested as heat; even Francis Bacon concluded by his inductive method that heat is associated with a rapid motion of the invisible parts of bodies.

Several 17th-century scientists proposed that the most appropriate definition of "quantity of motion" is not mv or $m\bar{v}$ but mv^2. This quantity was called by the Latin name *vis viva*, "living force"; that term was used up to the middle of the 19th century, when it acquired the familiar factor of $\frac{1}{2}$ and became "kinetic energy." In some of the 19th-century literature the phrase "*vis viva* theory" was used to refer to the kinetic theory of gases or more generally to the theory that the heat contained in a body is the kinetic energy of the motion of its molecules.

Huygens proved that total *vis viva* is conserved in elastic collisions, and used a similar conservation law in his theory of clockwork mechanisms (1673). G. W. Leibniz and his followers in the early 18th century advanced the *vis viva* concept as a measure of a more general "force" associated with moving objects; this was a step toward the modern concept of "energy," but no one as yet knew how to measure the other kinds of energy (thermal, electrical, magnetic) into which *vis viva* could be transformed. Only mechanical work (force multiplied by the distance through which it acts) could be included in the conservation equation.

During the first part of the 18th century the "*vis viva* controversy" raged between proponents of $m\bar{v}$ (mostly Cartesians) and of mv^2 (mostly Leibnizians). The controversy seems a little silly to modern physicists, since according to Newton's second law both quantities are equally valid ways of measuring the motion produced by a force. It is only a matter of convenience whether one chooses time or distance as a variable in the particular problem under consideration: a force acting for a definite time interval will produce a definite increment of $m\bar{v}$, but a force acting through a definite distance interval will produce a definite increment of mv^2.

But this way of dismissing the problem, though appropriate for a 20th-century mechanics textbook, conceals a fundamental issue in early modern science. One can get some understanding of that issue by recalling the elementary kinetic-theory derivation of the pressure exerted by a gas of billiard-ball atoms. Pressure is defined as the average force per unit area on a surface; but what is the force exerted by a single atom when it bounces off the side of the container? In the ideal case of perfectly elastic (but incompressible) spherical atoms striking a perfectly smooth wall, we would say that there is an infinite force acting for an infinitesimal time interval, defined in such a way that it has the *effect* of instantaneously reversing the component of the atom's momentum perpendicular to the wall. We call this an *impulsive* force, and we know that the product $F \Delta t$ has a well-defined finite value, equal to the change in momentum of the atom $\Delta(mv_\perp)$; we do not worry about assigning values to F and Δt separately. Yet we still think of this F as the same kind of physical entity as a finite, continuously acting force such as gravity.

In the 17th century it was not at all obvious that the impulsive forces acting in collisions are similar to continuous forces acting at a distance, or that they can be treated within the same mathematical framework. It was Newton who showed that this is so. In the *Principia* (1687) he stated his second law of motion in the form $F = \Delta(mv)$, which pertains to impulsive forces (suppressing any reference to the time interval, which one may assume to have been set equal to one). Nowhere in Newton's writings can one find the modern form of "Newton's second law," $F = ma$. (That was first written explicitly by Leonhard Euler in 1750.) But when he derived the effect of a gravitational force on a planet moving in its orbit, Newton used the fiction that the planet is pushed toward the sun by hammer blows acting at successive time intervals. In the limit as the strength of each individual blow goes to zero and the number of blows in a given time goes to infinity, Newton arrived at the correct result for a continuous force.

One may conclude from this that Newton viewed impulsive and continuous forces as equally valid concepts in the description of natural phenomena; at least that is the view of his followers. Alternatively one may conclude that Newton agreed with the Cartesians that only impulsive contact forces really exist in nature but that such forces appear to be continuous when the impulses are small and frequent. Or, with less justification, one could attribute to Newton the Leibnizian view that only continuous forces exist in nature, but when they come into play at atomic distances they appear to be impulsive.

One of Newton's early manuscripts, published by John Herivel (1965), provides an interesting sidelight on this point. Newton needed to find the centrifugal force exerted by a mass m moving in a circle of radius r with linear speed v, or rather the centripetal force that would have to be exerted on the moving mass toward the center of the circle to keep it from flying off on a tangent. The motion is the same as that of a ball bouncing around inside a hollow globe, in the limit where the number of sides in the polygonal path becomes a circle (see Fig. 1.2.1). It is clear[3] that the force for a square path is mv^2/r, and this result remains unchanged when one takes the limit. Thus the formula for centripetal acceleration, which we usually attribute to a continuously acting force exerted by an attracting

---

[3] The change of momentum is proportional to the vector difference between $\overrightarrow{bc}$ and $\overrightarrow{ab}$, that is, $\overrightarrow{bd}$, whose magnitude is $2r$, divided by the length of $\overrightarrow{ab}$, that is, $\sqrt{2}r$. Thus $\Delta p = mv(2r/\sqrt{2}r) = \sqrt{2}\,mv$. The time interval associated with each change of momentum is that needed to go $1/4$ of the perimeter of the square: $\Delta t = (1/4)(4ab/v) = \sqrt{2}r/v$. Thus $F = (\Delta p/\Delta t) = mv^2/r$.

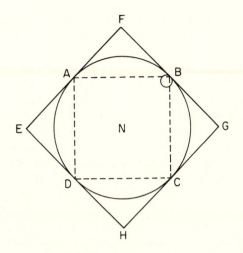

F$_{IG}$. 1.2.1. Motion of a ball inside a globe, from Newton's *Waste Book* (after Herivel 1965:130).

center, can just as well be derived from suitably arranged impacts. Even more remarkable, perhaps, is the analogy thereby demonstrated between gravitational force and kinetic gas pressure: for the mv^2 in both formulae results from multiplying a factor mv, proportional to the momentum change at a single impact between ball and container, by a factor v, proportional to the frequency of such collisions. Newton obviously should have discovered the kinetic theory of gases!

1.3 GAS PRESSURE

One of the most familiar examples of an attractive force is the *suction* which pulls a liquid up through a pump or straw. Should we call this "action at a distance," or attribute the effect to nature's "abhorrence of a vacuum"? Or can it be explained by impulsive forces acting only at contact?

In Aristotle's theory of the physical world, a vacuum cannot exist. It *must* not exist, for if it did various absurd things might happen; for example, light and heavy objects would fall at the same (infinite) speed.[4] In order to prevent such absurdities (note the implication of *purpose* here) the surrounding fluid will always rush into any space that one tries to

---

[4] If, taking some license, one may attribute to Aristotle the equation $v = F/R$, where F = force (proportional to weight) and R = resistance, then $v = \infty$ for any weight if $R = 0$. This conclusion is properly speaking not Aristotle's but that of his later interpreters.

evacuate, even if this means moving upward against the force of gravity. Thus the "abhorrence of a vacuum" becomes a force which can pull water up a tube.

But water can only be sucked up about 32 feet by a simple pump, as Galileo pointed out in 1638. That fact was presumably well known to people who had to pump water for a living, but scientists did not seem to appreciate its significance before the 17th century. It is just a bit implausible to say that nature's abhorrence of a vacuum is only strong enough to lift water 32 feet but no higher. The alternative explanation— that suction is due to the mechanical pressure of the air surrounding the pump—was at first even less plausible, for one would have to assume that we are all submerged in a fluid which exerts the incredibly large force of 15 pounds on every square inch of our skins. Nevertheless the mechanical philosophy which favored the air-pressure type of explanation was so influential at this time that scientists preferred this theory to either the "abhorrence of a vacuum" or any kind of attractive force.

There was considerable interest in experiments on suction and vacuum phenomena around the middle of the 17th century. Torricelli, in Italy, found that if water is replaced by mercury (which has a density 14 times as great) the length of the column which can be supported by whatever-it-is becomes about 30 inches rather than 32 feet; he invented a simple barometer by filling a glass tube (sealed at one end) with mercury, then inverting it in a bowl of mercury. The mercury would run down, creating a vacuum at the sealed end, until the difference in levels inside and outside the tube was 30 inches (Fig. 1.3.1). With a bulb at the sealed end, a few experiments could be performed in this "Torricelli vacuum."

In France, Blaise Pascal proposed an experiment to show that the height of mercury in a barometer would decrease at the top of a mountain,

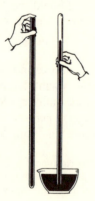

FIG. 1.3.1. Torricelli's experiment with a column of mercury in a tube longer than 30 inches (after Conant 1950:5).

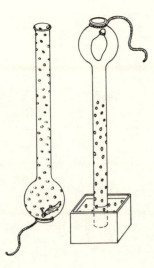

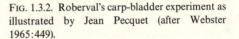

FIG. 1.3.2. Roberval's carp-bladder experiment as illustrated by Jean Pecquet (after Webster 1965:449).

where air pressure is presumably less; the experiment was successfully performed in 1648 by his brother-in-law Florin Perrier. It is often called the Puy de Dôme experiment.

Giles Persone de Roberval performed another experiment about the same time (1647) which was frequently quoted and repeated: he removed part of the swim-bladder from a carp, squeezed as much air out of it as possible and tied up the opening, and then inserted it in a Torricellian vacuum (Fig. 1.3.2). The bladder could be seen to inflate, convincing most observers that the small amount of residual air, previously compressed into a small space by atmospheric pressure, would expand to a greater volume when that pressure was removed.

Jean Pecquet publicized Roberval's carp-bladder experiment in his book on physiology (1651, English translation 1653), and introduced the term *elater* (Greek, "that which or one who drives") for the tendency of air to expand. (This was later modified to "elasticity.") Pecquet also suggested that air is composed of spongy or woolly particles, and that air near the surface of the earth is compressed by the weight of the atmosphere on top of it. Its elater resists this pressure and makes it expand when the pressure is reduced.

A more spectacular experiment was conducted by Otto von Guericke in Germany, in 1654: when the air was pumped out of two iron hemispheres, teams of horses could not pull them apart (Fig. 1.3.3). Guericke's experiment demonstrated the enormous strength of the force involved in suction but, unlike Pascal's and Roberval's, did not suggest the origin of that force.

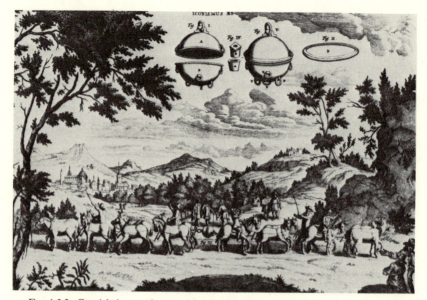

FIG. 1.3.3. Guericke's experiment with Magdeburg hemispheres (from I. B. Cohen, *Album of Science: From Leonardo to Lavoisier, 1450–1880* [New York: Scribner's, 1980]).

The culmination of this international inquiry into air pressure was the work of Robert Boyle, the British scientist already mentioned as a proponent of the clockwork universe and mechanical philosophy. Boyle was rich enough to be able to buy or have made for him the best possible scientific equipment, and also supported some less affluent scientists who worked on topics within his wide range of interests. In addition to "Boyle's law" he is known for giving a definition of chemical element, though the context indicates that he rejected the scientific value of this concept.[5]

Using an improved vacuum pump built for him by Robert Hooke in Oxford, Boyle conducted a series of investigations published in 1660 under the title *New Experiments Physico-Mechanicall, touching the Spring of the Air, and its Effects*. He argued that the column of mercury in a Torricelli barometer is supported by the mechanical pressure of the air acting on the mercury in the dish (Fig. 1.3.1); perhaps his most convincing

[5] "... I see not why we must needs believe that there are any primogeneal and simple bodies, of which, as of pre-existent elements, nature is obliged to compound all others. Nor do I see why we may not conceive that she may produce the bodies accounted mixt out of one another by variously altering and contriving their minute parts. . . ." See Boas (1952:498) for this and other quotations indicating Boyle's attitude toward the existence of elements.

evidence of this was his Experiment No. 17, in which he enclosed the barometer in a larger container from which the air could be pumped out (Fig. 1.3.4). As the air is removed, the level of the mercury in the tube gradually falls to that of the mercury in the dish. (The only possible objection is that it is not quite clear what the other theories of suction predict should happen under these circumstances.)

Boyle also discussed the atomic nature of air pressure. He professed neutrality between two alternative theories: (1) the Cartesian explanation based on whirling particles which push each other away; (2) Pecquet's idea that the particles of air are like coiled-up pieces of wool or springs that have an elasticity allowing them to push away their neighbors. On the second theory, which Boyle really preferred, one could imagine that the particles of air near the earth's surface are compressed by the weight of the atmosphere on top of them; this accounts at least qualitatively for their tendency to expand into any available space, and for the fact that the pressure of the atmosphere is greater at sea level than it is at greater heights.

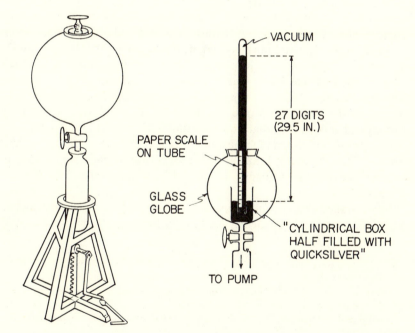

Fig. 1.3.4. Boyle's air pump as shown in his *New Experiments Physico-Mechanicall* (1660), and his experiment of removing the air above the reservoir of a barometer (after Conant 1950:18, 21).

Boyle's advocacy of the air-pressure theory of suction was quickly challenged by Franciscus Linus, a Jesuit scientist from Belgium. Linus proposed a new theory: when a vacant space is created above the mercury in the Torricelli barometer, an invisible threadlike object called the *funiculus* comes into existence, apparently drawn out of the mercury. The funiculus resists being stretched out by pulling in all the objects around it, and in particular it grabs the mercury column and prevents it from falling any farther down the tube.

The idea that suction is the effect of an attractive force exerted by an unseen object is not as silly as it may seem to the modern reader. In fact, you can "feel" the funiculus pulling your flesh whenever you put your finger on the open tube of a vacuum cleaner! And when you suck a liquid up a straw you probably think of yourself as an *active* power pulling in, rather than a passive receptacle letting air pressure do the work.

If we were interested only in those scientists who took positive steps toward our present scientific theories, we would ignore Linus or mention him only as an obstacle to progress toward truth. That is the "Whig" interpretation of the history of science: each person or event in the past is to be judged from the standpoint of the present.[6] In one sense Whiggism is unavoidable: if the kinetic theory of gases were not an important part of modern physics, a book about its history would find few readers.[7]

Boyle did not ignore Linus; instead, he published another book in 1662, in which he attempted to justify his theory of air pressure by further experiments and reasoning, and to reply to the objections of Linus and Thomas Hobbes. (Hobbes thought a Cartesian ether rather than a vacuum is in the space above the mercury in Torricelli's barometer.) It is this second publication that contains the first statement of the law for which Boyle has become famous; one might conjecture that, had it not been for the criticism of Linus and Hobbes, Boyle would have moved on to other subjects and someone else would have gotten the credit for the PV law.

Boyle argued that the funiculus theory was refuted by the experiments already mentioned, in particular his own Experiment No. 17 and Pascal's

---

[6] Historians of science have labeled this approach, which is often followed by scientists who write the history of their own subject, by analogy with Herbert Butterfield's *The Whig Interpretation of History* (1931). As described by Butterfield, Whig historians saw history in terms of progress toward modern liberal democracy, ignoring contemporary social context and personal motivations that are essential to understanding human behavior. In later writings Butterfield saw Whig historiography itself as a positive force in the movement toward democracy.

[7] Nevertheless books about obsolete scientific theories have recently been published: *The Caloric Theory of Gases* by Robert Fox, and *The Vortex Theory of Planetary Motion* by E. J. Aiton. This may be taken as a sign of the maturity of the history of science as a discipline independent of the contemporary sciences.

Puy de Dôme experiment. He pointed out that the latter had been re-peated in England, by Richard Townley of Lancashire and others, with similar results, showing that the column of mercury is supported by an atmospheric pressure whose strength varies with height above sea level.

Another objection to Linus' theory was that it seems implausible that the funiculus can hold up the entire mercury column simply by grabbing its top layer; only solids, not liquids, can be lifted this way.

To illustrate the springiness (*elater*) of air, Boyle described an experi-ment in which a certain amount of air was trapped in a J-shaped tube by a column of mercury whose height could be varied up to nearly 120 inches (Fig. 1.3.5). He noted that when the pressure of the atmosphere was effectively doubled by adding 29 inches of mercury, the volume of the enclosed air was reduced by half its original value. He generalized this to to the statement that the spring of the air is proportional to its density.

While this statement is equivalent to what we know as Boyle's law—the product of pressure and volume is constant (ignoring the possibility of variations in temperature)—Boyle had so far investigated only com-pressions and did not yet realize that a similar law should hold for expansions. At least that is what he tells us himself in his *Defence of the Doctrine touching the Spring and Weight of the Air* (1662), where he states that he had not reduced his data on expansion to "any certain hypothesis" when Richard Townley informed him that air when dilated loses its spring in proportion to its expansion.

The story of Townley's contribution to the discovery of "Boyle's law" was uncovered just a few years ago by the British historian of science Charles Webster (1965). It appears that an equally important role in the story was played by Townley's friend Henry Power, who started his experiments on air pressure in 1653. Power and Townley were stimulated to resume their research by the publication of Boyle's first book in 1660.

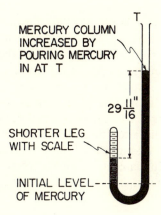

FIG. 1.3.5. Boyle's experiment on the relation between pressure and volume of air, as shown in his *Defence* (1662) and as described by Conant (1950:51).

It is not certain whether they discussed air pressure with Boyle at meetings of the Royal Society, but there does exist a report on their work which was sent to William Croune at the Royal Society and forwarded to Boyle. The title of the report was "Additional Experiments made at Townley Hall, in the years 1660 and 1661, by the advice and assistance of that Heroick and Worthy Gentleman, Richard Townley, Esqr. and those Ingenious Gentlemen Mr. John, and Charles Townley, and Mr. George Kemp." Unfortunately Power, who wrote the report, did not put his own name on the title page, and Boyle apparently did not see any letter of transmittal or other indication of Power's authorship, so he mentioned only Townley when he was giving credit for the hypothesis in 1662. The report was published a year later by Power himself as part of his book *Experimental Philosophy*, but that was too late to establish his priority.

Power, though he adopted Pecquet's conception of springy air particles, could not believe that air could expand entirely by its own elasticity and therefore wanted to retain some semblance of an ethereal substratum that could fill up the empty spaces in a rarefied gas. Many scientists in the 18th and 19th centuries held similar views: even though they were willing to attribute an elastic power to air particles, they retained an ethereal medium in which the particles "swim" rather than allowing the particles to move freely through completely empty space. The ether had to be there for other purposes, too: the transmission of electric and magnetic force, and of light if one adopted a wave theory.

While there is fairly good evidence that Power and Townley suggested the PV law, it is not historically incorrect to continue calling it Boyle's law, since Boyle did much more to publicize and establish it than anyone else. (During the 18th and 19th centuries the French called it Mariotte's law, but this ascription has now been completely abandoned, since there is no doubt not only that Mariotte's publication (1679) was later than Boyle's but that if he hadn't heard of Boyle's well-known books, he certainly should have.)

In any case, the discovery of a *quantitative* relation between pressure and volume, however important it might have been to later developments in physics, was not really as significant at the time as the proof of the *qualitative* fact that suction phenomena are due to air pressure. Only after explanations based on the abhorrence of a vacuum, attractive forces, ethers, and the funiculus had been rejected did scientists turn their attention to the mathematical formula relating the two variables that had been shown to be relevant. Thus if Boyle gets a little too much credit for the discovery of the PV law (despite his own willingness to share it with Townley), his contribution to the more fundamental problem of the *existence* of air pressure and its role in various physical phenomena has generally been underrated by those who take air pressure for granted and can't imagine how its existence could ever have been in doubt.

This distinction, between qualitative facts that have to be established first and quantitative laws that subsequently overshadow them, is illustrated by another well-known 17th-century discovery: the speed of light. The Danish astronomer Ole Rømer is universally credited with the "first determination of the speed of light," and it is often stated that he obtained a result equivalent to about two-thirds of the modern value. Yet his 1676 paper, the only one ever cited in this connection, does not mention any definite numerical value for the speed at all; Rømer only states the time that light would take to cross the earth's orbit. Since the earth-sun distance was not a well-established parameter at that time (in fact the data needed for its first accurate determination were just then being analyzed), one can hardly say that Rømer determined the speed of light in a quantitative sense. (Apparently it was Huygens who first published in 1690 the value usually attributed to Rømer.) But what he did do was far more important in the context of 17th-century physics: he showed that light is not propagated instantaneously from one place to another but moves at a finite speed. Descartes had asserted the opposite, for he thought that light, being an impulse transmitted through a rigid medium (his ethereal particles could not be expanded or compressed), would have to have infinite speed.[8] Rømer's demonstration that light moves at finite speed, a qualitative fact that now seems trivial, not only scored a point against the Cartesian system but ruled out one possible theory of the nature of light.

In 1663, when Henry Power published his *Experimental Philosophy*, the main opponents of the Cartesians were not yet the forward-looking Newtonians but the backward-looking Aristotelians and believers in occult forces. Thus Power was confident that he was addressing the true scientific "moderns," the new generation of mechanical philosophers who would appreciate his allusion to the theory of elastic air particles, when he wrote:

> You are the enlarged and elastical Souls of the world, who, removing all former rubbish, and prejudicial resistances, do make way for the Springy Intellect to flye out into its desired Expansion. . . .
> This is the Age wherein (me-thinks) Philosophy comes in with a Spring-tide. . . . I see how all the old Rubbish must be thrown away,

[8] Notwithstanding the dogmatic assertions he made on this point in his *Optics* and his correspondence in 1634 (Burke 1966), Descartes introduced a fictional model of light particles moving at different (finite) speeds in different media to derive Snell's law of refraction.

and carried away with so powerful an inundation. These are the days that must lay a new Foundation of a more magnificent Philosophy, never to be overthrown; that will Empirically and Sensibly canvass the *Phaenomena* of Nature, deducing the causes of things from such Originals in Nature, as we observe are producible by Art, and the infallible demonstration of Mechanicks; and certainly, this is the way, and no other, to build a true and permanent Philosophy. (Power 1663:191–92)

1.4 NEWTON AND THE THEORY OF GASES

According to current views of the history of science, as articulated by Alexandre Koyré, Thomas Kuhn, Imre Lakatos, and others, the origin and the fate of hypotheses depend not so much on an objective comparison with experimental data as on their relationship to established theory, the hard core of a research program, the paradigm, or the metaphysical worldview current at the time. Since the word "paradigm" has been widely adopted in this connection (in spite of or perhaps because of its vagueness and flexibility), I will use it here; this does not imply acceptance of Kuhn's theory of scientific revolutions, since Kuhn himself (1974) has now abandoned his original use of "paradigm" in favor of other, more restricted terms. I will thus use "paradigm" as a primitive undefined concept, assuming the reader will have gleaned some idea of what it means from the extensive discussions of the last decade.

In the normal course of scientific work, the paradigm provides not only the fundamental principles such as conservation laws for matter and energy but also guidance as to what kinds of questions are to be considered scientific problems, and examples of acceptable solutions. Scientists who share a paradigm form a community; they learn their subject from the same set of essentially interchangeable textbooks, go to the same meetings, read the same journals, and give jobs to each other's students. For them, the paradigm itself is not really subject to test; the change from one paradigm to another is a more or less discontinuous process—sometimes called a "revolution"—which is determined by factors that are partly rational, partly psychological or sociological. For this reason, successive paradigms in a science are sometimes called "incommensurable"—their proponents would not agree on a "crucial experiment" to decide between them, or on objective criteria by which the decision could be made.

During most of the 17th century there was no generally accepted paradigm in physics to replace the crumbling Aristotelian system. The Cartesian system very nearly succeeded in becoming the next paradigm:

Descartes provided an attractive philosophical framework, and Huygens supplied valuable and consistent solutions of some of the major problems in mechanics and optics. I suspect that a Cartesian paradigm in physics would have been more nearly like the *ideal* paradigm (as described by Kuhn) than any of which we have actual historical experience. Real paradigms must usually incorporate some contradictory elements in order to deal with a large range of imperfectly known phenomena, and cannot be so monolithic.

Newtonian mechanics was a paradigm for most scientists during the greater part of the last two centuries and is therefore the paradigm of paradigms for historians of science. To "unpack" that sentence (as the philosophers of science say), one may recall that until around 1900 it was assumed that any problem in physics could be solved, at least "in principle," by applying Newton's laws of mechanics; it was only necessary to determine the forces and mechanical properties of the parts of the system and then compute a solution for the appropriate set of differential equations. Since this was the most successful theory in *any* science, theorists in chemistry, geology, biology, psychology, and the social sciences tried to imitate it; but that meant adopting what was thought to be Newton's philosophy of nature as well as his scientific method. If we think that the sciences have now rejected the Newtonian paradigm, we can nevertheless use this historical case to understand what it means to be dominated by a paradigm.

One of the contradictory elements of the Newtonian paradigm was the concept of force, and this contradiction affected the development of the kinetic theory of gases. Newton agreed with the Cartesians in rejecting action at a distance as an innate property of matter, but was nevertheless willing to postulate both long- and short-range forces in order to achieve quantitative explanations of observed phenomena. If such explanations proved to be successful—as they often did—it was not clear what the next step should be: to go back and explain the force itself in terms of contact actions, or to continue using and elaborating the assumptions about forces to explain other phenomena. In Chapter Five we will see how the second strategy was maintained as a research program into the 20th century. The first alternative, despite its Cartesian aspect, was also pursued by those who considered themselves Newtonians, and this resulted in curiosities such as the kinetic theory of gravity, the vortex atom, and the virtual photon-exchange concept of modern quantum electrodynamics.

Newton himself chose the second alternative when he wrote the section on air pressure in the *Principia*. He suggested that air acts as if it were composed of particles that exert repulsive forces on neighboring particles, the magnitude of the force being inversely proportional to the distance between them. This force law is the simplest mathematical representation of the springiness of air particles postulated by Pecquet and Boyle, and it leads directly to Boyle's law.[9] It is not surprising that Newton adopted Boyle's conception of air pressure in preference to Descartes' conception of whirling particles, since one of the main goals of the *Principia* was to show the inadequacy of Descartes' theory of the universe. But the result of his choice was to establish a static model of gases in which pressure is attributed to interatomic repulsive forces, and thus to delay the development of the kinetic theory.

Newton made it quite clear that he was not claiming to have shown that gases really are composed of such repelling particles; only *if* one assumes that they are, *then* one can derive Boyle's law by making the force vary inversely as the distance rather than by some other power such as the inverse square or cube of the distance (which might have been thought more likely for other reasons). "Whether elastic fluids do really consist of particles so repelling each other," he wrote, "is a physical question. We have here demonstrated mathematically the property of fluids consisting of particles of this kind, [so] that hence philosophers may take occasion to discuss that question" (Newton 1687, Motte-Cajori translation, p. 302).

In the 18th century it was generally believed that Newton had *proved* that gases consist of repelling particles. This was in part a misunderstanding; few people bothered to read the *Principia* in detail or check its derviations, but simply relied on textbook summaries of the results. Even if they had read this section carefully they might still have concluded that any supposition published by the great Newton was probably true. To the growing band of Newtonians, every sentence of the *Principia* was stamped with the authority of the world's greatest scientist—especially since Newton had stated explicitly that he disdained hypotheses that were not well grounded in reality.

One indication that few of his followers read Newton's own statement of his theory of gas pressure is the fact that his careful limitation of the repulsive force to nearest neighbors was often ignored. Even the great Laplace thought he was authorized to assume that atoms exert repulsive

---

[9] Imagine the particles sitting in a cubic array with distance d between neighbors. If each linear dimension of the container is changed from L to xL (where x is a number that may be more or less than 1), the distances between neighbors change from d to xd. The force between them changes from F to F/x. This force when transmitted to the side of the container will produce a pressure equal to the force divided by the area of the square whose side is the distance between particles; hence the pressure changes from P to P/x^3. The total volume of the container has changed from L^3 to x^3L^3; hence the product, pressure × volume, remains constant.

forces at large distances but attractive forces at short distances, thus combining Newton's explanation of Boyle's law with his own theory of surface tension into a single force law. But Newton was well aware that if each atom repelled all the others with a force varying inversely as the distance, the pressure of the gas would no longer be a simple function of the density but would also depend on the total number of particles in the system, contrary to Boyle's results. (It would also vary with the shape of the container; the sum or integral of the forces on one atom exerted by the others does not converge to a limit as the number of particles goes to infinity.)

As with the law of gravity, Newton's prestige was thus thrown on the side of action at a distance, contrary to his own opinion. In limiting the repulsive force to nearest neighbors he seems to have had in mind a contact interaction between springy particles as in the Pecquet-Boyle model, not a force law in the modern sense—though the example he gives to justify his assumption is the attraction of magnetic bodies, which he says "is terminated nearby in bodies of their own kind that are next them" or by an interposed iron plate.

In another book, *Opticks*, Newton gave a different model of the atom. The following passage was often quoted by his followers:

> All these things being considered, it seems probable to me that God in the Beginning formed Matter in solid, massy, hard, impenetrable, movable Particles, of such Sizes and Figures, and with such other Properties, and in such Proportion to space, as most conduced to the end for which he formed them; and as these primitive Particles being Solids, are incomparably harder than any porous Bodies compounded of them; even so very hard as never to wear or break in pieces; no ordinary Power being able to divide what God himself made one in the first Creation. . . . And therefore that Nature may be lasting, the Changes of corporeal Things are to be placed only in the various Separations and new Associations, and Motions of these permanent Particles; compound Bodies being apt to break, not in the midst of solid Particles, but where those Particles are laid together. . . . (Newton 1730:400)

Elsewhere, however, Newton suggested that there may also be short-range *attractive* forces between atoms (Hall 1970:323–33).

It should be clear now that while the Newtonian paradigm encouraged assumptions about forces between elastic atoms, it did not prescribe a single force law, and it also retained the old idea of the ultimate atom as a hard indestructible building block of the universe. If one attempted to combine elasticity and absolute hardness in the same conceptual model, one would encounter a paradox: suppose two such atoms collide head-on with equal and opposite velocities. At some instant of time they must both come to a stop, since they are supposed to be impenetrable. At that instant their kinetic energy of motion has disappeared, but if they are *elastic* it must reappear an instant later so they can bounce back with the same velocities in opposite directions. But if they are *hard*, that is, undeformable, there is no way that energy can be stored momentarily inside the atoms. The usual process of compressing a springy ball, thereby storing potential energy which can be quickly converted back to kinetic energy (Fig. 1.4.1), is forbidden here; to say that an atom could change its shape or volume would imply that it is composed of smaller parts and thus is not an ultimate atom. (That implication could be avoided by postulating atoms made of a continuous deformable substance, but only at the cost of abandoning Newton's concept of a God-given hard atom quoted above.) The paradox is like the old problem, "What happens when an irresistible force meets an immovable object?" In this case it is not the force itself that is irresistible but the conservation of the total energy of the system. (Wilson Scott 1970 has given a detailed account of the historical conflict between atomism and conservation laws arising from this type of paradox.)

There are two possible resolutions of the paradox. The first is based on Huygens' modified Cartesian theory of collisions: one simply refuses to worry about instantaneous forces between colliding particles but applies the law of conservation of total momentum before and after the collision. The adjective "elastic" is interpreted operationally to mean that the atoms have the same total kinetic energy before and after the collision, but one does not inquire too closely into their kinetic energy *during* the collision, since that is difficult to measure anyway. (This approach, incidentally, is the one adopted by modern physicists in analyzing collisions of elementary particles.)

The alternative resolution is to replace the model of the hard atom occupying a finite volume of space by a point-center of repulsive force, such that two atoms would never come in contact unless they approach with infinite velocity. That was the suggestion of Roger Boscovich (1758), and it has often been adopted by scientists in the last 200 years.

Both solutions have disadvantages. In Huygens' theory one has to abandon any definite description of a force acting between two atoms when they collide. In Boscovich's theory the tangible atom is replaced by a system of forces, and it is difficult to see how a mathematical point can have all the properties we attribute to atoms.

These difficulties in finding a consistent and plausible model of the atom did not prevent scientists from accepting Newton's theory of gas pressure.

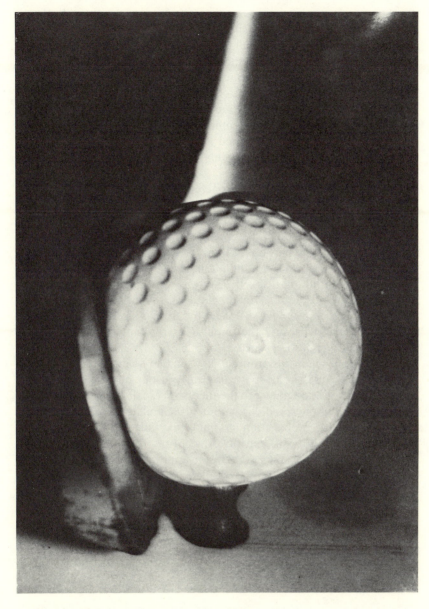

FIG. 1.4.1. High-speed photograph of iron hitting golf ball (courtesy of Professor Harold Edgerton, Massachusetts Institute of Technology).

A more serious problem was the discrepancy between theoretical and experimental values of the velocity of sound in air. Unlike the velocity of light, whose numerical value could not be estimated from theory and was therefore of little importance in the 17th century (see §1.3), the velocity of sound in a gas should be determined by the relation between pressure and volume of the gas, according to Newton's theory. In modern terminology the velocity is equal to the reciprocal of the square root of the *compressibility*,[10] which can be computed from the constant in Boyle's law. But when Newton substituted observed values in his formula he found that the measured speed was about 20% faster than the theoretical speed.

This discrepancy was clearly unacceptable, so Newton invoked an additional hypothesis to account for it. He proposed that the sonic impulse is passed along from one air particle to another more rapidly than one might expect because of the presence of solid particles, or because of the "crassitude" of the particles. I think Newton meant that the particles have a diameter σ which is not infinitesimal compared with the distance d between neighboring particles. If σ were zero, the time required for the impulse to travel distance d would be at least as long as the time it takes for a particle to move that distance, but if σ is not zero, the time is somewhat reduced because the impulse is transmitted instantaneously through the inside of the particle.[11]

At the time he made this proposal Newton had no independent method to determine the crassitude, so it was clearly just an ad hoc hypothesis or "fudge factor" inserted to rescue the theory (Westfall 1973).

The problem of the speed of sound in air stimulated further theoretical research on wave propagation in fluids by Euler and Lagrange (see Truesdell 1956), but the discrepancy was not resolved until the beginning of the 19th century; even then the solution appeared to depend on a theory of heat that was soon to become obsolete (see below §1.8).

1.5 Daniel Bernoulli and the First Kinetic Theory of Gases

A large part of what we call Newtonian mechanics was developed by a group of Swiss mathematicians in the 18th century; Leonhard Euler, the brothers James and John Bernoulli, and John's son Daniel Bernoulli. It was Euler who first put Newton's second law into the form $F = ma$ in

---

[10] The formula is $v = (d\rho/dp)^{-1/2}$, where ρ = density, p = pressure. As will be seen later (§1.8), the adiabatic rather then the isothermal compressibility should be used in this formula.

[11] He did not realize that σ/d is in fact quite small for gases at ordinary densities; the crassitude effect does not become important until the gas is so highly compressed that it is nearly a liquid (Enskog 1922).

1750; and it was James Bernoulli, also known for his work in probability theory, who first stated something like the principle of conservation of angular momentum in 1686. Euler and John Bernoulli showed how a wide range of mechanical problems could be formulated and solved using the techniques of integral and differential calculus (Leibniz's version of the calculus rather than Newton's). After the French mathematician Jean LeRond d'Alembert discovered the differential equations for a vibrating string, Daniel Bernoulli proposed the solution of this equation in the form of a sum of characteristic modes of vibration. The science of hydrodynamics was founded by John and Daniel Bernoulli, d'Alembert, and Euler. The contributions of these men to elasticity theory, celestial mechanics, and other disciplines established a unified and systematic approach to all problems in physics, astronomy, and engineering, going far beyond what one could learn from Newton's *Principia* (Truesdell 1968).

The "first kinetic theory of gases" has sometimes been attributed to Euler (Hooykaas 1948; cf. Truesdell 1968:274–76), but his paper of 1727 is based on a model of whirling particles suspended in an ether. Thus while Euler did derive here a result equivalent to $pV \propto Nmv^2$, the v is a rotational rather than a translational velocity. His theory was based on ideas about the structure of gases prevalent in the 17th and 18th centuries, its merit being to put these ideas in mathematical form. Daniel Bernoulli, on the other hand, must get the credit for the first kinetic theory in the modern sense even though he could not furnish enough evidence to convince his contemporaries to adopt it.

Bernoulli's theory occupies a few pages in the tenth chapter of his book on hydrodynamics, published in 1738. He first considers the force exerted by a gas of an infinite number of very small particles on a movable piston in a cylindrical container (Fig. 1.5.1). He *assumes* that the particles have the same velocity before and after the gas is compressed by the piston, an important point (as will appear later) that many writers pass over too

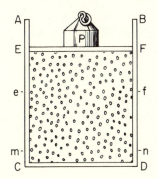

FIG. 1.5.1. Daniel Bernoulli's gas model.

quickly even today. "Now when the piston EF is moved to ef, it is subjected to a greater force by the fluid in two ways: first because the number of particles is now greater in proportion to the smaller space in which they are confined, and secondly because any given particle makes more frequent impacts" (from translation in Brush 1965:59).

If the motion of the piston from EF to ef reduces the volume to a fraction s of its previous value ($V_s = sV$), the surface area is reduced to a fraction $s^{2/3}$ of its previous value; hence the pressure increases by a factor $s^{-2/3}$ by the first cause. The effect of the second cause is estimated in a way that reveals a difference between Bernoulli's conception of gases and the modern one: he states that "impacts occur more frequently in proportion to the reduction in the mutual distances of the particles; evidently the number of impacts will be inversely proportional to the mean distances between the surfaces of the particles." That sounds reasonable enough if the distances are what we would call the *mean free path* traveled by a particle between successive collisions.[12] But Bernoulli did not make a clear distinction between the nearest neighbor of a particle and the one it actually hits next. As it turns out, this distinction does not affect the result. If the distance between centers of particles is D and their diameters are d, the compression increases the number of impacts by a factor

$$(D - d)/(Ds^{1/3} - d).$$

Combining the two factors Bernoulli finds that the pressure is increased by a factor

$$(P_s/P) = \frac{D - d}{(Ds^{1/3} - d)s^{2/3}} = \frac{1 - d/D}{s - s^{2/3}d/D} \qquad \text{where} \quad s = V_s/V.$$

Bernoulli now notes that if we put $d/D = 0$, we get $P_s/P = V/V_s$:

so that the force of compression is approximately inversely proportional to the volume occupied by the air. This is confirmed by a variety of experiments. This law can certainly be safely applied to air less dense than normal; though I have not adequately examined whether it also applies to air very much more dense; experiments of the necessary degree of accuracy have not yet been put in hand. An experiment is needed to fix the value of the quantity m [$= d/D$] but it must be made with very great accuracy, even when the air is strongly

[12] Bernoulli's formula is not quantitatively correct even then; the problem was discussed by van der Waals and others in the last part of the 19th century (§1.12).

compressed; the temperature of the air should be carefully kept constant while it is being compressed.

The last sentence suggests that Bernoulli was aware that compression often increases the temperature of a gas, but that it is possible to arrange an experiment so as to avoid this.

The next section of Bernoulli's discussion deals with the effect of temperature on gas pressure. This effect corresponds to what is called "Charles' law" in many textbooks, perhaps more appropriately "Gay-Lussac's law." As Bernoulli indicates in the following, the effect was already known even if not definitely established by the experiments of the French scientist Guillaume Amontons at the beginning of the 18th century, but somewhat awkward to express because of the lack of a standard temperature scale.

The pressure of the air is increased not only by reduction in volume but also by rise in temperature. As it is well known that heat is intensified as the internal motion of the particles increases, it follows that any increase in the pressure of air that has not changed its volume indicates more intense motion of its particles, which is in agreement with our hypothesis; for it is clear that the weight P needs to be greater to contain the air in the volume ECDF, according as the particles are in more violent motion. Moreover it is not difficult to see that the weight P will vary as the square of the particle velocity, on account of the fact that through this increased velocity the number of impacts and their intensity will increase equally at the same time. . . .

. . . The theorem set out in the preceding paragraph—namely, that in any air of whatever density, but at a given temperature, the pressure varies as the density, and furthermore, that increases of pressure arising from equal increases of temperature are proportional to the density—this theorem was discovered by experiment by Dr. Amontons; he has given an account of it in the Memoires of the Royal Academy of Sciences of Paris for 1702. The implication of the theorem is that if, for example, ordinary air at normal temperature can hold a weight of 100 lb distributed over a given surface area, and then its temperature is increased until it can hold 120 lb over the same area and at the same volume, then when the same air is compressed to half its volume and held at the same temperatures, it can support 200 and 240 lb respectively; thus the increases of 20 and 40 lb produced by the increase of temperature are proportional to the density. . . .

. . . From the known ratio between the different pressures of a given amount of air enclosed in a given volume, it is easy to deduce a measure of the temperature of the air, if only we agree in defining what we mean by "twice the temperature, 3 times the temperature," and so on; such a definition is arbitrary and not inherent in nature; in fact, it seems reasonable to me to determine the temperature of air by reference to its pressure, provided it is of standard density. The standard temperature from which the rest are measured should be obtained from boiling rainwater, because this no doubt has very nearly the same temperature all over the earth.

Accepting this, the temperatures of boiling water, of the warmest air in summer, and of the coldest air in winter in this country are as 6:4:3.

To summarize: Daniel Bernoulli introduced a gas model in which an indefinitely large number of similar particles of diameter d and speed v move in a closed container of volume V supporting a piston with pressure P by their impacts. The average distance between particles is D. The temperature T is a function of v. He showed that

(1) for variations of P and V at constant T, Boyle's law $PV = $ constant holds in the limit $d/D \to 0$;

(2) corrections to Boyle's law can be estimated for finite d/D, and might be checked against experimental data at high pressures if such were available;

(3) if the speed of the particles is varied at constant volume, the pressure is proportional to v^2;

(4) it is possible to define a temperature scale by using the equation $PV = CT$, where $C = $ constant. In that case T would be proportional to v^2.

Note that Bernoulli did not give his particles any properties other than size; in particular he did not mention their mass. By implication the mass of a single particle is infinitesimal, since he assumed there are an infinite number of particles in a finite volume; but the momentum transfer in a single particle impact does not play a role in the derivation of the pressure equation (except perhaps for the vague reference to "intensity of the motion"). In fact there is considerable similarity between Bernoulli's derivation of Boyle's law and Newton's. In both cases the increase in pressure due to compression is the product of two factors, one of them being inversely proportional to the distance between particles (the force in Newton's model, the number of impacts in Bernoulli's) and the other being inversely proportional to the square of that distance. For both derivations the second factor arises from the fact that the force has to be divided by the surface area to obtain the pressure. So the difference between the two theories reduces to the difference between a force acting by frequent impacts and a force acting continuously; and, as was pointed out earlier, one can easily go back and forth between these kinds of forces in Newtonian mechanics.

1.6 Chemical Atomic Theory

After Bernoulli there was little progress in the physical theory of gases for several decades. The next significant advances came in chemistry with the discovery that there are several kinds of gas, each with distinct properties. Here one may recall the work of Joseph Black, Henry Cavendish, and Joseph Priestley in Britain, and Carl Wilhelm Scheele in Sweden. Toward the end of the 18th century Lavoisier replaced the "phlogiston" theory of combustion with his theory based on the existence of a new gas, oxygen, and established the modern theory of chemical nomenclature; and the German chemist J. B. Richter proposed a new science, stoichiometry, to deal with quantitative aspects of chemical reactions. These developments are detailed in the standard histories of chemistry (for example Partington 1962, Ihde 1964); here they serve only as background for the work of Dalton, Gay-Lussac, and Avogadro.

John Dalton was an English scientist whose first major publication was a book on meteorology (1793), and it was a meteorological problem that first started him on the path to his chemical atomic theory. By the end of the 18th century it was known that the atmosphere consists primarily of oxygen and nitrogen and that these gases when separated have different densities at the same temperature and pressure. Dalton wondered why they stay mixed in the same proportions even at different heights above sea level, instead of spontaneously separating with the denser gas (oxygen) on the bottom.

Dalton speculated about differences between oxygen and nitrogen particles that might prevent them from separating into pure gases. First he tried the hypothesis that they have different sizes and shapes, so that a particle of one kind would tend to disrupt the regular lattice array formed by particles of the other kind. (It is clear from this hypothesis that he imagined the particles to be almost in contact, with little free space to move around.) Then he decided to extend the Newtonian model to mixtures of gases by assuming that particles repel others of the same kind but not those of different kinds. Thus each kind of gas acts like a vacuum to the others, and each behaves as if it were the only kind present.

The most important outcome of this early stage in Dalton's thinking was the "law of partial pressures": the total pressure of a gas mixture is simply the sum of the pressures each kind of gas would exert if it were occupying the space by itself. Though Dalton arrived at this law by using the Newtonian idea that gaseous particles exert pressure simply by the repulsive forces between neighbors, it is also derivable from the kinetic theory.

At some point Dalton decided that the weight of a particle is even more important than its size or shape, and began to study the weight relations of chemical compounds. If a particle of oxygen weighs eight times as much as a particle of hydrogen, then one might expect that 1 gram of hydrogen could react with exactly 8 grams of oxygen, forming 9 grams of water with no hydrogen or oxygen left over. (Of course the conclusion may be correct even if the premise is wrong.) This assumes that every particle of hydrogen has the same characteristic "atomic weight" and that every particle of oxygen has the same weight, different from that of hydrogen. The problem was then to find a consistent set of weights for the particles of every known element, from which the weight relations of all reactions could be calculated.

What is the connection between Dalton's atomic theory and the experimental data which he had available? In some chemistry textbooks and popular works on science it is stated that Dalton started from the empirical discovery that nitrogen and oxygen can combine in two different ratios, leading to the formation of two different nitrogen-oxygen compounds. In one of these, according to this account, exactly twice as much oxygen reacts with a given amount of nitrogen as in the other reaction. From this fact Dalton supposedly arrived at the hypothesis that the reactions involve combinations of individual atoms: one nitrogen plus one oxygen, or one nitrogen plus two oxygen.

Unfortunately for the advocates of inductive methods in science, it is now known that Dalton could not have arrived at his theory in this way, even though he may have tried to give that impression in his published writings. His laboratory notebooks show that he did not find a 2:1 ratio for the nitrogen-oxygen reaction before he worked out his theory; the closest he could get was about 2.7:1.7. Moreover, 20th-century chemists with historical interests (J. R. Partington and L. K. Nash) have tried to repeat Dalton's experiment and found that it is impossible to get a 2:1 ratio with anything like the conditions he described. The historical evidence indicates that Dalton first devised his atomic theory and later "rectified" his experimental results so as to illustrate his theory. Such behavior is not rare in science; we know that other scientists such as the biologist Gregor Mendel supported their theories with data of dubious authenticity (van der Waerden 1968).

Dalton's theory, worked out around 1804, assigned the following weights to atoms of the more common elements (1805:287):

Element	Dalton's At. Wt.	Modern At. Wt.
Hydrogen	1	1.008
Nitrogen	4.2	14.007
Carbon	4.3	12.011
Oxygen	5.5	15.999
Phosphorus	7.2	30.974
Sulfur	14.4	32.064

These are of course only *relative* weights, taking hydrogen = 1 by definition (the modern values take the carbon-12 isotope = 12.000 by definition); Dalton had no way of finding the absolute weight of a single atom.

The most important reason for the discrepancy between Dalton's atomic weights and modern values is not the inaccuracy of his experimental data but the theoretical assumptions he made about how atoms combine. In the case where there is only one compound of two elements, such as hydrogen and oxygen, he postulated that this compound contains one atom of each kind. Thus his formula for water is HO rather than H_2O, and oxygen would be assigned an atomic weight of 8 (if modern data are used) rather than 16. We will see shortly what his other assumptions were, but first we must look at the related work of another scientist.

The French chemist J. L. Gay-Lussac comes into our story because of three different discoveries concerning gases. One has already been mentioned, the establishment of the law of thermal expansion of gases (sometimes known as Charles' law). That law pertains to the expansion of a gas restrained by constant external pressure. Gay-Lussac also studied the free expansion of gases into a vacuum; he found that when a gas is suddenly allowed to occupy a much larger space by removing the external pressure, there is practically no change in its temperature. This fact contradicts some theories of gases and (at least in retrospect) is generally held to support the kinetic theory, (see below, §1.9).

Gay-Lussac's third discovery was the "law of combining volumes." He found that in gaseous reactions the volumes of the reactants and products are also related to each other by ratios of small integers. Thus for example 2 liters of hydrogen combine with 1 liter of oxygen to form 2 liters of water.

The most plausible interpretation of the law of combining volumes was that the volume of each gas is proportional to the number of particles it contains (assuming the same pressure and temperature). Then two particles of hydrogen would combine with one particle of oxygen to form two particles of water. That would of course imply that a "particle" of oxygen can split into two parts, one for each of the resulting water particles; thus gases may be composed of particles each of which is made of two or more atoms. It would also mean that an oxygen atom has weight 16 rather than 8 relative to hydrogen taken as 1, with similar changes in other atomic weights.

Dalton rejected this interpretation, even before it was fully articulated by Avogadro. While he based his objections partly on a criticism of the accuracy of Gay-Lussac's measurements, it is clear that his own conception of atoms was the main factor. His pictures of atoms show them nearly in contact, and since atoms of different elements have different sizes one would not expect to find the same number in a given volume. Even less could this be true for particles compounded of different numbers of particles.

The assumption that two atoms of the same element could combine to form a single particle in a gas was even more contrary to Dalton's ideas, for he believed that Newton had proved two similar atoms repel each other. Compound particles could only be formed out of different kinds of atoms, since according to Dalton's theory of mixed gases such atoms don't repel each other.

Even if one does assume that some special short-range attractive force acts between similar atoms, it is not at all clear why this force would not cause three or four atoms to cluster together instead of only two. Here was another paradox about atomic forces, usually referred to as the "saturation property" of chemical binding. The force by which one oxygen atom attracts other oxygen atoms is "used up" by the process of binding together two atoms and is not able to act on a third atom.[13] That is even more mysterious than Newton's force that acts or doesn't act between two atoms depending on whether they are nearest neighbors. The saturation property of chemical forces was not explained until after the invention of quantum mechanics in the 1920s.

The assumption that every gas contains the same number of *molecules* at the same pressure, volume, and temperature became known as Avogadro's hypothesis after the Italian scientist Amedeo Avogadro, who proposed it in 1811. As we have seen, it is tied to the assumption that in certain gases two or more atoms are combined to form separate particles, which we now call molecules, the number being chosen in accordance with Gay-Lussac's law of combining volumes. With this modification of

---

[13] Ozone (O_3) is the "exception that proves the rule"—it was discovered in 1840 and generally admitted to be triatomic by 1872 (Scott 1970:161).

Dalton's theory and modern data on chemical reactions one can derive the familiar molecular formulae (H_2O for water, H_2 for hydrogen gas, O_2 for oxygen gas, NH_3 for ammonia, CO_2 for carbon dioxide, etc.) and the familiar set of atomic weights ($H = 1$, $C = 12$, $O = 16$, etc.). This system was not generally accepted until after the Italian chemist Stanislao Cannizzaro presented it to the Karlsruhe Congress in 1860.

1.7 JOHN HERAPATH AND THE SECOND KINETIC THEORY OF GASES

Avogadro's hypothesis is favorable to the kinetic theory of gases insofar as it implies that the volume occupied by a certain number of molecules is independent of their size and shape, which suggests they are not ordinarily in contact. But it was not until 1859 that Maxwell showed that the hypothesis could be derived from the kinetic theory; the earlier version of kinetic theory we are now going to consider is actually inconsistent with it. That proves nothing except perhaps that the relation between physics and chemistry in the 19th century was not as close as it might have been.

John Herapath (1790–1868) was an Englishman who might be called an amateur scientist except that the distinction between amateur and professional was not very significant at the time. Instead, let's call him an outsider, a clever eccentric who never enjoyed any recognition from the scientific community but lived to see his ideas vindicated. He was concerned with fundamental aspects of mechanical explanation at a time when Bristish science was dominated by Baconian empiricism and honored Newton in name but not in spirit. The calculus was still being taught with the clumsy technique of dots and fluxions, ignoring the powerful methods of continental mathematicians—though that situation was soon to be corrected by a group of reformers at Cambridge University. The Royal Society had degenerated to little more than a fashionable London club, many of whose members had only a dilettante interest in science; there too, reforms were to come in the following decades.

Herapath was initially interested in developing an explanation of gravity in terms of the impacts of particles of an ethereal fluid, somewhat along the lines of the kinetic theory of gravity proposed by G. S. LeSage and many others. In the usual version of this theory one attempts to explain the attractive force between two bodies as an indirect effect of the way each shields the other from the stream of gravific particles. Severe difficulties arise when one tries to work out this theory in detail, especially if one asks what happens to the particles *after* they strike the bodies. The remarkable aspect of the kinetic theory of gravity is that it seems to occur spontaneously to almost anyone who wants to reduce action at a distance to contact action. Herapath's version was somewhat different: he proposed to take account of the effect on the gravific particles of the high temperatures in the space near the sun. In this way he came to consider the relation between temperature and particle velocity.

After worrying for some time about the paradox of collisions between particles that are both hard and elastic (see above, §1.4), Herapath decided to adopt the Huygens solution—except that he failed to grasp the way Huygens' theory differed from Descartes'. Herapath simply assumed that the scalar momentum mv of a particle is a measure of its temperature and that the total momentum of a system is conserved in collisions while individual momenta tend to be equalized. Thus he ignored both the vector nature of momentum and the conservation of kinetic energy in elastic collisions.

Herapath arrived at the correct relation between pressure, volume, and particle velocity (aside from constant factors) by a method too complicated to describe here; the important point is that he wrote the result in the form

$$PV \propto T^2,$$

where T, which he called "true temperature," is proportional to mv; and he claimed that if two fluids at different temperatures are mixed, the temperature of the mixture must be calculated by averaging the original "true temperatures" of the components.

Here at last was a definite prediction from the kinetic theory differing from the predictions of other theories, which could apparently be tested by a simple experiment. The test could even be done with water: Herapath predicted that the temperature of a mixture of equal amounts of water at the freezing and boiling points, 32° and 212°F, would be 118.4° rather than $(32 + 212)/2 = 122°$.

In fact this experiment had been done before, by de Luc and Crawford, with different results. Herapath argued that de Luc's result of 119°, which was quite close to Herapath's prediction, was more likely to be valid precisely because it was significantly different from what would have been expected; hence de Luc must have checked it fairly carefully before publishing it. Crawford's result of 122° could be disregarded for the opposite reason. But it was evident that further experiments were needed to settle the question.

Herapath submitted his first paper on kinetic theory to the Royal Society of London in 1820, hoping to get it published in the *Philosophical Transactions*. That would have given his views wide circulation in the international scientific community; moreover, as Herapath himself frankly admitted, it would have enhanced his personal reputation so that he could embark on a career of teaching and scientific research.

Humphry Davy, a well-known chemist, became president of the Royal Society shortly after the submission of Herapath's paper and was mainly responsible for its fate. Davy had earlier supported the general idea that heat is molecular motion rather than a substance, and thus he might have been expected to be receptive to a theory that gave this idea a precise mathematical formulation. But his reaction was negative, for several reasons. The only one he made explicit to Herapath was his reluctance to consider heat as a simple quantity that could be completely extracted from a body by annihilating the motion of its molecules, thus implying the existence of a lowest temperature ("absolute zero"). It is also evident that he found Herapath's derivations difficult to follow and would have preferred a less abstract, less mathematical approach emphasizing instead the correspondence between concepts and observations at each step; that seems to be a persistent difference in the attitudes of chemists and physicists. Finally, we may conjecture that Davy had a metaphysical repugnance for the basic assumption of kinetic theory—that particles move through empty space with no interactions except when they collide. By this time Davy had adopted some of the sentiments of Romantic nature philosophy; for example, seeing the world as an interconnected system dominated by all-pervading forces rather than by push-pull mechanisms.

Herapath was told that his paper would not be published in the *Philosophical Transactions* and he was advised to withdraw it, since according to the custom of the Royal Society once a paper was "read" (formally presented, if only by title or abstract, at a meeting) it became the property of the Society and could not be returned to its author. Herapath did this and sent his paper to an independent scientific journal, the *Annals of Philosophy*, where it was published in 1821. But he did not abandon his efforts to get the Royal Society to endorse his work. Another paper describing his proposed experiment on mixing water at different temperatures remained in the Royal Society Archives; Herapath wanted the Society to sponsor this experiment, and when this was refused he decided that Davy and the other Fellows were conspiring to persecute him. He wrote a series of letters to the *Times* (London) attacking Davy and challenging the entire Royal Society to solve a series of problems in mathematical physics, for a stake of £500. There were no takers, nor did anyone reply to Herapath in public. Presumably he was regarded as just another crank.

Denied recognition by the scientific establishment, Herapath nevertheless did have an opportunity to present his ideas to a larger public. The

Annals of Philosophy, though now forgotten, was not by any means an obscure journal in the early 19th century. Michael Faraday and other reputable scientists published there. It was similar in content and circulation to the *Philosophical Magazine* (with which it merged in 1826), the *Annalen der Physik und der Chemie*, or the *American Journal of Science*. So if Herapath's theory was ignored by most scientists in the 1820s, we cannot simply blame Davy and the Royal Society but must recognize that the theory was out of harmony with prevailing ideas about the nature of gases and heat, and failed to convince its readers that those ideas should be revised.

Before discussing the changes in physics that had to occur before the kinetic theory of gases could be accepted, I want to sketch briefly the rest of Herapath's career. We recall that the most important event in England around 1830 was the introduction of steam locomotives for hauling goods and passengers in railway trains. This development attracted Herapath's interest, and by 1836 he had settled into a position as editor of the *Railway Magazine*, later known as *Herapath's Railway Journal*. In addition to giving him an adequate income and some influence on the booming railway industry, control of this magazine allowed him to publish his own articles on any subject whatever, including the kinetic theory of gases and his earlier dispute with Davy.

One of the first scientific articles to appear in the *Railway Magazine* in 1836 was a calculation of the speed of sound, which Herapath said he had completed four years earlier. This was in fact the first calculation of the average speed of a molecule from the kinetic theory of gases. (J. P. Joule, who is usually credited with this accomplishment, was simply following Herapath's method.)

In the 1840s, stimulated by the publications of Thomas Graham on gas diffusion and of Regnault on compressibility, Herapath revised and elaborated his kinetic theory and published the two-volume treatise *Mathematical Physics* (1847). He claimed that he had proved from his theory in 1844 that the time required for a given volume (V) of gas to pass through a small hole into a vacuum should be proportional to $V\sqrt{G/T_H}$, where G = specific gravity and T_H is Herapath's "true temperature." (Recall that T_H is proportional to molecular velocity.) This was before he learned that Graham had established a similar result by experiment.

Herapath also claimed to have predicted in advance Regnault's result (1846) that the pressure of very dense gases is greater than that given by Boyle's law. This is indeed what the model of atoms as impenetrable spheres would lead one to expect if one ignores short-range attractive forces (see Bernoulli's derivation, §1.5). However, earlier experiments had indicated deviations in the opposite direction, suggesting that attractive forces are more important than repulsive. (Later kinetic theories managed to account for the fact that attractive forces dominate at low temperatures, repulsive at high.)

James Prescott Joule is the only scientist known to have read Herapath's *Mathematical Physics* during the first few years after its publication, but William Thomson and James Clerk Maxwell later mentioned it. Herapath lived long enough to see the kinetic theory revived by others. In 1860, after Maxwell's first paper was reported in a British magazine, Herapath published a letter calling attention to his own earlier work, and thus helped to ensure that he would be remembered as one of the pioneers of kinetic theory, though by this time he could not be credited with the *first* kinetic theory, since Daniel Bernoulli's chapter in *Hydrodynamica* had also been rediscovered.

As I mentioned at the beginning of this section, Herapath's version of kinetic theory is inconsistent with Avogadro's hypothesis, a fact that passed unnoticed in the 1820s but would have called for revision if anyone had tried to integrate it with chemical atomic theory after 1860.[14]

1.8 THE NATURE OF HEAT

Between 1820 and 1850 there was a major change in physical science. From the viewpoint of this book the most obvious effect of the change was that the response to any proposed kinetic theory of gases became very favorable in 1850, whereas it had been very unfavorable in 1820. For physics as a whole, the change meant the linking together of previously separate disciplines—mechanics, electricity, magnetism, light and heat— and thus in a sense the birth of physics itself as we now understand it. The kinetic theory was soon to provide one additional, and ultimately dominating, link: the link to atomistics.

The change I refer to is usually ascribed to the discovery of the law of transformation and conservation of energy in the 1840s by Julius Robert Mayer, James Prescott Joule, Ludvig Colding, Hermann von Helmholtz, and others.[15] This discovery was so fundamental that one could also

---

[14] In Herapath's theory $P \propto Nmv^2 \propto NT_H^2/m$, where N = number of particles in unit volume, m = mass of one particle, T_H = "true temperature" = mv. The specific gravity of the gas is proportional to the density $D = Nm$. Thus substituting $m = D/N$ in the equation for P, one gets $P \propto N^2T_H^2/D$. Herapath claims that if two gases have the same pressure and temperature they must have the same ratio N^2/D; hence N is proportional to the square root of the specific gravity. In particular, "if oxygen is 16 times heavier then hydrogen, there is quadruple the number of particles of oxygen in a given volume that there is of hydrogen" (Herapath 1847, 1:260).

[15] The multiplicity of names involved has stimulated a century-long debate on the history of this "simultaneous discovery": see articles by P. G. Tait and J. Tyndall in *Philosophical Magazine*, 1863; Lindsay (1973:11–13); Elkana (1974); Knott (1911:209–13); Kuhn (1959).

define physics since 1850 as the science of different forms of energy and their transformations. (That would leave chemistry as the science of different forms of *matter* and their transformations, though as it happened the subsequent development of science did not permit such a clear-cut distinction and the two sciences are unavoidably mixed together.)

The discovery of energy conservation did not constitute a "revolution" as that term is usually understood by historians of science; rather it was an act of imperial conquest. The Newtonian paradigm was not over-thrown; instead it was extended to cover a wider domain of science and technology. Yet at the same time one can argue that energy conservation was an early phase of a long revolution that did eventually dethrone the Newtonian paradigm. This is true in the obvious sense that a generalized energy concept was prerequisite to the formulation of the quantum and relativity theories. But it is somewhat misleading to stress the existence of energy as an independent entity, as if none of its forms—mechanical, electrical, thermal, etc.—were more fundamental than the others. For many 19th-century physicists (in spite of the arguments of Mayer, Ernst Mach, and Wilhelm Ostwald) *kinetic* energy was the basic quantity to which other forms of energy should in principle be reduced. Their goal was to replace explanations of phenomena based on postulating different *substances* (including ether, caloric, etc.) to explanations based on postulating different *kinds of motion*. This "kinetic worldview" eventually proved to be too simplistic as a basis for all physics, but it did influence the development of several 19th-century theories.

To understand how energy and the kinetic worldview got into physics in the first half of the 19th century, we must first grant that scientists in this period had a strong desire to unify their theories of different phenomena and were fascinated by experiments showing the interrelations of diverse forces in nature. The desire for unity may originally have been associated, as has been suggested by Kuhn (1959) and others, with Romantic nature philosophy and its postulate that all phenomena are manifestations of a single underlying force or an antagonism of two opposing forces in an organismic universe. But convictions about the unity of nature survived the death of nature philosophy as a respectable scientific doctrine.

One of the earliest examples of unification was the doctrine of the "identity of light and heat," stemming from discoveries about radiant heat around 1800. William Herschel detected heat-producing rays just beyond the red end of the visible spectrum of sunlight, suggesting a continuous transition from light to radiant heat. In other experiments it was shown that radiant heat obeys the same laws of reflection, refraction, and interference as does light. Evidently light and radiant heat are the same *qualitative* phenomenon, differing only in the value of some *quantitative* parameter. It was also believed that radiant heat is essentially the same as other kinds of heat; hence it followed that light and heat in general are the same. A definite answer to the question "what is light?" would determine the answer to "what is heat?" and conversely.

There were two possible answers to such questions: "substance" or "quality." One could postulate either that light (heat) is a special kind of matter or that it is a property—most likely a mode of motion—of some kind of matter. Whether the matter is fluid, has weight, is composed of discrete particles, etc. are subsidiary problems that can be worked out once the primary issue of substance versus quality has been settled.[16]

At the beginning of the 19th century it was widely accepted, in part on Newton's authority, that light is a substance—in particular, a stream of particles. Thomas Young failed to change this situation, either by performing his famous two-slit interference experiment or by quoting statements of Newton in favor of a wave theory.

Heat was also considered a substance, a fluid called "caloric." Chemical reactions, changes of temperature, and changes of state (melting, vaporization) were associated with the gain or loss of definite amounts of caloric, and many scientists thought that a body in a particular state (for example 1 gram of water at 15°C and atmospheric pressure) *contains* a definite amount of caloric. (For that matter many people today think the same thing, perhaps misled by terms like "heat capacity," "latent heat," "specific heat," and "calories," which are left over from the caloric theory.)

One could accept the caloric theory of heat and still believe that the heat inside a body has some connection with the motion of its molecules—at least one finds a number of statements to this effect in the scientific literature around 1800. But the phenomenon of heat *radiation* seemed to show that heat itself is something other than just molecular motion, for it can pass through a vacuum or even from the sun to the earth.

Pierre Simon de Laplace, the French mathematician best known for his achievements in theoretical astronomy, applied the caloric theory to calculate the speed of sound in gases (1816). Newton had shown that this speed is related to compressibility, but the value obtained by taking the compressibility from Boyle's law was about 20% less than the experimental value. Laplace suggested that the compressions and expansions that occur in the passage of a sound wave are *adiabatic*—they do not involve any exchange of heat with the surroundings. The temperature of

---

[16] There are some pitfalls associated with this terminology. Newton (1672), after proposing his theory that white light is a mixture of rays of different colors, argued that light could not be a quality because colored rays had different refrangibilities (were bent different amounts by a prism), and qualities cannot themselves have qualities!

the compressed gas would momentarily rise as it is compressed, then fall as it expands. In this case the compressibility cannot be that given by Boyle's law, which is valid only for constant temperature. Laplace and his colleague S. D. Poisson showed that Newton's formula for the speed of sound has to be multiplied by a factor $\sqrt{c_p/c_v}$, where c_p and c_v are the specific heats at constant pressure and constant volume respectively. For any gas c_p will be greater than c_v because the amount of heat needed to raise the temperature includes the extra amount needed to expand the gas to the larger volume it will occupy at the higher temperature. For air, oxygen, nitrogen, and many other common gases c_p/c_v is about 1.4, so $\sqrt{c_p/c_v}$ gives just the right correction to eliminate the 20% discrepancy (Finn 1964).

In addition to this triumph the caloric theory was the basis for two other major advances in heat theory: Joseph Fourier's mathematical theory of heat conduction (1807, 1822) and Sadi Carnot's analysis of the motive power of heat in steam engines (1824). In both cases, as with the Laplace-Poisson theory of sound propagation, the main results turned out to be independent of the assumption that heat is a substance, but before 1850 they had to be counted as successful applications of the caloric theory. (See Truesdell 1980 for a detailed analysis of the caloric theory.)

The caloric theory was also adapted to interpret Newton's model of repelling gas particles. It was generally held that the caloric fluid itself consists of particles that repel each other but are attracted to atoms of ordinary matter and cluster around them to form atmospheres. (Some 20th-century readers, coming upon such descriptions without knowledge of their historical context, have called this model a precursor of the modern "electron-cloud" model of the atom.) Laplace proposed a somewhat different model in which caloric is continually radiated and absorbed by each atom in equilibrium with its neighbors. It any case it was agreed that caloric is a "repulsive agent" in gases; the more you add, the more the atoms push each other apart. (The various modifications and applications of the caloric theory of gases have been discussed in detail by Robert Fox 1971.)

In the 1820s most scientists would probably have agreed that some form of the caloric theory offered a satisfactory explanation of nearly every thermal phenomenon, and no experiments involving heat gave decisive evidence against it. The experiments of Rumford and Davy, later thought to have disproved the caloric theory, were well known but not considered crucial; the heat generated by boring a cannon or rubbing together pieces of ice could be ascribed to caloric squeezed out of the bodies. Nevertheless the caloric theory was about to be rejected in favor of another theory that, by any objective standards, could not explain as many facts about heat as well as the caloric theory. The new theory that dominated discussions of heat in the 1830s was *not* the kinetic theory or thermodynamics but a theory that most 20th-century scientists have never heard of: the *wave* theory of heat.

To understand this curious turn of events, we must recall that the "identity of heat and light" required one to adopt the same theory for both. Before 1820 that principle simply reinforced the caloric theory, since light was thought to be a substance. But in 1818 the French Academy of Sciences held a prize competition on the subject of the diffraction of light, and one of the entries was a brilliant paper by Augustin Fresnel based on the wave theory of light. The panel of five judges included three supporters of the particle theory of light. One of them, Poisson, detected a consequence of Fresnel's equations that Fresnel himself had overlooked: they predicted that a bright spot of light should appear in the center of the shadow cast by a circular disk placed in a beam of light. (The spot is produced, theoretically, by constructive interference of waves diffracted around the edge of the disk.) Poisson seized upon this prediction as a means of disposing of the wave theory, once and for all, since he assumed that no such bright spot could be found. However, Fresnel and his friend François Arago did the experiment and found that the bright spot does indeed appear just where the wave theory predicts.

The unexpected evidence of the Poisson bright spot, together with Fresnel's success in accounting for many other properties of light, forced the judges to award him the prize of the Paris Academy (though they did not completely abandon their personal preferences for the particle theory). With the help of this endorsement from the world's most prestigious scientific society, the wave theory of light gradually won the allegiance of most European and American physicists and had become the established doctrine by 1830.

With the triumph of the wave theory of *light*, physicists naturally turned their thoughts to the wave theory of *heat*. Such a theory had been discussed, though only rather vaguely, by several scientists including Newton. The French physicist André Marie Ampère was the first to present a definite formulation in two short articles (1832, 1835). He considered various theoretical problems such as the conservation of mechanical energy in interactions between vibrating atoms and a vibrating ether, and the description of thermal motion by differential equations. In effect he identified heat with the *vis viva* of vibrations and postulated that heat will flow between vibrating atoms by means of ether vibrations, in such a way as to make the *vis viva* of each atom nearly the same. While he assumed that the *difference* of *vis viva* of two atoms is proportional to the *difference* of temperatures, he did not go so far as to say that the absolute amount of *vis viva* is proportional to an absolute temperature.

A remarkable feature of the wave theory of heat is that it provided a fairly smooth transition from the caloric theory of the early 1800s to the kinetic theory of the 1850s. It was not necessary to reject the caloric theory completely, at first; one could simply say that the caloric fluid was being identified with the ether, so that the same fluid could transmit both heat and light. Then one could say that it doesn't really make any difference if we replace the *quanity* of ether by *quantity of motion* (*vis viva*) of the ether—if you have twice as much ether you will have twice as much motion. Thus, as Ampère put it (1845:348),

> We find manifestly the same result by considering the subject ... according to the system of emission [caloric theory] or according to that of vibrations, substituting for the quantity of caloric in the first system, the *vis viva* of the vibrating molecules in the second.

By analogy with the transmission of sound from one solid body to another, Ampère invoked the transfer of vibrations from an atom to an intervening medium, then back to another atom. Once the reader has followed this far, he will be prepared for the next step (not taken by Ampère but implicit in his statement of the theory) that heat or temperature in material bodies is measured simply by the energy of motion of their atoms; the ether need be brought in only when heat has to be transmitted through empty spaces.

Later, especially after the development of Maxwell's theory of electromagnetic waves in the 1860s, it was realized that radiant heat cannot simply be equated to heat but must be treated as a separate form of energy that can be transformed into ordinary heat. The wave theory of heat was then replaced by two independent theories: kinetic theory, which dealt with heat as molecular motion, and radiation theory, which treated radiant heat as electromagnetic waves with frequencies somewhat lower than those of visible light. Thermodynamics encompassed both but only on a macroscopic level; indeed there was no adequate and comprehensive microscopic theory of matter-radiation interactions until quantum theory was developed.

Perhaps the main point of this excursion into the forgotten wave theory of heat is that the kinetic theory was not strong enough by itself to vanquish the caloric theory before 1850, but could have sneaked in under the wing of the wave theory if anyone had taken advantage of the opportunity. It is easy to imagine that Ampère might have developed a kinetic theory of gases out of the tentative hints given in his 1835 paper. He had already advanced the cause of the kinetic worldview in the previous decade with his proposal that magnetism is not a separate entity but rather a form of motion of electricity. Unfortunately Ampère could not carry on this program—he died in 1836.

My suggestion that the kinetic theory could have been brought back along with the wave theory of heat is not completely hypothetical, for that is just what J. J. Waterston tried to do. Waterston, a Scotsman who applied his talents to many areas of science and technology, has often been mentioned along with Herapath as having developed ideas that were precursors to the kinetic theory, and indeed the experiences of the two men were remarkably similar. Waterston, like Herapath, started out by speculating on a mechanical explanation of gravity, then developed a theory of gases which he submitted to the Royal Society of London. Writing in 1845, he was able to invoke the wave theory of heat in support of his views; but as in Herapath's case 25 years earlier, the response of the Royal Society was negative. Worse, Waterston (who was stationed in Bombay as a navy instructor at the time) did not learn about the Royal Society's policy on retention of manuscripts in time to withdraw his paper before it was officially read at a meeting, and he had not kept a complete copy of it. The result was that only a brief abstract was published at the time, while the paper itself remained in the Royal Society Archives and was not discovered until after Waterston's death.

The story of J. J. Waterston, which has been told in detail elsewhere (Haldane 1928; Brush 1976a, chap. 3), is not really an important part of the history of the kinetic theory, because his major work was not published when it could have had any influence on other scientists. I have mentioned it as an example of the connection between the wave theory of heat and the doctrine that heat is molecular motion; one might argue, with Edward Daub (1971), that even the minimal publicity given to Waterston's ideas did make some impression on August Krönig, who revived the kinetic theory in 1856. For those who worry about "priority" issues it should be noted that Waterston's published abstract (1846) stated a restricted version of what later became known as the "equipartition theorem": in a mixture of particles of different masses, the average *vis viva* or kinetic energy of each kind of particle will be the same. Among his later publications, a paper on the theory of sound and a note which includes an estimate of molecular sizes are of some interest, though they passed largely unnoticed in the 19th century. During his own lifetime Waterston's name was known to the scientific community only for his work on the sun's heat, and in particular for his suggestion—adopted temporarily by William Thomson—that the source of this heat is the continual infall of meteors.

From a modern perspective Waterston's case illustrates the degeneration of institutions like the Royal Society, which scientists had established in earlier centuries. Such institutions can sometimes be beneficial to the progress of science, as we have seen in connection with the discovery of momentum conservation and the shift from the particle to the wave theory of light. But when scientists and the public have too much respect for the authority of an institution, its leaders can stifle unorthodox ideas. This is rarely an absolute power; an individual scientist can evade institutional censorship if he is persistent enough (like Herapath, or, better, Galileo). It is only when expensive projects requiring the cooperation of many people are needed that the power of instiutions becomes dominant. Fortunately science depends more on the ideas and persistence of individuals than on expensive collaborative projects.

1.9 Clausius: "The Kind of Motion we call Heat"

Soon after the discovery of energy conservation by Mayer, Joule, Helmholtz, and others in the 1840s, Rudolf Clausius began to develop its application to the interconversions of heat and mechnical work. His series of papers beginning in 1850 established the basic equations of the science that soon became known as "thermodynamics."

Willism Thomson[17] (later "Lord Kelvin") and W.J.M. Rankine in Scotland independently published similar theories a little later than Clausius and are sometimes listed with him as cofounders of thermodynamics. Thomson used Sadi Carnot's (1824) theory to develop a scale of absolute temperature (1848), now called the "Kelvin" scale, and was the first to give a *general* statement of the principle of dissipation of energy in 1852 (see Chapter Two). Rankine encumbered his heat equations with the details of a theory of molecular vortices, which though interesting in itself did not help to establish thermodynamics.[18]

The first law of thermodynamics was merely a special case of the law of conservation of energy but brought out explicitly some important

[17] Thomson is an interesting exception to the generalization of the previous section, that physicists shifted their allegiance from the caloric theory to the wave theory of heat before adopting thermodynamics; working more within the engineering tradition, Thomson learned about heat from Fourier, Clapeyron, and Regnault, and accepted the caloric theory until around 1850. For more extensive discussion see Sharlin (1979) and Crosbie Smith (1976, 1977, 1978).

[18] See however the papers by Hutchison (1973, 1981a, 1981b).

properties of matter.[19] It states that the energy E of a body may be changed by adding heat (q) or allowing the body to do mechancial work (w) on its surroundings. The only kind of work we consider here is expansion against an external pressure P, changing the volume by the increment ΔV. (Imagine a hot gas pushing up a piston.) Since mechanical work is defined as (force) $\times$ (distance through which the force acts), and pressure is defined as force per unit area, we see that the work done by the substance is equal to $P \Delta V$ (ΔV = cross section area of piston $\times$ distance it moves).

Conversely if the substance is compressed (by pushing the piston in), ΔV is negative and thus the work done *by* the substance is negative; equivalently, a positive amount of work is done *on* the substance. By convention we will use w to denote the work done by the substance; it may be either a positive or negative quantity. Similarly if the substance *loses* heat, we would say that q is a negative quantity.

With these defnitions the first law may be written

$$\Delta E = q - w = q - P \Delta V.$$

The quantities denoted by capital letters are properties of the substance itself and are determined by its *thermodynamic state*, that is, by its temperature, pressure, volume, and perhaps other variables such as magnetic field. Quantities denoted by small letters pertain to a particular kind of *change* in the state of the system but are not uniquely determined by the thermodynamic state or even by the energy difference ΔE between two states. In particular a substance does not "contain" any definite amount of *heat*; the word "heat" only describes a way it can gain or lose energy, but once the energy is gained or lost one cannot identify it as "heat energy" as distinct from some other kind of energy.

A familiar illustration is "heating by compression." If you rapidly compress a gas, for example in a tire pump, it gets "hotter"—its temperature rises from T_1 to T_2—though you have not added any heat to it but have simply done mechanical work on it. In this case $\Delta E = -w = -P_2(V_2 - V_1)$, a positive quantity, since $V_2 < V_1$. (If the pressure changes during this process, one has to compute $-\int P \, dV$.) Alternatively you could have added enough heat at constant volume to raise the temperature to T_2, then slowly compressed it (keeping the temperature constant at T_2) to the same final volume V_2. Then according to the first law of thermodynamics the net energy change must be the same in the two cases even though different amounts of heat have been added to the gas; the

---

[19] This and the following five paragraphs are a digression from the historical account, inserted for readers who have not studied thermodynamics.

energy depends only on the thermodynamic state of the system, not on how it got to that state.

A more dramatic demonstration of this point is sometimes used in introductory physics courses. One has a cylinder of compressed carbon dioxide gas which has been kept at room temperature. When the gas is squirted into the atmosphere, it can be seen to condense into dry-ice crystals. The expansion of the gas has cost it so much energy that it is cooled below its freezing point, though no *heat* has been lost. (We are relying on the fact that heat will only flow spontaneously from a hotter body to a cooler one; see Chapter Two.) Of course the crystals evaporate in a few seconds as they gain heat from the surrounding air.

The last example shows that the phrase "latent heat," still used in modern textbooks, is quite misleading. It is not true that a certain amount of heat must be supplied to change a solid to a liquid or gas. The amount of heat depends on the pressure (or density) at which the transition occurs, and may sometimes be zero. What *is* constant is the amount of energy that must be supplied to go from a state at (P_1, T_1) to another state at (P_2, T_2).

At this point one might ask whether such facts were not already known at the beginning of the 19th century, and if so how could they have been explained by the caloric theory? Heating by compression was certainly known, and as noted earlier Laplace and Poisson understood this phenomenon well enough to base a correct calculation of the speed of sound on it. In the caloric theory one could reasonably assume that the *temperature* depends on the *density* of caloric, that is, the quantity per unit volume. Thus compression would raise the temperature without adding any heat from outside ("adiabatic" compression), and expansion would lower the temperature. One objection to this explanation—though this was apparently not clearly understood until around 1850—is that in the free expansion of a gas into a vacuum there is practically no cooling effect (Gay-Lussac 1807). The change of temperature depends on the presence of an external pressure, for example a piston which the gas must push back. This implies that the heating or cooling effect is linked to the performance of mechanical work.

In addition to suggesting that temperature is not simply determined by the density of a caloric fluid, the result of the free-expansion experiment also implies that no "internal" work is involved in increasing the average distance of molecules in a gas. That would contradict the hypothesis that there are long-range repulsive forces between molecules; if there were such forces, they would accelerate the molecules as they moved apart, and if temperature depended on molecular speed, the expansion would *raise* the temperature. This argument has sometimes been used to show that the Boyle-Newton theory of gas pressure is invalid; it is not a very good refutation, since it depends on an extra postulate (temperature increases with molecular velocity) which adherents of that theory did not usually accept.

After the kinetic theory had been accepted, the free-expansion experiment could be used to draw conclusions about long-range forces between gas molecules. Joule and William Thomson did the experiment more accurately in the 1850s and found that there is usually a small cooling effect; this result can be interpreted as showing that there are weak long-range *attractive* forces between molecules.

Now a few words about words. According to the viewpoint presented in the last few paragraphs the statement that "heat is a form of molecular motion" is not very meaningful. Nevertheless that is how 19th-century scientists generally described the basic postulate of the kinetic theory; the more restricted definition of heat given above was adopted only after the principles of thermodynamics had become familiar. In a historical account it is preferable to employ the terminology that was in general use in the period under discussion rather than translate it into modern terms that may be misleadingly precise. I deviated slightly from this policy in the last section when I used the expression "wave theory of heat" instead of "undulatory theory of heat," but this can be justified on the grounds that English-speaking historians of science always refer now to the "wave theory of light" even though it was called the "undualtory theory of light" in the 19th century. I will continue to use the term "thermodynamics" since it was actually adopted in the 19th century, even though Clausius originally called it "the mechanical theory of heat." Similarly I use "kinetic theory of gases" throughout this chapter, although that phrase does not seem to have emerged until the 1870s. Maxwell's original phrase, "the dynamical theory of gases," is unsuitable even for an historical discussion because the word dynamic (or "dynamism") implied an emphasis on *force* rather than atomic motion in the 19th-century German literature. For the same reason it would be confusing to refer to the *general* idea that heat is molecular motion (in solids and liquids as well as gases) as the "dynamical theory of heat." The best compromise seems to be "*vis viva* theory of heat," with the understanding that this is gradually to be replaced by "kinetic theory of heat" and eventually "statistical mechanics," with increasing levels of precision and mathematical sophistication.

Having adopted these definitions, I will now make the historical assertion that the *vis viva* theory of heat, available since the 17th century

or earlier, had to be given serious consideration as soon as energy conservation and thermodynamics had been introduced in the middle of the
19th century. It was still logically possible to reject it, as did J. R. Mayer
and later Ernst Mach, by denying the need to reduce heat to any other
form of energy. This antireductionist or positivist stance was the basis
for the "energetics" movement at the end of the 19th century, but it was
uncongenial to most scientists.

Having accepted the *vis viva* theory of heat, one still had several possible
hypotheses to choose from. The molecular motion might be translational,
rotational, or vibrational, or a combination of all three; the molecules
might be small relative to the space in which they move, or large and thus
crowded together; the motion might be similar for each molecule in the
system or differ according to a definite pattern. One must not forget that
physicists still believed that an ether is needed to transmit energy between
bodies in the form of light or radiant heat; if the ether also fills the space
between molecules inside a body, it should have some effect on their motion. The old idea that molecules "swim" in the ether, or are suspended
by it at definite equilibrium points around which they may vibrate, was
not yet dead.

Among these possibilities the kinetic theory of gases was perhaps the
simplest but by no means the most plausible. It required a certain amount
of boldness to ignore the ether and assert that molecules move through
space at constant velocity, encountering no resistance except when they
collide with each other or a boundary surface. Waterston, in his 1845
manuscript, worried about the ether problem but set it aside. Herapath
(1847), Joule (1848), and A. K. Krönig (1856) did not mention it, but one
may suspect that they did not feel responsible for giving a complete theory
of matter, consistent with contemporary knowledge of all forms of energy.
Clausius did.

In his first paper on kinetic theory, "The Kind of Motion we call Heat"
(1857), Clausius stated that he had been thinking about molecular motion
even before writing his first article on thermodynamics (1850), but had
abstained from publishing his ideas because he wanted to establish the
empirical laws of heat without making them appear to depend on any
molecular hypothesis. Now that Krönig had taken the lead with his paper
on gases (1856), there was no question of priority, but the time seemed
auspicious to attempt a unified description of several phenomena from
the kinetic viewpoint. Krönig had assumed that the molecules have only
translatory motion (and, as Clausius was perhaps too polite to point out,
had not even given the correct numerical factor in the pressure equation
for that simple case). Clausius concluded that one must also include other
kinds of molecular motion, such as rotation, and showed how one could

estimate the fraction of the total energy which is translational by using specific heat data.[20]

By including rotational motion in his kinetic theory, Clausius was compromising not with alternative theories but with empirical knowledge of gas properties. But the result of this compromise was damaging to the kinetic theory all the same: the ratio of translational energy to total energy came out to be 0.6315 for the common gases whose ratio of specific heats is 1.421. Now 0.6315, as Maxwell and others intuitively realized, is not a very nice number. It is unlikely (though not impossible) that a direct calculation based on a plausible molecular model would lead to such a number. Perhaps the best that can be said for 0.6315 is that, despite the accuracy implied by its four significant figures, it is not too far away from 3/5, and we will see later there is some hope of making sense out of 3/5.

Clausius did take one step in this direction by reviving Avogadro's proposal that gaseous molecules may contain two or more atoms of the same kind. Some chemists had already come to the same conclusion, but Clausius was probably the first to introduce the idea into mid-19th century physics.

Another assumption dictated by experimental data was the extremely small size of molecules: Clausius stipulated that "the space actually filled by the molecules of the gas must be infinitesimal in comparison to the whole space occupied by the gas itself." Moreover, "the influence of the molecular forces must be infinitesimal." This means not only that the forces between molecules at their *average* distances are negligible but also that the short-range repulsive forces that cause molecules to rebound at collisions must act over a very small portion of the path of the molecule. If these conditions were not satisfied the gas would not obey the ideal gas laws. Of course by this time it was well known from Regnault's experiments that real gases do *not* obey the ideal gas laws, but Clausius was

---

[20] The basic principle of his calculation is that the heat needed to raise the temperature at constant pressure can be written as the sum of two terms.

$$c_p \, dT = c_v \, dT + AP \, dV.$$

The first term is the amount of heat needed to raise the temperature at constant volume; the second is the work done in expanding the gas at constant pressure (A = mechanical equivalent of heat). For an ideal gas, $PV = RT$, and so $P \, dV = R \, dT$. If the specific heats are independent of temperature, the equation can be integrated from $T = 0$ to T, and the equation becomes

$$c_p T = c_v T + ART = c_v T + APV.$$

The term $c_v T$ is what Clausius calls the total heat H in the gas; the second term is related to the kinetic energy of translatory motion K, by the equation $K = (3/2)PV$, since $PV = (1/3)Nmv^2$. Hence $K/H = (3/2)(c_p/c_v - 1)$.

unable in 1857 to carry out the complex calculations needed to compute the deviations using a molecular model, so he limited his theory to ideal gases.

While the strict mathematical deductions of his theory were thus limited to gases obeying the laws of Boyle and Gay-Lussac—that is, temperatures and pressures not too far from those of the atmosphere—Clausius did not hesitate to propose a quantitative description of molecular motion in other states of matter:

> In the *solid* state, the motion is such that the molecules move about certain positions of equilibrium without ever forsaking the same, unless acted on by external forces. In solid bodies, therefore, the motion may be characterized as a vibrating one, which may, however, be of a very complicated kind. In the first place, the constituents of a molecule may vibrate among themselves; and secondly, the molecule may vibrate as a whole; again, the latter vibrations may consist in oscillations to and fro of the centre of gravity, as well as in rotatory oscillations about this centre of gravity. In cases where external forces act on the body, as in concussions, the molecules may also be permanently displaced.
>
> In the *liquid* state the molecules have no longer any defnite position of equilibrium. They can turn completely around their centres of gravity; and the latter, too, may be moved completely out of place. The separating action of the motion is not, however, sufficiently strong, in comparison to the mutual attraction between the molecules, to be able to separate the latter entirely. Although a molecule no longer adheres to definite neighbouring molecules, still it does not spontaneously forsake its neighbours, but only under the united actions of forces proceeding from other molecules. . . . In liquids, therefore, an oscillatory, a rotatory, and a translatory motion of the molecules takes place, but in such a manner that these molecules are not thereby separated from each other, but, even in the absence of external forces, remain within a certain volume.
>
> Lastly, in the *gaseous* state the motion of the molecules entirely transports them beyond the spheres of their mutual attraction, causing them to recede in straight lines according to the ordinary laws of motion. If two such molecules come into collision during their motion, they will in general fly asunder again with the same vehemence with which they moved toward each other. . . .

From this qualitative picture Clausius was able to develop a theory of changes of state. Thus, the *evaporation* of a liquid can be explained by assuming that even though the average motion of its molecules may not be sufficient to carry them beyond the range of the attractive forces of their neighbors, "we must assume that the velocities of the several molecules deviate within wide limits on both sides of the average value," and therefore a few molecules will be moving fast enough to escape from a liquid surface even at temperatures below the boiling point.

The phenomena of *latent heat* could also be explained by the kinetic theory, if one adopted Clausius' description of the three states of aggregation:

> In the passage from the solid to the liquid state the molecules do not, indeed, recede beyond the spheres of their mutual action; but, according to the above hypothesis, they pass from a definite and, with respect to the molecular forces, suitable [ordered] position, to other irregular positions, in doing which the forces which tend to retain the molecules in the former position have to be overcome.

Whenever a body is moved against the action of a force, mechanical work must be done, and therefore, according to the law of conservation of energy, heat must be supplied.

> In evaporation, the complete separation which takes place between the several molecules and the remaining mass evidently again necessitates the overcoming of opposing forces

and so heat must again be provided (latent heat of vaporization).

Near the end of his 1857 paper Clausius calculated the average speeds of molecules of oxygen, nitrogen, and hydrogen at the temperature of melting ice and found them to be 461 m/sec, 492 m/sec, and 1,844 m/sec respectively. A Dutch meteorologist, C.H.D. Buys-Ballot, looked at these numbers and realized a consequence that had escaped the notice of Herapath, Joule, Waterston, and Clausius: if the molecules of gases really move that fast, the mixing of gases by diffusion should be much more rapid than we observe it to be. For example, if you release an odorous gas like ammonia or hydrogen sulfide at one end of a room it may take a minute or so before it is noticed at the other end; yet according to the kinetic theory all the molecules should have traversed the length of the room several times by then.

Here was almost the makings of another "Poisson bright spot" situation (cf. §1.8) The hostile critic, Buys-Ballot, thought he had refuted the new theory by pointing out an obvious contradiction between its predictions and the real world. But this time a new experiment could not save the theory. Instead, Clausius had to make an important change in the theory itself. Abandoning his earlier postulate that the gas molecules have infinitesimal size, he now assumed that they have a large enough diameter or "sphere of action" so that a molecule cannot move very far without hitting another one.

Clausius now defined a new parameter: the *mean free path* (L) of a gas molecule, to be computed as the average distance a molecule may travel before interacting with another molecule. He argued that L may be large enough compared with molecular diameters so that the basic concepts of kinetic theory used in deriving the ideal gas law are unimpaired, yet small enough so that a molecule must change its direction many times every second, and may take a fairly long time to escape from a given macroscopic region of space. In this way the *slowness* of ordinary gas diffusion, compared with molecular speeds, could be explained.

The mean free path is inversely proportional to the probability that a molecule will collide with another molecule as it moves through the gas. For spheres of diameter d this probability is proportional to the collision cross section (πd^2) and to the number of molecules per unit volume (N/V).[21] Thus the mean free path is determined by the formula

$$L = kV/Nd^2,$$

where k is a numerical constant of order of magnitude 1 (its precise value was a matter of dispute for some time).

When Clausius introduced the mean free path in 1858 it may have looked like only an ad hoc hypothesis invented to save the theory, since he did not have any independent method for estimating the parameters N and d in the above formula. But before anyone had a chance to criticize it on those grounds, Maxwell incorporated the mean free path into his own kinetic theory and showed that it could be related to gas properties such as viscosity. As a result it soon became a valuable concept, not only for interpreting experimental data, but also for determining the size of molecules and thus justifying its own existence.

1.10 TRANSPORT THEORY AND THE SIZE OF ATOMS

According to one currently popular view, progress in science depends on proposing hypotheses that can be refuted by experimental observations (Popper 1962). Paradoxically, a theory that can explain all the facts is considered not only worse than a theory that is disprovable; it is not even regarded as a proper scientific theory at all. On this view, "refutability" is the criterion for demarcation between science and nonscience.

---

[21] Collision will occur if the *centers* of two molecules come within distance d, so the result is the same as if one shot a point-particle at a circular target of *radius d*.

In May 1859 James Clerk Maxwell, a brilliant young Scottish physicist who had just begun to establish his reputation with a prize essay on the rings of Saturn, wrote a letter to George Gabriel Stokes, the expert on fluid dynamics. In a few succinct sentences he indicated his current interest in the kinetic theory of gases, his deduction of an important prediction from it, and his expectation that experimental data would refute that prediction and thereby finally dispose of the theory:

I saw in the Philosophical Magazine of February, '59, a paper by Clausius on the 'mean length of path of a particle of air or gas between consecutive collisions', on the hypothesis of the elasticity of gas being due to the velocity of its particles and of their paths being rectilinear except when they come into close proximity to each other, which event may be called a collision. . . . I thought it might be worthwhile examining the hypothesis of free particles acting by impact and comparing it with phenomena which seem to depend on this 'mean path'. I have therefore begun at the beginning and drawn up the theory of the motions and collisions of free particles acting only by impact, applying it to internal friction of gases, diffusion of gases, and conduction of heat through a gas (without radiation). Here is the theory of gaseous friction with its results. . . .

. . . I do not know how far such speculations may be found to agree with facts, even if they do not it is well to know that Clausius' (or rather Herapath's) theory is wrong [footnote: i.e., inadequate] and at any rate as I found myself able and willing to deduce the laws of motion of systems of particles acting on each other only by impact, I have done so as an exercise in mechanics. Now do you think there is any so complete a refutation of this theory of gases as would make it absurd to investigate it further so as to found arguments upon measurements of strictly 'molecular' quantities before we know where there be any molecules? One curious result is that μ [coefficient of viscosity or "internal friction"] is independent of the density, for

$$\mu = MNlv = Mv/\sqrt{2}\pi s^2.$$

This is certainly very unexpected, that the friction should be as great in a rare as in a dense gas. The reason is, that in the rare gas the mean path is greater, so that the frictional action extends to greater distances.

Have you the means of refuting this result of the hypothesis? (Brush 1965:26–27)

Stokes apparently replied—his letter has not survived—that experiments on the damping of a pendulum swinging in air showed that gas viscosity does go to zero at low densities, as one would intuitively expect but contrary to the kinetic-theory prediction. So Maxwell, in his first published paper on "Illustrations of the Dynamical Theory of Gases" (1860), pointed out that his theoretical result—that the viscosity of a gas is independent of density and increases with temperature—disagreed with the only observations he knew about.

Maxwell's 1860 paper was nevertheless a major contribution to the theory it seemed to be refuting. One feature of the paper, the introduction of a statistical velocity-distribution law and a more rigorous approach to the partition of energy among different kinds of atomic motion, will be discussed in the next section. Here I want to mention Maxwell's work on "transport properties," the behavior of a gas not in thermal equilibrium but subject to variations of pressure, temperature, or composition.

Maxwell was the first to formulate a general framework in which diffusion, viscosity, and heat conduction can be treated as special cases of a general process in which a quantity such as momentum or energy is transported by molecular motion. Previously only diffusion had been intensively studied as a gas property; viscosity and heat conduction are quantitatively so small in gases that they are easily masked by other effects (turbulence, convection), except under specially designed experimental conditions. Thus the publication of Maxwell's theory of these processes proved to be a powerful stimulus to experimental work by providing a definite prediction to test (and in this sense Popper's emphasis on the refutability of the hypothesis is justified).

Maxwell himself initiated the experimental test of his predictions for the effect of temperature and density on viscosity. He found that the viscosity coefficient of air is indeed constant over a wide range of densities, contrary to the alleged experimental facts mentioned above. Later it was pointed out by O. E. Meyer and others that the analysis of the pendulum data (which Stokes transmitted to Maxwell) had been based on the *assumption*-natural enough before Maxwell's theory was known—that the viscosity coefficient does go smoothly to zero as the density goes to zero. While it is true that its value must be zero *at* zero density (if there is no gas left, it can't exert any viscous resistance to the swing of the pendulum), the drop occurs quite suddenly at a density lower than that reached in most experiments before 1850.

Crudely speaking, the theory refuted the experiment in this case. One might argue that it was not the experiment itself but only its *interpretation* that was refuted (Shimony 1977). But in ordinary scientific discourse the term "experimental fact" is commonly applied to data inferred from experiments with the help of some kind of interpretation or theoretical assumption; the "neutral observation language" or "pointer reading" of the philosopher of science has no counterpart in the real world. As Einstein observed (Heisenberg 1971:63), our theories determine what we can observe; hence experiments can never furnish a completely impartial test of a theory.

The investigation of the temperature dependence of viscosity did not yield such a clear-cut result. According to the original "billiard ball" model (elastic spheres with no forces except at contact) the viscosity coefficient ought to be proportional to the square root of the absolute temperature ($\mu \propto \sqrt{T}$). But Maxwell and others found a stronger temperature variation, $\mu \propto T^x$, where x ranges from about .75 to 1.0.

At about the same time (early 1860s) Maxwell developed a much better formulation of transport theory which avoided the mean-free-path approximation. All the results of the new theory depended on the velocity-distribution function for a gas not in thermal equilibrium—a function that Maxwell was unable to determine—except in the special case of repulsive forces inversely proportional to the 5th power of the distance between two particles. For this case the velocity-distribution function did not have to be known, and Maxwell found that the viscosity coefficient is *directly* proportional to the absolute temperature, $\mu \propto T$. Maxwell's own experimental results agreed with this, and so he concluded by 1866 that the kinetic theory gave completely accurate predictions for gas viscosity.[22]

Fifty years later, when David Enskog and Sydney Chapman worked out formulae for the transport coefficients for many other force laws, the temperature dependence of viscosity could be used to draw conclusions about which law most nearly represents the actual force between atoms. But in the 1860s, when the mere existence of an atomic structure of matter was no more than a plausible hypothesis, Maxwell's theory was used to accomplish a major advance: the first reliable estimate of the *size* of an atom. For this purpose the earlier result relating μ to the particle diameter d was most useful:

$$\mu \propto L \propto V/Nd^2.$$

(L = Clausius' mean free path, N = number of molecules in volume V.) Josef Loschmidt, an Austrain physicist and chemist, pointed out in 1865 that this relation could be used to determine d if one other equation for N and d were known. In particular, he suggested that the volume occupied

---

[22] The viscosity coefficient is still independent of density for the inverse fifth-power force law, as long as the density is not so high or so low that the gas properties do not depend mainly on binary collisions of particles; at very low densities interactions with the surface of the container dominate the flow behavior, whereas at very high densities simultaneous interactions of three or more particles must be taken into account.

by the gas molecules themselves, if they were closely packed, should be approximately the volume of the substance condensed to the liquid state,

$$V_l \approx Nd^3.$$

If the density of a substance is known in both the liquid and gaseous states, the ratio or "condensation coefficient" $V/V_l = V/Nd^3$ could be combined with the mean free path ($L \approx V/Nd^2$) to obtain a value for d.

In this way Loschmidt concluded that the diameter of an "air molecule" is about $d = 10^{-7}$ cm. This value is about four times too large according to modern data, but considerably better than any other estimate available at the time.

The corresponding value of the number of molecules in a cubic centimeter of an ideal gas at standard conditions (0°C, 1 atm pressure) would be

$$N_L \approx 2 \times 10^{18}.$$

Although Loschmidt himself did not give this result explicitly in his 1865 paper, it can easily be deduced from his formula, and so this number is now sometimes called "Loschmidt's number." Its modern value is 2.687×10^{19}. It should not be confused with the related constant, "Avogadro's number," defined as the number of molecules per gram-mole, equal to

$$N_A = N_L/V_0 = 6.02 \times 10^{23},$$

where $V_0 = 22420.7$ cm^3 atm mole^{-1}. Avogadro himself of course did not give any estimate of this number, but only postulated that it should have the same value for all gases (see §1.6).

During the next few years, other scientists (the most influential being William Thomson) made similar estimates of atomic sizes and other parameters with the help of the kinetic theory of gases. As a result, the atom came to be regarded as no longer a merely hypothetical concept but a real physical entity, subject to quantitative measurement, even though it could not be "seen." This was perhaps the most important contribution of the kinetic theory to 19th-century science; yet it was carelessly brushed aside by skeptics like Ernst Mach and Wilhelm Ostwald, who argued at the end of the century that we still have no convincing evidence for the existence of the atom and should therefore banish it from the elite company of established physical theories.[23]

[23] See §2.7 for the later history of this problem.

1.11 The Maxwell-Boltzmann Statistical Approach

In June 1850 the *Edinburgh Review*, a magazine widely read by scientists and other intellectuals, published a long article on Quetelet's recent works on statistics. The author, John Herschel, did not sign his name to the review (following the 19th-century British tradition of anonymous reviewing), but his identity was generally known, especially after he reprinted it in a book of his essays in 1857. Herschel's review of Quetelet may be said to mark the introduction of continental statistical theory into British science.[24] It made a strong impression on James Clerk Maxwell, at that time a 19-year-old student in Edinburgh, soon to enter Cambridge University. Echoing some of Herschel's phrases, Maxwell wrote to his friend Lewis Campbell:

the true logic for this world is the Calculus of Probabilities . . . as human knowledge comes by the senses in such a way that the existence of things external is only inferred from the harmonious (not similar) testimony of the different senses, understanding, acting by the laws of right reason, will assign to different truths (or facts, or testimonies, or what shall I call them) different degrees of probability. . . . (Campbell & Garnett 1969:143)

This is a clear enough statement of the reason for using probability theory, except that it leaves out what is generally thought to be the basic axiom of 19th-century physical scientists: that the phenomena themselves—at least inorganic phenomena—are rigorously determined by the laws of nature together with a complete specification of the state of the system at one instant of time. Each molecule has a definite position and velocity at any given time, and will change to a new position and velocity in a way that depends on its interactions with the other molecules. It is only our lack of knowledge about these details that compels us to use a statistical approach, just as we have to use statistics in setting insurance rates or in treating errors of observation. We will discuss in the next chapter how this presumption of determinism gradually weakened, partly as a result of the development of kinetic theory.

Clausius used probability concepts in his derivation of the mean-free-path formula (1858), but it was Maxwell who converted the kinetic theory of gases into a fully statistical doctrine. The crucial step was his translation of the normal distribution law or law of errors (Fig. 1.11.1), discovered by Adrain and Gauss and extensively applied by Quetelet, into a distribution law for molecular velocities.

[24] Gillispie (1963), Garber (1972), Porter (1981).

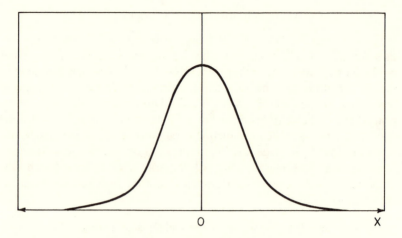

FIG. 1.11.1. Normal statistical distribution.

I will first state the Maxwell distribution law and then explain it. The number of molecules whose speed (scalar magnitude of the velocity vector) lies between v and $v + dv$ is proportional to $v^2 e^{-v^2/\alpha^2}\, dv$, where α is a constant to be determined later. For a given value of α the curve looks like Fig. 1.11.2.

The Maxwell curve, unlike the original law of errors, is unsymmetrical—first because the v^2 factor multiplying the exponential gives greater weight to higher values of v, and second because negative values of v are not allowed (by definition of magnitude of a vector).

On integrating the function $v^2 e^{-v^2/\alpha^2}$ over all values of v from 0 to ∞, we obtain the result[25] $\alpha^3 \sqrt{\pi}/4$; thus if we want the probability distribution $f(v)$ for a single molecule, defined such that its integral is 1, we must take

$$f(v) = (4/\alpha^3\sqrt{\pi})v^2 e^{-v^2/\alpha^2}.$$

The average speed is then found to be[26]

$$\bar{v} = 2\alpha/\sqrt{\pi},$$

and the average kinetic energy is[27]

$$\tfrac{1}{2}mv^2 = (\tfrac{3}{4})m\alpha^2.$$

[25] $\int_0^\infty v^2 e^{-hv^2} = \pi^{1/2}/4h^{3/2}$ is a well-known definite integral.
[26] $\int_0^\infty v^3 e^{-hv^2}\, dv = 1/2h^2$.
[27] $\int_0^\infty v^2 e^{-hv^2}\, dv = (3/8h^2)(\sqrt{\Pi/h})$.

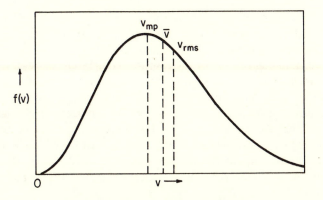

FIG. 1.11.2. Maxwell velocity distribution.

From the last result it can be seen that the parameter $h = 1/\alpha^2$ in Maxwell's distribution law is inversely proportional to the absolute temperature T. In modern notation $\alpha^2 = 2kT/m$, where k is known as "Boltzmann's constant."

The effect of temperature on molecular velocity can now be described more accurately in terms of Maxwell's distribution law. At any temperature there will be molecules of all different speeds in a gas, but as the temperature increases the number of fast molecules increases relative to the number of slow ones (Fig. 1.11.3).

The importance of the Maxwell distribution law is due to the fact that some properties of gases depend not on the *average* speed of molecules but on the number that have exceptionally high speeds. For example, the

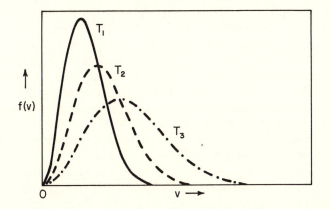

FIG. 1.11.3. Maxwell distribution for three different temperatures.

rate of loss of atmospheric gas from a planet into space depends on how often an individual molecule near the top of the atmosphere acquires an upward velocity that exceeds the escape velocity.

Maxwell's first proof of his distribution law (1860) was not very persuasive to other physicists. Following the reasoning Herschel had used in his 1850 article to derive the law of errors, Maxwell simply asserted that the distribution function must satisfy certain abstract mathematical properties, such as spatial symmetry. These properties could be expressed as functional equations whose solution must have the exponential form. Nothing was said about the properties of the molecules or their collisions, which presumably would have the effect of bringing about the distribution.

In another paper (1867) Maxwell tried to justify his distribution law by showing that if the molecules had once attained such a distribution, collisions would not disturb it. This proof was also unsatisfactory: one of the steps is erroneous, and the result itself is not strictly true except as an average over long periods of time.

The problem was taken up from a different viewpoint by the Austrian physicist Ludwig Boltzmann (1844–1906). Boltzmann's original concern was to derive the second law of thermodynamics from the principles of mechanics. This attempt led him to inquire into the molecular properties of thermodynamic states and to develop a general treatment of thermal equilibrium.

Boltzmann's first major achievement (1868) was to extend Maxwell's distribution law to the case when an external force field such as gravity is present. He found that one can still have thermal equilibrium with constant temperature in a vertical column of gas; the density and pressure vary exponentially with height, that is, with gravitational potential energy. In modern notation, the relative probability that a molecule will be found at a place where the potential energy is V is $e^{-V/kT}$. This is the so-called *Boltzmann factor*. Its widespread use in molecular physics is due to the fact that V can be the potential energy function for *all* forces acting on the molecule, including those of other molecules; thus the Boltzmann factor, combined with Maxwell's velocity-distribution law, gives the probability of any molecular state, in liquids and solids as well as gases. This may be stated briefly as the *basic principle of statistical mechanics* (Maxwell-Boltzmann distribution law): the relative probability of a molecular state with total energy $E = \Sigma(\frac{1}{2}mv^2) + \Sigma V$ is $e^{-E/kT}$.

Boltzmann did not succeed in proving this principle except in some special cases, but his attempts to justify it were extremely illuminating. They fall into two general categories: (1) kinetics of approach to equilibrium; (2) equal a priori probability postulate.

The kinetics argument was first presented in a long paper published in 1872. Boltzmann considered first a uniform gas of identical spherical particles, with no external forces, and computed the effect of collisions on the velocity-distribution function f. In order to do this he assumed that the probability that two colliding molecules have velocities $\vec{v}_1$ and $\vec{v}_2$ is equal to the *product* of the probabilities of these two velocities. This may be written

$$f_2(v_1, v_2) = f_1(v_1)f_1(v_2),$$

where the subscripts on f indicate the number of molecules. In other words there is no statistical correlation between the velocities of two molecules before they collide. Each collision changes the values of f for four particular values of v—decreasing it for the velocities of the two molecules before the collision and increasing it for the values they acquire after the collision. (Notice that in this step of the derivation f is being treated not as a fixed probability but as a function giving the *number* of molecules having each speed, and these numbers will change with each collision.) The probability of each such collision depends on f itself, so the result of the kinetic argument is an integro-differential equation for f which looks something like

$$\frac{\partial f_1}{\partial t} = \int [\, f_1(v_1')f_1(v_2') - f_1(v_1)f_1(v_2)]g(v_1, v_2, v_1', v_2'),$$

where v_1' and v_2' are velocities after the collision. The function $g(\ )$ depends on the nature of the forces between the molecules.

The above equation, with other terms added to take account of gradients in temperature, fluid velocity, and external forces, is known as *Boltzmann's transport equation* or in some circles simply as "the Boltzmann equation." It determines f for a general class of physical situations, not limited to thermal equilibrium.

Boltzmann then showed that collisions always push f toward the equilibrium Maxwell distribution. In particular, the quantity $H = \int f \log f$ always decreases with time unless f is the Maxwell distribution, in which case H maintains a fixed minimum value. This statement is now known as Boltzmann's H-theorem.[28]

---

[28] The letter H was introduced by S. H. Burbury in 1890. It does not seem to stand for anything in particular, though it has occasionally been suggested that it was meant to be a Greek capital eta, following Gibb's use of η for entropy. Boltzmann originally used the letter E and made the connection with entropy quite explicit.

For a gas in thermal equilibrium, Boltzmann's H is proportional to minus the entropy as defined by Clausius (1865). While the entropy in thermodynamics is defined only for equilibrium states, Boltzmann suggested that his H-function could be considered a generalized entropy having a value for any state. Then the H-theorem is equivalent to the statement that the entropy always increases or remains constant, which is one version of the second law of thermodynamics. The justification for Maxwell's distribution law is then based on the assertion of a general tendency for systems to pass irreversibly toward thermal equilibrium.

The second argument, based on a priori probabilities, makes it a little clearer why a system in equilibrium should obey the Maxwell distribution law. It also explains the paradoxical fact that *each* molecule has kinetic energies ranging from zero to infinity, with probabilities independent of the energies of the other molecules, in spite of the fact that the total energy of all the molecules is supposed to be fixed. Boltzmann (1868) considered a gas of N particles with total (kinetic) energy E, and imagined that the energy is divided into J discrete pieces e such that $E = Je$. He then postulated that every microstate of the system, defined by assigning J_1 pieces of energy to particle 1, J_2 to particle 2..., J_N to N, such that $\Sigma J_i = J$, has equal probability. The probability that a given particle has energy $J_i e$ is then equal to the number of microstates for which particle i has this amount of energy (summing over all possible ways of distributing $J - J_i$ pieces to the other $N - 1$ particles) divided by the total number of microstates. A correction factor, proportional to the square root of the energy, must be inserted when averaging over microstates with different energies, to conform to the mathematical structure of mechanical systems in three-dimensional space.

Although the intermediate stages of this calculation involve some rather complicated formulas, the Maxwell velocity distribution comes out at the end when one takes the limit $J \to \infty$, $e \to 0$, $N \to \infty$, keeping E/N fixed. It is worth noting that for a finite number of particles the velocity distribution for a single particle does depend on the velocities of the other particles and has a finite maximum value, but this constraint is removed in the limit $N \to \infty$.

According to the a priori probability argument, the Maxwell distribution is the one most likely to be found in thermal equilibrium because it corresponds to the largest number of microstates. Other distributions can occur, but they have a probability that is relatively insignificant when the number of particles is on the order of 10^{22} (as it is for a macroscopic sample of gas). From this viewpoint it is irrelevant how the system got to thermal equilibrium; no knowledge of collision dynamics is assumed (in contrast to the kinetic argument); one merely assumes that all micro-states are equally likely.

By comparing the two approaches, Boltzmann could conclude that the process of irreversible approach to equilibrium, which is a typical example of an entropy-increasing process, corresponds to a transition from less probable to more probable microstates. Entropy itself can therefore be interpreted as a measure of probability. In particular, let W be the probability of a macrostate, that is, proportional to the number of microstates corresponding to that macrostate. Then the entropy of a macrostate, according to Boltzmann, is proportional to the logarithm[29] of W. In modern notation,

$$S = k \log W.$$

The fact that a macrostate can be assigned a certain "probability" does not necessarily mean that its existence results from a random process. On the contrary the use of probabilities here is perfectly compatible with the assumption that each microstate of a system is rigorously determined by its previous state and the forces acting on it. We need to use probability measures because we must deal with macrostates corresponding to large numbers of possible microstates. Boltzmann might have avoided the connotations of the word probability by using a neutral term such as "weighting factor." On the other hand the fact that the word "probability" *was* used, and the fact that one would get similar results in a more direct way by starting from the assumption that molecular states *are* chosen by a random process, may have had some influence on the subsequent history of atomic physics (see Chapter Two).

Another conclusion reached by Maxwell is that the energy of a system must be equally shared among all its parts, on the average. This is now called the "*equipartition theorem,*" and it includes as a special case Waterston's proposition that in a mixture of particles of different masses, each has the same average kinetic energy. Maxwell insisted that the theorem should also apply to internal motions of molecules such as rotation and vibration. In modern terminology each mechanical "degree of freedom" has the same energy; thus in a system of mass-points each point has three degrees of freedom corresponding to three directions of motion in space. If two mass-points are bound together by some kind of force,

---

[29] The logarithmic relation was chosen because when two independent systems are combined and considered as a single system, the probability of a state of the combined system is the *product* of the separate probabilities. Their entropies will therefore be additive, as is the case in thermodynamics ($\log W_1 W_2 = \log W_1 + \log W_2$).

there are still a total of six degrees of freedom even though they may be described as three degrees of translational motion of the center of mass, two of rotation (about the two axes perpendicular to the line between the points and to each other), and one of vibration. A rigid nonspherical solid nonspherical solid would also have six degrees of freedom—it can rotate about three perpendicular axes.

According to Maxwell, diatomic molecules must be regarded as having at least six degrees of freedom. Since the ratio of specific heats computed from kinetic theory is[30] $c_p/c_v = (n + 2)/n$, where n = number of degrees of freedom per molecule, this ratio should be no greater than 1.33 for diatomic molecules. Yet it was well known that the experimental value of the ratio was about 1.4 for the gases believed to be diatomic. Another way of expressing this discrepancy is Clausius' result (see §1.9) that the ratio of translational to total energy (K/H) is about 0.6; according to the equipartition theorem it should be 0.5 for diatomic molecules.

Here was a clear-cut refutation of a prediction from the kinetic theory. Maxwell concluded in 1860 that the theory could not be correct for real gases, because of the specific-heats discrepancy and the implausible prediction it made about gas viscosity (see §1.10). That did not keep him from doing further work on the theory but it did serve to maintain in his mind a certain degree of skepticism about the kinetic theory. Boltzmann (1876) suggested that a diatomic molecule might be described by a model having only five "effective" degrees of freedom, the sixth (rotation around the line between the atoms) being unaffected by collisions and therefore not contributing to the specific heat. Maxwell (1877) would not accept that solution, and the specific-heats problem remained a mystery until after the introduction of quantum theory.

One other way in which the Maxwell-Boltzmann distribution law, and thus the equipartition theorem, might be justified, was through the "ergodic hypothesis."[31] This hypothesis (Boltzmann 1871, Maxwell 1879) asserts that the a priori probability assumption can be *derived* from the kinetic argument: a mechanical system will eventually pass through *all* microstates before returning to any microstate a second time. If that is true, then the average value of any property of the system, taken over a sufficiently long time, will be equal to the average value taken over all microstates. Since for a large system the overwhelming majority of micro-

[30] If there are n degrees of freedom per molecule of which three correspond to translational kinetic energy of the center of mass, then the ratio K/H (see footnote 20) will be $3/n = (3/2)(c_p/c_v - 1)$, which is equivalent to the equation given in the text.

[31] This term was given its present meaning by P. T. Ehrenfest (1911). The word "ergodic" was first used by Boltzmann, but in a different sense.

states belong to the "equilibrium" microstate, the result will be the same as if one used the Maxwell-Boltzmann distribution law.

Maxwell and Boltzmann never tried to prove that the ergodic hypothesis is correct; they used it only as a heuristic justification for the equipartition theorem. In the 1890s Lord Kelvin and others suggested that this theorem may not be valid for certain kinds of mechanical models, especially those invented to describe interactions between atoms and ether. In fact if the ether itself is regarded as a mechanical system to which the equipartition theorem applies, it would seem that the specific heat of any gas should be very large or infinite because energy must be supplied to keep the numerous modes of ether vibration in thermal equilibrium with atomic motions.

This difficulty in accounting for the apparent failure of some degrees of freedom to take up their proper share of energy was eventually resolved by the quantum theory. In modern textbook accounts of the subject this is usually presented as a cause-and-effect relationship: the quantum theory is supposed to have been invented as a direct response to the equipartition problem. The problem has been dramatically characterized as the "ultraviolet catastrophe": if one tries to express the total energy of ether vibrations as an integral over all possible frequencies, giving each mode of vibration the same average energy, the integrand diverges at the high frequency end.[32] Max Planck's distribution law for radiation supplies an exponential factor that drastically reduces the share of energy in the high-frequency modes and makes the integral converge. Thus the anomaly resulting from applying Newtonian mechanics to the ether (that is, to black-body radiation) is resolved by the quantum hypothesis.

If the conventional account of the origin of Planck's quantum theory, sketched in the previous paragraph, were true, it would imply an important role for the equipartition problem in the development of modern physics. But in fact no one in 1900 really believed that Newtonian mechanics entailed the ultraviolet catastrophe—certainly not Lord Rayleigh, who is supposed to have discovered the anomaly, or Planck, who resolved it. It was only *after* quantum theory had been proposed that the "crisis" of the ultraviolet catastrophe was invented to show why Newtonian mechanics was faulty.

We have here an excellent example of the distinction between what philosophers of science call the "logic of discovery" and the "logic of justification." The logic of discovery is the process (not always logical) by which a scientist actually arrived at his discovery—in Planck's case,

---

[32] The number of vibration modes at frequency v is proportional to v^2 for an elastic solid; thus the total energy would be given by an integral of the form $\int_0^\infty v^2 \, dv$.

analysis of black-body radiation by thermodynamics and electromagnetic theory, with some semiempirical curve-fitting in the final stage (see below, §3.1). The logic of justification is the reasoning used to establish the validity of the discovery afterward; arguments that convinced the original discoverer may not be persuasive to other scientists, and new evidence may have been accumulated to strengthen the case. But confusion arises when, as sometimes happens, the two logics are combined in telling the history of the discovery—there is a natural tendency to assume that the discovery was made for the reasons that *now* seem to provide its best support. (Thus we preserve the myth that the replacement of the Ptolemaic by the Copernican system of astronomy was due to the desire to avoid a proliferation of "epicycles on epicycles"; historical analysis shows that there was no such "crisis" at the time of Copernicus, and that his own system (1543) achieved no significant reduction in the number of circular motions needed to represent planetary motions. Whatever may have been his reasons for preferring a heliocentric system, they could not have included most of the ones we now use to justify it.)

1.12 The Van der Waals Equation and the Critical Point

Most of the quantitative results of the kinetic theory mentioned so far are valid only for gases that are approximately "ideal"—that is, the density is low enough so that the size of each molecule is small compared with the space between molecules. In this case the laws of Boyle and Gay-Lussac, usually combined in the "ideal gas equation of state" $PV = RT$, will be valid. Moreover, in deriving the Maxwell-Boltzmann transport equations and the formula for transport coefficients, one needs to include only binary collisions.

Daniel Bernoulli (1738) did realize that modifications to the ideal gas law would be expected if the ratio of particle diameter to average interparticle distance, d/D, is not vanishingly small (§1.5). But it was not until 1873 that the Dutch physicist Johannes Diderik van der Waals gave the correct first approximation to the theoretical correction involved, and showed how, by including as well an estimate of the effects of long-range attractive forces, one could account for some of the peculiar properties of gases at high densities.

Van der Waals argued that when the molecules occupy a finite space, the volume V in the equation of state should be replaced by $V - b$, where b depends on the molecular diameter d. But, contrary to what one might think, b is not simply the total molecular volume $N(\pi d^3/6)$, but, for low densities, should be taken as four times that volume: $b = (2\pi d^3/3)N$. (For the justification of the factor of 4, which involves a detailed analysis of collisions, the reader is referred to any elementary text on kinetic theory.)

Interatomic attractive forces had previously been postulated by Laplace and others to explain capillarity and surface tension in liquids, but before 1850 it was assumed that such forces were effective primarily at short distances; it was thought that in gases there must be long-range *repulsive* forces to account for the resistance of the gas to the external pressure (§1.4). In the van der Waals theory, the attractive forces were supposed to dominate at large distances between atoms, whereas repulsive forces were effective only at distances where the atoms are nearly in contact. Van der Waals did not specify a particular form for this force law, but argued that the net effect of the attraction would be to increase the external pressure in the equation of state by an amount inversely proportional to the square of the volume, a/V^2.[33] (Thus one can imagine that an atom leaving the surface of the system will be pushed back by the pressure exerted on the outside and pulled back by the attractive forces of other atoms inside the system.)

Combining these two corrections, van der Waals arrived at the equation of state

$$[P + (a/v^2)](v - b) = RT,$$

where a and b are constants characteristic of each gas.

The most interesting feature of the van der Waals equation is not its ability to fit small deviations from the ideal gas law, but the fact that it gives a remarkably good qualitative description of the behavior of the system near the gas-liquid critical point. The critical point had been discovered by 1822 by a French scientist, Cagniard de la Tour. (See also Schmidt 1823.) He found that when liquid alcohol was sealed in a glass tube with its own vapor (no air present) and heated, the liquid-gas meniscus eventually disappears. Cagniard de la Tour and others showed that for every substance there is a particular pressure, volume, and temperature, called the critical point, at which the distinction between liquid and gas vanishes. (For water, $P_c = 218$ atm; $V_c = 3.2$ cc/g; $T_c = 374°$C.) It was initially assumed that the liquid simply changes to a gas above the critical point, but in 1863 the Irish physical chemist Thomas Andrews demonstrated that the supercritical substance can be changed continuously into either gas or liquid by appropriate variations of temperature and pressure.

---

[33] Since the pressure correction is thus assumed to vary as the square of the number of particles in a unit volume, it can be attributed to the sum of contributions from all pairs of particles, regardless of distance. This implies a rather peculiar force law; for example it might be the limit of the function $F(r) = -c$ for $r \leq R$, as $R \to \infty$ while $c \to 0$, keeping the product Rc constant.

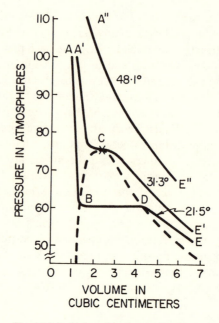

FIG. 1.12.1. Andrews' isotherms for CO_2.

Andrews' isotherms for carbon dioxide are shown in Fig. 1.12.1. At low temperatures, a gas (E) when compressed will condense (D) and change its volume discontinuously to a much smaller value (B), characteristic of the liquid state. Further compression of the liquid requires much higher pressure (A). At the critical temperature (about 31°C) the volume change (BD) becomes zero and the isotherm has only a point of inflection at the critical point C. At higher temperatures the isotherms approach the hyperbolic shape of Boyle's law (PV = constant). One can go continuously from liquid to gas or conversely by following a path above C in the PV plane.

In 1871 James Thomson (brother of William Thomson) suggested that there might be a continuous transition between gas and liquid even at temperatures below the critical temperature (Fig. 1.12.2), along an S-shaped path (bcd), although most of this path would not correspond to stable states of the system. (For example, the state just to the left of D might be like a supercooled gas.)

Van der Waals' theory, which was developed in part specifically to account for the critical-point phenomena described by Andrews, confirmed James Thomson's hypothesis. The van der Waals equation of state is a cubic equation in V; it yields three real roots for temperatures below the critical temperature ($T_c = 8a/27bR$), corresponding to the points B, c,

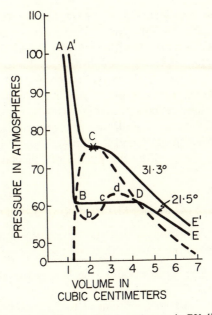

F$_{IG}$. 1.12.2. James Thomson's S-shaped curve in PV diagram.

and D. As the temperature increases to T_c, the three roots merge into one, and above T_c two of them are imaginary. Thermodynamic analysis shows that the dotted part of the isotherm, bcd, is indeed unstable, so the system must jump discontinuously from B to D; Maxwell showed that the line BD must be drawn in such a way that it cuts off equal areas above and below the dotted curve.

As the first successful explanation of phase transitions, van der Waals' theory demonstrated the fertility of the atomistic approach and stimulated much research on liquid-gas critical phenomena in the last quarter of the 19th century, especially in Holland. It was a major breakthrough to show that the same atomic model could be used to explain two different states of matter, for some scientists had previously attributed different properties to "gas molecules" and "liquid molecules." Even though the van der Waals theory was eventually replaced by more sophisticated theories of the critical point in the 20th century, it played an important role by demonstrating that qualitative changes on the macroscopic level, such as changes from the liquid to the gaseous state, might be explained by quantitative changes on the microscopic level.

Unfortunately the substantial progress achieved by van der Waals and his Dutch colleagues was not given proper recognition by scientists in other countries until much later. Maxwell, Clausius, and Boltzmann were well aware of the significance of this pioneering work and attempted to build on it. But the positivist critics of "mechanistic science" in the 1880s and 1890s ignored it when they claimed that the kinetic-theory research program was no longer giving enough new results to keep the atomistic hypothesis alive. Although those critics represented only a vociferous minority of scientists at the time, they have misled a few modern philosophers of science into thinking that the kinetic program was "degenerating" at the end of the 19th century because of the equipartition problems mentioned in §1.11 and a supposed failure to generate fruitful new research (for the most recent version of this claim see Clark 1976, Gardner 1979).

1.13 The "Statistical Mechanics" of J. Willard Gibbs

In order to explain and predict the physical properties of bulk matter from the properties of atoms, three things are needed:

(1) a theory of the structure and especially the interactions of individual atoms;
(2) fundamental laws or "equations of motion" determining the states and changes of state of atoms in terms of their structure and interactions;
(3) a statistical technique for estimating properties of systems of large numbers of atoms.

At the beginning of the 20th century only some rather crude models were available for (1), such as billiard balls (elastic spheres), point-centers of attractive or repulsive force, or some combination thereof. Yet, despite considerable progress as a result of quantum theory, leading to much more elaborate theories of atomic structure and interactions, the simple 19th-century models continued to play an important role in 20th-century theories of the properties of matter, since they were the most convenient ones to employ in actual calculations (3) and "quantum effects" could often be separated out.[34]

By fundamental laws (2) I mean either Newtonian mechanics or quantum mechanics; in some cases relativity theory needs to be used (for example when dealing with systems at extremely high temperatures and/or densities).

The statistical technique (3) was actually in rather good shape, thanks to the work of Maxwell and Boltzmann in the last part of the 19th century; all that was needed was to put it into a mathematical form that was independent of (1) and (2). This task was accomplished by the American

---

[34] But see the interesting discussion by Lieb (1979) on the need to invoke quantum effects in order to explain the stability of matter, if all forces are electrical.

theoretical physicist Josiah Willard Gibbs (1839–1903), who also gave (3) its modern name, "statistical mechanics."

Gibbs was the son of a professor of sacred literature (specializing in philology) at Yale; he seems to have inherited his scientific talent from his mother's side of the family (Wheeler 1951:6–7). He received a Ph.D. in engineering from Yale in 1863, served as a tutor there for three years, then studied physics and mathematics for a year each at Paris, Berlin, and Heidelberg. He was appointed professor of mathematical physics at Yale in 1871, but without any salary (he had an independent income) until 1880, when Johns Hopkins tried to lure him away. Yale then decided to pay him (though only two thirds of what Hopkins had offered), so he stayed there the rest of his life.

Gibbs was almost completely unknown in the United States during his lifetime, but was recognized in the European scientific world for his series of papers on thermodynamics (1873–78). Although these were published in an obscure journal (*Transactions of the Connecticut Academy of Arts and Sciences*), he sent reprints to all the major physical scientists in Europe. Maxwell was so impressed that he made plaster casts of Gibbs' thermodynamic surfaces, which he sent to Gibbs and others, and recommended that his colleagues read Gibbs to straighten out their confusions. Wilhelm Ostwald, the German pioneer of physical chemistry, published a translation of Gibbs' major papers in his series "Classics of Exact Sciences" in 1892. His work had considerable influence, especially in Holland.

Gibbs' most famous single discovery is the "phase rule," the theorem stating that

$$f = n - r + 2,$$

where f is the number of thermodynamical degrees of freedom in a heterogeneous system in thermodynamic equilibrium, n the number of components, and r the number of phases. Thus for a one-component system, if there is only one phase, one can vary the temperature and pressure independently ($f = 2$), but if there are two phases (such as gas and liquid in equilibrium) there is only one independent variable. This means that there is a vapor-pressure curve, the pressure being determined by the temperature. If there are three phases (solid, liquid, gas) there are no degrees of freedom, so both pressure and temperature have unique values ("triple point").[35]

---

[35] In addition to the many applications of the phase rule in physics, chemistry, and engineering, one should note its application to history by the American historian Henry Adams in 1909. Taking human thought as the substance, he discussed transitions from a religious to a mechanical phase circa 1600, then to an electrical phase in 1900; successive phases occupied times equal to the square roots of previous ones.

During the 1880s Gibbs worked on Maxwell's electromagnetic theory; the main outcome was his system of vector analysis, which eventually displaced the Hamilton quaternion system advocated by P. G. Tait.

Gibbs then turned to the kinetic theory of gases, which Boltzmann was already beginning to transform into a more abstract mathematical theory by the use of what Gibbs call "ensembles." An ensemble is a collection of systems, similar in some respects and different in others, the differences being usually on a microscopic level only and not accessible to observations. For example, there is the "microcanonical ensemble" (introduced by Boltzmann in 1884 under the name *Ergode*) which consists of systems having the same (or nearly the same) value of the total energy, though differing in the positions and velocities of the individual particles.[36]

In his definitive treatise *Elementary Principles in Statistical Mechanics* (1902), Gibbs presents the microcanonical ensemble as one of a series of ensembles. The next generalization is the "canonical ensemble," which consists of many systems each containing the same N particles but with all possible values of the total energy ϵ; the number of systems with energy E is proportional to $e^{-\epsilon/\theta}$, where θ is called the "modulus" of the distribution (subsequently to be equated to kT, where k = Boltzmann's constant). Gibbs introduces this ensemble as being "the most simple case conceivable, since it has the property that when the system consists of parts with separate energies, the laws of the distribution in phase of the separate parts are of the same nature—a property which enormously simplifies the discussion, and is the foundation of extremely important relations to thermodynamics." He defines the "coefficient of probability" as

$$P = e^{(\psi - \epsilon)/\theta},$$

where θ and ψ are constants determined by the condition that the integral of P over all phases must be equal to 1. He also calls $\eta = \log P$ the "index of probability."

By comparing the properties of this ensemble with the equations of thermodynamics, Gibbs shows that η corresponds to the entropy and θ corresponds to the temperature. One may calculate the various quantities depending on the configuration by using the coefficient of probability for configurations, and in many cases these average values can be expressed as derivatives of ψ with respect to certain parameters

Gibbs then considers the "grand ensemble" in which different systems may have different numbers of particles. In order to define such an en-

---

[36] The term "ergodic" was misapplied by Paul Ehrenfest to a *single* system that passes through all sets of positions and velocities corresponding to the same total energy (Brush 1976a, chap. 10).

semble he has to answer the question, "If two phases differ only in that certain entirely similar particles have changed places with one another, are they to be regarded as identical or different phases? If the particles are regarded as indistinguishable, it seems in accordance with the spirit of the statistical method to regard the phases as identical. In fact, it might be urged that in such an ensemble of systems as we are considering no identity is possible between the particles of different systems except that of qualities, and if v particles of one system are described as entirely similar to one another and to v of another system, nothing remains on which to base the identification of any particular particle of the first system with any particular particle of the second. And this would be true, if the ensemble of systems had a simultaneous objective existence. But it hardly applies to the creations of the imagination. . . ." He defines *generic* phases as phases that are not altered by the exchange of identical particles and *specific* phases as phases that are altered by such an exchange. If v_1, $v_2, \ldots, v_n$ are the numbers of the different kinds of molecules in a system, then one generic phase will contain $v_1! v_2! \ldots v_h!$ specific phases. It is postulated that the number of systems in a certain region of phase space is proportional to

$$N e^{(\Omega + \mu_1 v_1 + \cdots + \mu_h v_h - \epsilon)/\theta},$$

where the constant Ω is determined by the condition that the total number of systems in the ensemble is equal to N. The coefficients μ may be identified with the constants that appear in thermodynamic expressions for the equilibrium of multicomponent systems; they were called "potentials" by Gibbs (now "chemical potentials"). The entropy may be defined with respect either to generic or specific phases. At the end of his book, Gibbs gives the following argument for choosing the former:

> . . . let us suppose that we have two identical fluid masses in contiguous chambers. The entropy of the whole is equal to the sum of the entropies of the parts, and double that of one part. Suppose a valve is now opened, making a communication between the chambers. We do not regard this as making any change in the entropy, although the masses of gas or liquid diffuse into one another, and although the same process of diffusion would increase the entropy, if the masses of fluid were different. It is evident, therefore, that it is equilibrium with respect to generic phases, and not with respect to specific, with which we have to do in the evaluation of entropy. . . .

This is known as the "Gibbs paradox" in statistical mechanics, and it is frequently stated that it could be resolved only by the introduction of quantum mechanics. Gibbs did not seem to regard it as a paradox at all, but simply as a convenient illustration of the reasons for choosing one of two alternatives, both of which were otherwise equally plausible as long as one did not consider the possibility of variations in the number of particles.

One of the two "clouds over the dynamical theory of heat and light" at the end of the 19th century, according to Lord Kelvin (1901), was the discrepancy between experimental specific heats and the predictions of kinetic theory based on the equipartition theorem. This led to discussions of whether the equipartition theorem—that is, the assumption that all degrees of freedom have the same average energy in thermal equilibrium—could be derived from pure (nonstatistical) mechanics, perhaps by the use of the "ergodic hypothesis," as it was later called. This hypothesis, suggested in a rather tentative and ambiguous way by Maxwell and Boltzmann, stated that a single system left to itself will eventually pass through all possible configurations (sets of position and velocity variables) consistent with the given total energy. If it passes through them in a time that is short compared with the time it takes to make observations, it would then be legitimate to use the microcanonical ensemble—to replace averages over the path of a single system by averages over the states of many systems.

As a result of a detailed critical discussion by Paul and Tatiana Ehrenfest in 1911, the question was raised whether the ergodic hypothesis is even mathematically possible for any dynamical system, and answered in the negative by Artur Rosenthal and Michel Plancherel in 1913 (see Brush 1971). Subsequently it was suggested that a weaker assumption, the "quasi-ergodic hypothesis"—that the system passes "as close as you like" to each configuration—might be sufficient to prove the equipartition theorem and to justify the use of statistical averaging. The discussion of this question generated a new branch of mathematics, ergodic theory, but turned out to have practically no influence on the development of physics in the 20th century. The reason, I think, was that the introduction of quantum mechanics made it seem legitimate, and more convenient, to start by *postulating* a statistical distribution rather than to try to derive it from a deterministic mechanical theory. For this purpose Gibbs' formulation of statistical mechanics turned out to be most useful.

Gibbs himself avoided the specific-heat problem by abstaining from detailed calculations for special molecular models. In the preface of his *Elementary Principles* he gives his rationale, which needs to be quoted at somewhat greater length than usual because one or two sentences taken out of context have sometimes been used to give the impression that Gibbs was an anti-atomist in the camp of energetics or positivism.

We may therefore confidently believe that nothing will more con-
duce to the clear apprehension of the relation of thermodynamics to
rational mechanics, and to the interpretation of observed phenomena
with reference to their evidence respecting the molecular constitution
of bodies, than the study of the fundamental notions and principles of
that department of mechanics to which thermodynamics is especially
related.

Moreover, we avoid the gravest difficulties when, giving up the
attempt to frame hypotheses concerning the constitution of material
bodies, we pursue statistical inquiries as a branch of rational mechanics. In the present state of science, it seems hardly possible to
frame a dynamic theory of molecular action which shall embrace the
phenomena of thermodynamics, of radiation, and of the electrical
manifestations which accompany the union of atoms. Yet any theory
is obviously inadequate which does not take account of all these
phenomena. Even if we confine our attention to the phenomena dis-
tinctively thermodynamic, we do not escape difficulties in as simple a
matter as the number of degrees of freedom of a diatomic gas. It is
well known that while theory would assign to the gas six degrees of
freedom per molecule, in our experiments on specific heat we cannot
account for more than five. Certainly, one is building on an insecure
foundation, who rests his work on hypotheses concerning the constitution of matter.

Difficulties of this kind have deterred the author from attempting
to explain the mysteries of nature, and have forced him to be con-
tented with the more modest aim of deducing some of the more obvious propositions relating to the statistical branch of mechanics. . . .

According to Gibbs' statistical mechanics, all the thermodynamic properties of the system can be determined from a single function, known as
the "phase integral" or integral over states,

$$Z = e^{-\psi/\theta} = \int e^{-\epsilon/\theta}$$

(the letter Z has come to be adopted from the German literature, where Z
stands for *Zustand*, state). For example, the average energy $\bar{\epsilon}$ of a system,
defined as the average of the energy of each state ϵ multiplied by its
probability P and integrated over all states,

$$\bar{\epsilon} = \int \epsilon P,$$

would be computed, using the Boltzmann factor $e^{-\epsilon/\theta}$ for the relative probability of a state, as

$$\bar{\epsilon} = \int \epsilon e^{-\epsilon/\theta} \bigg/ \int e^{-\epsilon/\theta} = \partial \ln z / \partial(-1/\theta),$$

since $P = e^{(\psi - \epsilon)/\theta} = Z^{-1} e^{-\epsilon/\theta}$.

This approach was made familiar to physicists in the early decades of the 20th century by Einstein, Planck, L. S. Ornstein, and Paul Ehrenfest. Thus in 1913 Ehrenfest showed how one could estimate the rotational contribution to the specific heat of a diatomic gas, replacing the integral by a sum over discrete terms representing the contributions of quantized states, and the calculation was further refined by E. Holm (1913) and Planck (1915). A mathematical study by H. Poincaré (1912) showed how, using the Fourier transform, one could invert the relation and prove that (under certain reasonable assumptions) the Planck distribution law for black-body radiation implies discrete energy states (McCormmach 1967). Planck's interest was stimulated by Poincaré's approach, and his first explicit use of the term "Zustandssumme" is found (as far as I know) in his 1921 paper on "Henri Poincaré und die Quantentheorie." His first use of the letter Z for this quantity seems to be in a 1924 paper which begins by introducing "die sogennante Zustandssumme $Z = \Sigma e^{-E/kT}$." We return to this topic in Chapter Four.

(Additional references for this section: Barker 1976, Daub 1976, Haas 1936.)

II. IRREVERSIBILITY AND INDETERMINISM

2.1 Introduction

I have left for a separate chapter one of the most significant aspects of the development of the kinetic theory of gases in the 19th century: the statistical explanation of irreversibility and its role in the growing acceptance of indeterminism in modern physics. The topic calls for a somewhat broader perspective than our earlier discussion of the properties of gases and heat and the motions of atoms.[1]

Historians of science generally recognize two major revolutions since the Renaissance. The First Scientific Revolution, dominated by the physical astronomy of Copernicus, Kepler, Galileo, and Newton, established the concept of a "clockwork universe" or "world-machine" in which all changes are cyclic and all motions are in principle determined by causal laws. The Second Scientific Revolution, associated with the theories of Darwin, Maxwell, Planck, Einstein, Heisenberg, and Schrödinger, substituted a world of process and chance whose ultimate philosophical meaning still remains obscure. How did the transition from fixed cycles to random processes occur?

The American physicist Kenneth Ford noted a few years ago that it is historically inaccurate to suppose that indeterminism was suddenly introduced into physical science as a result of the adoption of quantum mechanics:

> The idea that the fundamental processes of nature are governed by laws of probability should have hit the world of science like a bombshell. Oddly enough, it did not. It seems to have infiltrated science gradually over the first quarter of this century. Only after the quantum theory had become fully developed in about 1926 did physicists and philosophers sit up and take note of the fact that a revolution had occurred in our interpretation of natural laws. (Ford 1968:730)

But the roots of indeterminism can be traced back to a much earlier period without losing the thread of historical continuity. In 1932 the British astrophysicist Sir Arthur Eddington, surveying the "decline of determinism," suggested that the first inkling of the existence of

---

[1] For further details and references see Brush (1976b).

"secondary laws" having statistical rather than absolute validity had come from the observation that heat flows from hot to cold: the reverse is not impossible, only extremely improbable. Eddington was well aware that the theory of heat conduction, as well as the second law of thermodynamics, embodied a principle of irreversibility or directionality of time, and that the attempt to explain irreversibility by the atomic-kinetic theory of matter in the 19th century had made physicists familiar with probabilistic modes of thought, thereby helping to create a favorable climate for the introduction of quantum indeterminism in the 1920s. But later writers on the history of modern physics have been so impressed by the remarkable discoveries in atomic and nuclear physics at the beginning of this century that they tend to discount the radical significance of developments before 1900. The image of the complacent Victorian physicist, so certain that all the fundamental laws of nature had been discovered that he thought there was nothing left to do except determine the physical constants a little bit more accurately, has become a cliché. While a few 19th-century physicists did suggest that the foundations of their science were securely established, others were deeply skeptical of the adequacy of the worldview based on deterministic Newtonian mechanics, even if they had no satisfactory alternative to offer.

In reading the casual remarks, and even what seem to be carefully considered pronouncements by 19th-century scientists, the 20th-century reader often finds a puzzling ambiguity. Words like randomness, irregularity, and indeterminism may appear to imply that molecules or other entities do not move in paths that are completely determined by the positions, velocities, and forces of all particles in the system; yet the same author may express himself in a manner that is completely consistent with the view that these paths really are determined but we simply cannot obtain complete *knowledge* of them. What we might now call a crucial distinction between ontological and epistemological indeterminism is frequently blurred in these writings. This ambiguity can be used to argue that almost every one of the scientists quoted here believed only in epistemological, not ontological randomness—but such an argument would conceal (I claim) a gradual but extremely significant historical shift in the meaning of concepts. As Yehuda Elkana, an Israeli historian of science, has shown in the case of the German term *Kraft* (usually translated "force"), certain concepts may be in a state of flux during the time when a new theory is being created, and one misses something by trying to force a precise meaning on words like "statistical" or "probabilistic," which may acquire such meanings only after the appropriate theory has been developed. Indeed, this chapter might be considered the account of an episode in the history of the changing meaning of the word "statistical."

2.2 THE COOLING OF THE EARTH

According to one widely held viewpoint in the history of ideas, 18th-century thought was dominated by the concept of the "Newtonian world-machine." God had been assigned the role of master clockmaker who designed a universe so perfect that it could run indefinitely without any need for divine tinkering. Closer inspection of Newton's own writings shows that he was actually quite firmly opposed to this concept, which had been popularized by earlier writers such as Robert Boyle. Newton's objections were both physical and moral: he pointed to the existence of irreversible processes tending to dissipate motion and gravitational perturbations that seemed to threaten the stability of the solar system, and he warned that restricting God to the creation and design of the world while denying His continual supervision was a step on the road to atheism (Alexander 1956:13–14, 177–80). Nevertheless, in those areas of physical science where Newton's mathematical principles of natural philosophy could be applied with decisive results—planetary and lunar astronomy, fluid dynamics, elasticity, electrostatics, etc.—18th-century theorists scored numerous successes and developed increasing confidence in the mechanistic worldview.

During the same period a new type of problem began to attract considerable scientific attention: the internal heat of the earth and its connection with geological history. Attempts at quantitative treatment go back to the French scientists Dortous de Mairan (1719) and Buffon (1774), who argued that the present temperature near the surface is the result of a process of cooling from an initial hot molten state. Buffon, from a series of laboratory experiments with hot iron spheres, concluded that the time required for the earth to reach its present temperature was about 75,000 years. The astronomer J. S. Bailly (1777) suggested that progressive cooling is a property of all objects in the solar system, but each is at a different stage: Jupiter is still too hot to allow life to exist, while the moon is already too cold; but everything must eventually reach a state of inanimate cold.

Such speculations came into conflict with the "uniformitarian" view of the earth's history developed by British geologists—Hutton, Playfair, and later Lyell. While admitting that the inside of the earth is probably very hot, they did not agree that the internal heat was more intense in earlier epochs, since that would have meant allowing causes whose magnitude was substantially greater than those now observable to sneak back into geological theory. Instead, they proposed that the internal heat has remained constant, providing power for recurrent upheavals to counteract the effects of erosion; this cyclic scheme has been called the "Huttonian earth-machine" (Davies 1969). It was also necessary, in order to ensure that there had been no significant long-term change in physical conditions on the earth's surface, to invoke the conclusion of the mathematical astronomers that the earth's average distance from the sun could only oscillate between certain minimum and maximum values.

The critics of uniformitarian geology—such forgotten men as Richard Kirwan, John Hunter, John Murray, and George Greenough—protested that internal heat must, like all other heat, flow from hot to cold and therefore could not be assumed to remain constant unless it was being continually replenished somehow. But these qualitative objections carried little weight until after J.B.J. Fourier developed his mathematical theory of heat conduction at the beginning of the 19th century. By his own account (1890, 2:114) it was the problem of terrestrial temperatures that led him to develop his theory, and one of its first applications was the derivation of a simple formula relating the time required for a homogeneous sphere to cool from a specified initial temperature, the present temperature gradient at the surface, and the conductivity and heat capacity of the substance. The temperature gradient (rate of change of the average temperature as one goes down into the earth) was a frequently measured quantity at this time, since it could be used to find the central temperature by extrapolation, and Fourier could give rough estimates of the other parameters in his equation. The result was such a long period of time for the cooling of the earth (about 200 million years) that Fourier did not even bother to write it down explicitly. He concluded (1819) that the rate of heat flow through the surface was so slow that it could have had no significant effect on the temperature during the epochs of importance to geologists at this time. Thus the initial effect of Fourier's work was to confirm the uniformitarian view; it would cause difficulty only when the geologists later expanded their own time scale.

Quite apart from its geological application Fourier's theory of heat conduction marked a major turning point in the history of physics, for two distinct reasons. First, it established a new methodology for the formulation and solution of problems, based on partial differential equations. Although such equations had been developed extensively by 18th-century mathematicians, Fourier's exploitation of trigonometric series expansions (though at first without adequate mathematical justification) greatly enhanced the power and scope of this mode of analysis. During the past 150 years physicists have followed Fourier by expressing their theories of electricity, magnetism, optics, gravity, gases, and continua in terms of differential equations; the trigonometric series expansion has grown from a mathematical convenience to a new language—spectrum or harmonic analysis—which is especially suited to the description of physical processes in terms of frequencies. (The availability of this mathematical technique also reinforced the tendency to describe processes as forms of periodic, wavelike motion.) Fourier's influence can be seen in both of the forms of quantum mechanics which were developed in 1925–26: Schrödinger's wave mechanics, based on a partial differential equation for the "wave function," and Heisenberg's matrix mechanics, based on the manipulation of "Fourier coefficients" for transitions involving radiation of various frequencies.

The second notable feature of Fourier's heat conduction theory is that it is explicitly based on a postulate of irreversibility (1807:33):

> When heat is unequally distributed among the different points [parts] of a solid body, it tends to come to equilibrium and pass successively from hotter to colder parts. At the same time the heat dissipates itself at the surface and loses itself in the surroundings or the vacuum. This tendency toward a uniform distribution, and this spontaneous cooling which takes place at the surface of the body, are the two causes which change at every instant the temperature of the different points.

Irreversibility is expressed mathematically by the fact that the time variable t appears in a first-derivative term. If we followed the system "backward in time" by reversing the sign of the time variable (replacing t by $-t$), we would see heat flow from cold to hot, a process which is not allowed by the theory. Newton's second law, by contrast, contains the time variable in a second derivative,

$$F = ma = d^2x/dt^2,$$

so the "time-reversal" transformation, $t \rightarrow -t$, leaves the right-hand side unchanged. Thus if we see an object falling toward the earth with velocity v, we will observe that it has an acceleration a toward the earth due to a gravitational force F. An observer whose time direction is reversed (for example by running a movie film of this process backward) will see the object moving *away* from the earth—the velocity $v = dx/dx$ has been *reversed* by the transformation $dt \rightarrow -dt$—a process which *is* allowed by the theory; but he will still observe the same acceleration *toward* the earth, because $d^2x/dt^2 = d^2x/d(-t)^2$.

For this reason Newton's laws are said to be *time-reversible*. Strictly speaking this is true only if one deals only with *forces* that are themselves time-reversible. (If we had introduced air resistance, a force acting always in a direction opposite to the velocity v, into the problem in the previous

paragraph, the motion would no longer be time-reversible.) It is sometimes asked why Fourier and other scientists in the first half of the
19th century did not discuss the apparent contradiction between reversible Newtonian mechanics and irreversible heat conduction; but in fact
there is no contradiction at the phenomenological level, since Newtonian
mechanics had already been successfully applied to problems involving
dissipative forces (such as air resistance or friction), which are not time-
reversible. The contradiction arises only when one assumes that all forces
at the atomic level must be reversible.

2.3 THE SECOND LAW OF THERMODYNAMICS

A few years after Fourier presented his theory of heat conduction to the
Académie des Sciences, Sadi Carnot (1824) published his essay on the
motive power of fire and the efficiency of steam engines. Carnot's essay
is the source of the second law of thermodynamics, established in its
modern form by Rudolf Clausius in 1850, and thus is generally regarded
as the origin of the principle of irreversibility in physics. Perhaps Carnot
has been treated a little too generously in this respect by later writers,
overcompensating for his neglect by the scientific community in his own
lifetime. Irreversibility is mentioned rather casually in Carnot's memoir,
not as a new scientific law but as a common-sense basis for the pre-
scription that contact between bodies of different temperatures should
be avoided in operating steam engines; such contact produces a flow of
heat which could otherwise have been utilized in expanding the steam
and performing mechanical work, hence entails a loss of motive power.
The well-known tendency of heat to flow from hot to cold is now identified as the cause of inefficiency in steam engines, and is the reason
why real engines can never attain their theoretical maximum efficiency
according to Carnot's theory.

In the geological context there was nothing regrettable about the flow
of heat from hot to cold; if the earth had not cooled it would never have
become hospitable to life. The unfavorable consequences of irreversibility
were still in the distant future, when the continued cooling of the earth
and sun would presumably freeze us to death. But when Carnot pointed
out that the production of useful work in steam engines depends on the
controlled use of heat flow, it became evident that all *uncontrolled* heat
flow must be, in human terms, a *waste* or *dissipation* of potential motive
power. Though Carnot in 1824 had not yet decided that heat is actually
transformed into mechanical work in the steam engine—he still considered heat or "caloric" to be a conserved substance—he did succeed in
attaching negative connotations to its natural behavior.

William Thomson (Lord Kelvin) combined the theory of terrestrial refrigeration and the thermodynamic analysis of steam engines into a grand statement of a new law of nature: a "universal tendency toward dissipation of energy." This was in 1852, after Thomson had written several papers on Fourier's theory of heat conduction, and had used it to extrapolate the earlier thermal history of a sphere with specified present temperature distribution and uniform conductivity. His statement included the assertion that the earth had been too hot in the past for human habitation, and would be too cold in the future. Thomson had also studied the properties of steam, proposed an absolute temperature scale based on Carnot's theory, and accepted the interconvertibility of heat and mechanical work just a few months after Rudolf Clausius used this interconvertibility as the basis for the first law of thermodynamics (1850).

The second law of thermodynamics was formulated in various ways by Clausius, Thomson, and later writers. For some reason the negative versions seem to be favored by physicists: "it is impossible to obtain mechanical work by cooling a substance below the temperature of its surroundings," or "it is impossible to cause heat to flow from a low temperature to a high temperature without some equivalent compensation." It is fairly obvious that the basis for excluding such processes is that they are incompatible with the assumed natural tendency of heat to flow from hot to cold.

Clausius, who talked about "equivalent compensations," was forced to introduce a quantitative measure of the "equivalence value of a transformation" in order to be able to compare heat flows with heat conversions. His measure, proposed in 1854, was the function.

$$dQ/T,$$

where dQ is the differential heat absorbed by a substance at [absolute] temperature T. Note that if the same quantity of heat dQ is emitted by a substance at temperature T_1 and absorbed by another substance at a *lower* temperature T_2, the net change in equivalence value $(-dQ/T_1 + dQ/T_2)$ is necessarily positive. Thus the statement "heat always flows from hot to cold unless another compensating process occurs" becomes "the equivalence value of an uncompensated transformation is always positive."

The function dQ/T turned out to be so useful in thermodynamic analysis that Clausius (1865) decided to give it a short, snappy name; and thus "entropy" was baptized eleven years after its birth. Since dQ/T was defined as the *change* in entropy (dS), the new statement of the second law was: entropy always increases (or remains constant in the special case of reversible processes). But the question "What *is* entropy?" remained unanswered within the macroscopic science of thermodynamics.

2.4 MAXWELL AND HIS DEMON

During those eleven years when the concept of entropy was percolating in the mind of Clausius, the concept of disordered molecular motion was being revived and systematically developed by James Clerk Maxwell (§1.11). It is generally recognized that statistical ideas and methods were first introduced into physics in connection with the kinetic theory of gases, in a somewhat limited way by Clausius but more explicitly by Maxwell. At the same time, modern writers invariably point out that in this case statistics was used only as a matter of convenience in dealing with large numbers of particles whose precise positions and velocities at any instant are unknown, or, even if known, could not be used in practice to calculate the gross behavior of the gas. Nevertheless, it is claimed, 19th-century physicists always assumed that a gas is really a deterministic mechanical system, so that if the superintelligence imagined by Laplace were supplied with complete information about all the individual atoms at one time he could compute their positions and motions at any other time as well as the macroscopic properties of the gas. This situation is to be sharply distinguished, according to the usual accounts of the history of modern physics, from the postulate of atomic randomness or indeterminism which was adopted only in the 1920s in connection with the development of quantum mechanics. Thus, part of the "scientific revolution" that occurred in the early 20th century is supposed to have been a discontinuous change from classical determinism to quantum indeterminism.

There is no doubt that some 19th-century thinkers did see determinism as the essence of science. Thus James Buchanan, writing on religion in 1857, asserted:

> It has been undeniably the effect of scientific inquiry to banish the idea of Chance, at least from as much of the domain as has been successfully explored, and to afford a strong presumption that the same result would follow were our researches extended beyond the limits within which they are yet confined. (1857:307)

Similarly W. Stanley Jevons, a philosopher of science, wrote in 1877:

> We may safely accept as a satisfactory scientific hypothesis the doctrine so grandly put forth by Laplace, who asserted that a perfect knowledge of the universe, as it existed at any given moment, would

give a perfect knowledge of what was to happen thenceforth and for ever after. Scientific inference is impossible, unless we may regard the present as the outcome of what is past, and the cause of what is to come. To the view of perfect intelligence nothing is uncertain. (1958:738–39)

Hence, as Laplace himself had remarked in 1783 (Gillispie 1972:10), there is really no such thing as "chance" in nature, regarded as a *cause* of events; it is merely an expression of our own ignorance, and *"probability belongs wholly to the mind"* (Jevons 1958:198).

But was this view really held by scientists themselves? I think that by the time Jevons wrote the words quoted above, support for absolute determinism was already beginning to collapse. In arguing for some degree of continuity between the 19th and 20th centuries, I do not want to overstate the case; if it were possible to quantify the causal factors responsible for the ultimate effect, I would guess that 20th-century events (including the discovery of radioactive decay, though it actually occurred just before 1900) accounted for perhaps 80% of the impetus toward atomic randomness, while the 19th-century background accounted for the remaining 20%. That 20% is still significant in view of the fact that most historians and scientists seem to give no weight at all to the role of the well-publicized debates on the statistical interpretation of thermodynamics during the 1890s, or to the well-established use of probability methods in kinetic theory.

The claim that 19th-century kinetic theory was based on molecular determinism must rely heavily on the evidence of the writings of James Clerk Maxwell and Ludwig Boltzmann; though in the absence of any explicit statements one might legitimately infer that they tacitly accepted the view of their contemporaries. In fact the situation is a little more complicated: their words were ambiguous but their equations pushed physical theory very definitely in the direction of indeterminism. As in other transformations of physical science—the cases of Kepler, Fresnel, Planck, and Heisenberg might be adduced here—mathematical calculation led to results that forced the acceptance of qualitatively different concepts.

Maxwell's earliest work in kinetic theory, in particular his introduction of the velocity-distribution law, seems to derive from the tradition of general probability theory and social statistics (as developed by Quetelet and interpreted by John Herschel) rather than from the mechanistic analysis of molecular motions (§1.11). Maxwell's law asserts that each component of the velocity of each molecule is a random variable which is statistically independent of every other component of the same and every other molecule. Only in his later papers did Maxwell attempt to justify the law by relating it to molecular collisions, and even then he needed to assume that the velocities of two colliding molecules are statistically independent. On the other hand, the computation of gas properties such as viscosity and thermal conductivity, whose comparison with experimental data provided the essential confirmation of the theory, did involve the precise dynamical analysis of collisions of particles with specified velocities, positions, and force laws. Without determinism in this part of the theory Maxwell could not have achieved his most striking successes in relating macroscopic properties to molecular parameters.

The debate on the statistical interpretation of irreversibility started in 1867, just after the publication of Maxwell's major work on kinetic theory. It was around the same time that Clausius, in a public lecture in Frankfurt, pointed out the cosmic consequences of his new version of the Second Law of Thermodynamics (1865), "The entropy of the universe tends to a maximum":

> The more the universe approaches this limiting condition in which the entropy is a maximum, the more do the occasions of further changes diminish; and supposing this condition to be at last comletely obtained, no further change could evermore take place, and the universe would be in a state of unchanging death. (1868:419)

In Scotland, Maxwell was responding to his friend P. G. Tait's request for assistance in drafting a textbook exposition (1871) of thermodynamics, by imagining a tiny gatekeeper who could produce violations of the second law. Maxwell's immortal Demon proved that Victorian whimsy could relieve some of the gloom of the Germanic Heat Death. The Demon is stationed at a frictionless sliding door between two chambers, one containing a hot gas, the other a cold one. According to Maxwell's distribution law, the molecules of the hot gas will have higher speeds on the average than those in the cold gas (assuming each has the same chemical constitution), but a few molecules in the hot gas will move more slowly than the average for the cold gas, while a few in the cold gas will travel faster than the average for the hot gas. The Demon identifies these exceptional molecules as they approach the door and lets them pass through to the other side, while blocking all others. In this way he gradually *increases* the average speed of molecules in the hot gas and *decreases* that in the cold, thereby in effect causing heat to flow from cold to hot (see also Knott 1911:213–15).

In addition to reversing the irreversible, Maxwell's Demon gives us a new model for the fundamental irreversible process: he translates heat flow into molecular mixing. The ordinary phenomenon, heat passing from a hot body to a cold one, is now seen to be equivalent (though not always identical) to the transition from a partly ordered state (most fast molecules in one place, most slow molecules in another place) to a less ordered state. The concept of *molecular order and disorder* now seems to be associated with heat flow and entropy, though Maxwell himself didn't make the connection explicit.

Maxwell's conclusion was that the validity of the second law is not absolute but depends on the nonexistence of a Demon who can sort out molecules; hence it is a *statistical* law appropriate only to macroscopic phenomena. That conclusion would have been less significant if the second law had originally been deduced, like other statistical laws, from analysis of large numbers of elementary events. But, having already been accepted as a member of the category of general "laws of nature," like the first law of thermodynamics, Newton's laws of motion, etc., the second law of thermodynamics now became a rotten apple in the barrel, though this danger was not apparent until much later. Only in the 1920s did the suspicion arise that, for example, the *first* law of thermodynamics, "like the second," might also have only statistical rather than absolute validity.

To call the second law a "statistical law" does not of course imply *logically* that it is based on random events—to the contrary. If Maxwell's Demon could not predict the future behavior of the molecules from the observations he makes as they approach the door, he could not do his job effectively. And in some of the later discussions it appeared that a relaxation of strict molecular determinism would make complete irreversibility *more* rather than less likely. (This was indeed the effect of the Burbury-Boltzmann "molecular disorder" hypothesis mentioned below.) Nevertheless at a more superficial level of discourse the characterization "statistical" conveyed the impression that an element of randomness or disorder is somehow involved.

Maxwell himself did not consistently maintain the assumption of determinism at the molecular level, though he occasionally supported that position, for example, in his lecture on "Molecules" at the British Association meeting in 1873. Yet in the same year, in private discussions and correspondence, he began to repudiate determinism as a philosophical doctrine. A detailed exposition of his views may be found in a paper titled "Does the progress of physical science tend to give any advantage to the opinion of necessity (or Determinism) over that of the contingency of events and the Freedom of the Will?" presented to an informal group at Cambridge University. The answer was *no*—based on arguments such as the existence of singular points in the trajectory of dynamical systems, where an infinitesimal force can produce a finite effect. The conclusion was that "the promotion of natural knowledge may tend to remove that prejudice in favor of determinism which seems to arise from assuming that the physical science of the future is a mere magnified image of that of the past" (Campbell & Garnett 1969:434).

By 1875 Maxwell was asserting that molecular motion is "perfectly irregular; that is to say, that the direction and magnitude of the velocity of a molecule at a given time cannot be expressed as depending on the present position of the molecule and the time" (1965:436). His article "Atom" in the *Britannica*, also published in 1875, indicated that this irregularity must be present in order for the system to behave irreversibly.

2.5 BOLTZMANN AND THE
STATISTICAL INTERPRETATION OF THERMODYNAMICS

Ludwig Boltzmann, whose publications and students transmitted the methods and results of kinetic theory from the 19th century to the 20th, proposed a quantitative measure of irreversibility in 1872. In what later became known as the *H*-theorem he generalized Clausius' entropy to a function defined in terms of the velocity distribution, and showed that collisions between uncorrelated molecules would push this function toward the value corresponding to the equilibrium Maxwell distribution.

When Boltzmann's Viennese colleague Josef Loschmidt pointed out (1876) that according to Newton's laws one should be able to return to any initial state by merely reversing the molecular velocities,[2] Boltzmann argued (1877a) that entropy is really a measure of the probability of a state, defined macroscopically. While each *microscopic* state (specified by giving all molecular positions and velocities) has equal probability by assumption, macroscopic states corresponding to "thermal equilibrium" are really collections of large numbers of microscopic states and thus have high probability, whereas macroscopic states that deviate significantly from equilibrium consist of only a few microscopic states and have very low probability. In a typical irreversible process the system passes from a nonequilibrium state (for example high temperature in one place, low in another) to an equilibrium state (uniform temperature); that is, from less probable (lower entropy) to more probable (higher entropy). To reverse this process it is not sufficient to start with an equilibrium state and reverse the velocities, for that will almost certainly lead only to

[2] One sometimes gets the impression that Loschmidt was a critic of the kinetic theory because of his role in this debate. But as we saw in §1.10, he was also responsible for one of the first major triumphs of kinetic theory, its application to a reliable estimate of the diameter of an atom. It should also be noted that William Thomson was the first, in 1874, to point out the "reversibility paradox" usually attributed to Loschmidt.

another equilibrium state; one must pick one of the handful of very special microscopic states (out of the immense number corresponding to macroscopic equilibrium) which has evolved from a nonequilibrium state, and reverse its velocities. Thus it is possible that entropy may decrease, but extremely improbable.

The distinction between macro- and microstates is crucial in Boltzmann's theory, though he failed to explain it as clearly as one might wish. Like Maxwell's Demon, an observer who could deal directly with microstates would not perceive irreversibility as an invariable property of natural phenomena. It is only when we decide to group together certain microstates and call them, collectively, "disordered" or "equilibrium" macrostates that we can talk about going from "less probable" to "more probable" states. This is an irreversible process in the same sense that shuffling the deck after dealing a grand-slam hand in bridge is an irreversible process; the rules of the game single out certain distributions of cards as "ordered" (all the same suit or all aces, kings and queens in the same hand), and we call these "rare" distributions although in fact *each* of the possible distributions of 52 cards among four hands of 13 each has exactly the same probability.

If you play bridge long enough you will eventually get that grand-slam hand, not once but several times. The same is true with mechanical systems governed by Newton's laws, as Henri Poincaré showed with his *recurrence theorem* in 1889: if the system has fixed total energy and is restricted to a finite volume, it will eventually return as closely as you like to any given initial set of molecular positions and velocities. If the entropy is determined by these variables, then it must also return to its original value, so if it increases during one period of time it must decrease during another. This apparent contradiction between the behavior of a deterministic mechanical system of particles and the second law of thermodynamics was used by Ernst Zermelo in 1896 to attack the mechanistic worldview. His position was that the second law is an absolute truth, so any theory that leads to predictions inconsistent with it must be false. This refutation would apply not only to the kinetic theory of gases but to any theory based on the assumption that matter is composed of particles moving in accordance with the laws of mechanics.

Boltzmann had previously denied the possibility of such recurrences and might have continued to deny their certainty by rejecting the determinism postulated in the Poincaré-Zermelo argument. (That was in fact Planck's strategy in 1908.) Instead, he admitted quite frankly that recurrences are completely consistent with the statistical viewpoint, as the card-game analogy suggests; they are *fluctuations* which are almost certain to occur if you wait long enough. So determinism leads to the same (qualitative) consequence that would be expected from a random sequence of molecular states! In either case the recurrence time is so inconceivably long that our failure to observe it cannot constitute an objection to the theory.

While Boltzmann sidestepped the issue of determinism in the debate on the recurrence paradox, maintaining a somewhat ambiguous "statistical" viewpoint, he had to face the issue more squarely in another debate that came to a head at almost the same time. E. P. Culverwell in Dublin had raised, in 1890, what might be called the "reversibility objection to the H-theorem," not to be confused with the "reversibility paradox" discussed by William Thomson, Loschmidt, and Boltzmann in the 1870s. Culverwell asked how the H-theorem could possibly be valid as long as it was based on the assumption that molecular motions and collisions are themselves reversible, and suggested that irreversibility might enter at the molecular level, perhaps as a result of interactions with the ether.

The ether was always available as a source and sink for properties of matter and energy that didn't quite fit into the framework of Newtonian physics, although some physicists were by this time quite suspicious of the tendency of their colleagues to resolve theoretical difficulties this way.

Culverwell's objection was discussed at meetings of the British Association and in the columns of *Nature* during the next few years. It was S. H. Burbury in London who pointed out, in 1894, that the proof of the H-theorem depends on the Maxwell-Boltzmann assumption that colliding molecules are uncorrelated. While this would seem a plausible assumption to make before the collision, one might suppose that the collision itself introduces a correlation between the molecules that have just collided, so that the assumption would not be valid for later collisions. Burbury thought that the assumption might be justified by invoking some kind of "disturbance from without [the system], coming at haphazard" (Burbury 1894:78).

Boltzmann, who participated in the British discussions of the H-theorem, accepted Burbury's conclusion that an additional assumption was needed, and called it the hypothesis of "molecular disorder." He thought it could be justified by assuming that the mean free path in a gas is large compared with the mean distance of two neighboring molecules, so that a given molecule would rarely encounter again a specific molecule with which it had collided, and thus become correlated (1964:40–41).

"Molecular disorder" is not merely the hypothesis that states of individual molecules occur completely at random; rather it amounts to an exclusion of special ordered states of the gas which would lead to violations of the second law. In fact such ordered states *would* be generated by a random process, as Boltzmann noted in his discussion of the recurrence paradox. In modern terminology, one makes a distinction between "random numbers" and "numbers generated by a random process"—in preparing a table of random numbers for use in statistical studies, one rejects certain subsets, for example pages on which the frequencies of digits depart too greatly from 10%, because they are inconveniently nonrandom products of a random process. Boltzmann recognized that the hypothesis of molecular disorder was needed to derive irreversibility, yet at the same time he admitted that the hypothesis itself may not always be valid in real gases, especially at high densities, and that recurrence may actually occur (1964:41–42).

2.6 MAX PLANCK ON INDETERMINISM

Max Planck was one of those scientists who, in the 1880s, believed that the principle of irreversibility must be regarded as an absolute law of nature, even at the cost of abandoning the atomic theory, and his student Ernst Zermelo carried this viewpoint into the 1896–97 debate on the recurrence paradox. Yet it was Planck who, in 1900 and afterward, insisted that Boltzmann's statistical interpretation of entropy is the key to the understanding of the quantum theory of radiation, and by his popular lectures promoted the thesis that the hypothesis of molecular disorder is necessary to the explanation of irreversibility. Planck's reversal of his earlier position furnishes a singular exception to his own oft-quoted dictum that new scientific ideas win out not by rational persuasion of the opponents but only by waiting for them to die out (1950:33–34).

In the 1890s Planck developed a "theory of irreversible radiation processes" in which he attempted to show that electromagnetic theory could provide an explanation of irreversibility where mechanics had failed. This would have been a victory for the electromagnetic worldview, which some physicists at that time expected to replace the mechanical worldview. But two circumstances turned the thrust of his efforts in an unexpected direction. The first was the task of editing Gustav Kirchhoff's lectures on heat, which happened to include an exposition of the kinetic theory of gases, and thus forced Planck (always a meticulous scholar) to become thoroughly familiar with the mathematical apparatus of kinetic theory even though he had little faith in its physical validity. The second was Boltzmann's critique (1897–98) of the technical details of Planck's derivation of irreversibility from radiation theory. The result: Planck conceded that his theory really required an assumption of elementary disorder ("natural" radiation) very similar to that which Burbury and

Boltzmann had used in kinetic theory; and by 1899 he was formulating an "H-theorem" for radiation which was mathematically analogous to the original one for gases.

Even without an atomistic or statistical interpretation, entropy is a crucial quantity that links equilibrium and nonequilibrium (time-dependent) properties of physical systems. Before 1900 only a handful of scientists—Willard Gibbs, Boltzmann, Planck—realized this. For Planck, the study of the entropy-energy relation for a radiating system was the key to obtaining a theoretical formula for the observable frequency distribution of black-body radiation, currently the object of great interest among his experimental colleagues in Berlin. Planck probably could have found the correct distribution by using only the analytical methods of macroscopic thermodynamics; his debt to Boltzmann is evident only in the step from that distribution to the quantum hypothesis. Not only did Planck adopt Boltzmann's relation between entropy and probability; he could also find in one of Boltzmann's early papers (1877b) an explicit calculation of the statistical properties of an assembly of particles with quantized energies. All of this reinforced Planck's belief in the validity of the statistical interpretation of entropy.

Since in his later years Planck became known as a supporter of causality and determinism, it is important to remember that he was an influential spokesman for indeterminism in the period of transition from classical to quantum mechanics. In 1898, introducing his hypothesis of "natural radiation" in order to derive irreversibility, he remarked that the disorder or indeterminacy [*Unbestimmtheit*] of the radiation components "lies in the nature of the subject"—not perhaps in nature itself but at least in any rational theory we can construct about nature. The following year, in a lecture to a scientific congress at Munich, he referred again to the indeterminacy associated with the fact that we cannot specify precisely the frequency of a beam of radiation but must assume that the energy is distributed irregularly over a small range of frequencies (1900c).

In 1900 Planck proposed a new frequency-distribution law for black-body radiation, which led directly to the introduction of the quantum hypothesis. It is frequently stated that this innovation was motivated by a crisis of classical physics: the "ultraviolet catastrophe" predicted by a native application of statistical equilibrium theory to the modes of motion of the ether. If a Victorian physicist had believed that all possible vibration frequencies of the ether must share equally in the total energy, then he would have concluded that all energy would be sucked into vibrations of indefinitely high frequency, contrary to observation as well as to common sense. As it happened, no respectable Victorian physicist did believe that, and certainly Max Planck did not. His step toward a quantum theory of radiation was not a rejection of 19th-century statistical physics but a more passionate embrace of that much-maligned beauty. For it was precisely Boltzmann's formula relating entropy to probability, $S = k \log W$, which Planck used to develop his quantum theory; and it was Planck who first wrote the formula in its modern form with a precise numerical value for "Boltzmann's constant," k.

The new physics was most definitely a legitimate offspring of the old, though adolescence brought defiant claims of independence and denials that anything useful could be inherited from a supposedly stiff and stuffy parent. Leaders of the younger generation, like Paul Ehrenfest, insisted that there was an unbridgeable gap between classical and quantum principles; whoever accepted the latter must abandon the former, though the calculational methods developed in the 19th century would still be of immense utility. (This was especially true of Willard Gibbs' statistical mechanics, which was found to be well suited to quantum calculations.) Planck, representing the older generation, was reluctant to admit that quantum theory entailed any real break with classical physics, especially the wave theory of light. This attitude led him to assert, in 1932, that the worldview of quantum physics is just as rigidly deterministic as that of classical physics.

Perhaps Planck was really a determinist at heart throughout his long career; it is not my purpose to convict him of inconsistency in his innermost beliefs. I claim only that by his scientific accomplishments and by his popular lectures and essays he helped to undermine the commitment of physics to determinism and thus helped to popularize the efficacy of probabilistic thinking. At Leiden in 1908 and again at Columbia University (New York) in 1909, he reminded his audience that the principle of irreversibility could be justified only by postulating microscopic disorder. A more extended statement of his views was given in a lecture at Berlin in 1914:

> Like the social sciences, physics has learnt to appreciate the great importance of a method completely different from the purely causal, and has applied it since the middle of the last century with continually increasing success. This is the statistical method, and the newest advances in theoretical physics have been bound up with its development. . . . [While this method might seem] unsympathetic to the scientific needs of many workers, who desire principally an elucidation of causal relations, yet it has become absolutely indispensable in practical physics. A renunciation of it would involve the abandonment of the most important of the more recent advances of physical

science. It must also be borne in mind that physics, in the exact
sense, does not deal with quantities that are absolutely determined;
for every number obtained by physical measurements is liable to a
certain possible error. (1960:57–58)

Yet this same lecture can be cited as evidence that on another level
Planck did not after all want to abandon determinism; having made the
distinction between dynamical and statistical laws, he rejected the possibility that *all* laws are statistical, and insisted that some laws must be
taken as dynamical (that is, exact) in order to provide some foundation
for deducing the statistical ones. For, after all, "the assumption of absolute determinism is a necessary basis for every scientific investigation"
(1960:67–68). (For further discussion of Planck's views see Goldberg
1976.)

2.7 THE NEW CENTURY: BROWNIAN MOVEMENT

At the first international physics congress, held at Paris in 1900, the
director of the Puy-de-Dôme Observatory stated that "The most important question, perhaps, of contemporary scientific philosophy is that of
the compatibility or incompatibility of thermodynamics and mechanism."
By "thermodynamics" Bernard Brunhes meant the irreversibility of the
second law, and by "mechanism" he intended to suggest the position that
all natural phenomena can be explained by the application of Newton's
laws. Some physicists, like H. A. Lorentz, believed that compatibility
had already been established by the postulate of molecular disorder.
But the debate continued with new arguments and examples. Max Born
(1882–1970) recalled in 1949 that when he first began to read the scientific literature there was a violent discussion raging about statistical
methods in physics and the validity of the *H*-theorem. According to Born,
it was Paul Ehrenfest and his wife Tatiana who finally cleared up the
matter beyond any doubt. The Ehrenfests published in 1911 an extensive
critical review of the foundations of the statistical approach in mechanics;
while it had little to say about quantum theory, this article was widely
read and quoted during the following decades and seems to have been regarded as the definitive treatment of the subject. In addition to discussing
the need for the hypothesis of molecular chaos and similar assumptions
in the kinetic theory of gases, the Ehrenfests remarked that "the last few
years have seen a sudden and wide dissemination of Boltzmann's ideas"
in connection with studies of electrons, colloids, and radiation (1959:68).
They concluded at one point that the "postulate of determinism" seems

to hold only for "visible" states; as soon as one includes a microscope among the instruments available to the observer, one encounters phenomena such as Brownian movement for which determinacy no longer holds (1959:36–37).

Brownian movement, the irregular motion of microscopic particles suspended in fluids, had been observed long before the botanist Robert Brown established its general character in 1828. Brown's achievement was to show that the motion could not be attributed to any supposed vitality of the particles themselves, since all kinds of inorganic as well as organic substances behaved similarly. While Brownian movement was often said to be an effect of the thermal molecular motion of the surrounding fluid, no quantitative theory of this effect was successful until the work of Einstein in 1905. Before this, as Henri Poincaré reminded his audience at the 1900 Paris congress, Léon Gouy (1888) had argued that the continual renewal of the particle's motion by contact with a medium at constant temperature could be interpreted as a violation of the second law of thermodynamics. Einstein, who had done extensive work in statistical physics before 1905, recognized that the motion of small particles in fluids might provide an illustration of the need for the atomistic approach to thermodynamics, a case in which *fluctuations* produce an observable and calculable effect. Whereas Boltzmann had argued against Zermelo that fluctuations of sufficient magnitude to return the system to an initial ordered state would be so rare as to escape human observation, Einstein's theory indicated that the combination of many random molecular impulses—which produce no net effect *on the average*—could account for the entire motion of the Brownian particles. Jean Perrin's verification of Einstein's quantitative predictions provided not only a reconfirmation of the validity of the classical atomic-statistical theory of matter, but a dramatic new proof of the existence of the atom itself.[3] From about 1910 on, Brownian movement could be cited as visible evidence of atomic randomness. Later it was to be cited for a contrary purpose, as an example of an *apparently* random phenomenon that could be explained by reduction to the deterministic motion of invisible entities, thus suggesting the possibility of escaping from the indeterminism of quantum mechanics by postulating causal "hidden variables." The paradox, like many others in this field, is rooted in the ambiguous use of the word "statistical" mentioned above.

[3] Thus Perrin may be credited with "proving" that atoms exist, in the sense that he persuaded the scientific community to accept their existence; the first person to "see" an atom was Erwin Mueller, in 1955, with the field-ion microscope which he perfected.

2.8 EINSTEIN'S THEORIES AND RANDOMNESS

Albert Einstein, even more than Planck, is notorious for his refusal to accept indeterminism as a necessary consequence of quantum mechanics: "God does not play dice" (Born 1971:82, 91, 149, 155, 158, 188, 199; Einstein 1949:86–87). His historical role in the fall of determinism is even more remarkable. He published three important papers in 1905; each of them could later be seen as an attack on determinism, though such was certainly not Einstein's intention. I have already mentioned the paper on Brownian movement; the others deal with the special theory of relativity and with the photon theory of light (often referred to as the "paper on the photoelectric effect").

Modern physicists and philosophers usually assert that relativity theory, unlike quantum mechanics, retains the causal determinism of Newtonian mechanics. While this assertion may be logically correct at present, it is historically misleading. Before about 1930, the two theories were often cited together as challenges to the entire structure of classical physics, and since relativity was the first to attain a definitive formulation it was available as a model in the development of quantum theory. The best example of this influence of relativity is to be found in Heisenberg's original paper on the indeterminacy principle, where he compares the impossibility of talking about the simultaneity of distant events with the impossibility of talking about the precise position and momentum of a particle (1927:179).

Einstein's photon theory of light has been specifically credited by Max Born as the inspiration for his own statistical interpretation of the wave function. In 1926, on first introducing this interpretation, Born wrote:

> I adhere to an observation of Einstein on the relationship of wave field and light quanta; he said, for example, that the waves are present only to show the corpuscular light quanta the way, and he spoke in this sense of a "ghost field." This determines the probability that a light quantum, the bearer of energy and momentum, takes a certain path . . . . (Ludwig 1968:207)

In his Nobel lecture, Born was even more generous: he said that Einstein

> had tried to make the duality of particles—light quanta or photons—and waves comprehensible by interpreting the square of the optical wave amplitudes as probability density for the occurrence of photons. This concept could be carried over at once to the ψ-function . . . . (Born 1964:262)

It is hard to find a concrete basis for these remarks in Einstein's published writings; the 1905 paper seems to me to indicate only that the statistical methods of classical gas theory are very useful in dealing with the behavior of systems of numerous entities whose precise nature (particle or wave) is not known. In further papers on quantum theory during the next ten years, Einstein showed the utility of the statistical approach without committing himself to any assumptions about fundamental randomness. The high point of his work in this period was a derivation (1916) of Planck's distribution law for black-body radiation from the assumption of spontaneous as well as stimulated emission of radiation by matter. Other physicists acclaimed this derivation as the final emancipation of quantum theory from any reliance on classical physics. But for Einstein himself,

> The weakness of the theory lies on the one hand in the fact that it does not get us any closer to making the connection with wave theory; on the other, that it leaves the duration and direction of the elementary processes to 'chance.' Nevertheless I am fully confident that the approach chosen is a reliable one. (van der Waerden 1967:7)

For Max Born, this derivation was the decisive step toward indeterminism, the step which "made it transparently clear that the classical concept of intensity of radiation must be replaced by the statistical concept of transition probability" (1964:258).

2.9 WHAT'S GOING ON INSIDE THE ATOM?

Another major discovery at the beginning of the 20th century seemed to offer direct evidence of indeterminism in nature. As Einstein noted in his 1916 paper on radiation, "the statistical law which we have assumed corresponds to that of radioactive reaction" (van der Waerden 1967:66), and indeed by that time radioactive decay had become the best-known example of a random process. In 1902 Rutherford and Soddy wrote that "The idea of the chemical atom in certain cases spontaneously breaking up with evolution of energy is not of itself contrary to anything that is known of the properties of atoms for the causes that bring about the disruption are not among those that are yet under our control."

Physicists seemed to have no compunctions about using the word "spontaneous" to describe an event which is at the same time determined by some cause. For the progress of physics the important fact was that the atoms behave *as if* they could decide by themselves, independently of all external influences, when to explode. From that fact would follow the observed exponential decay of radioactivity, and thus Egon von Schweidler and others, starting in 1905, could construct statistical theories of radio-active processes.

One could ask "was Rutherford a determinist?" and one could easily argue that Rutherford, like Maxwell, Boltzmann, Planck, and Einstein, would find it inconceivable that anything in nature could happen "by chance" without any cause at all. Hence whenever we find them using the words "probability," "chance," "statistical," or "spontaneous" we must assume that such terms refer only to our *lack of knowledge* of causes, not to the *absence of causes*.

The fallacy of that interpretation is that it could apply equally well to Born and Heisenberg, or to anyone who believes in an *"uncertainty* principle" as distinct from an *"indeterminacy* principle." Heisenberg himself emphasized that his principle applies to our *knowledge* about the world. The assertion that quantum mechanics has abandoned causality means that, in the proposition "if we know the present exactly we can predict the future," it is not only the conclusion but also the premise that is false, according to Heisenberg (1927:197):

> One might be led to the conjecture that under the perceptible sta-tistical world there is hidden a "real" world in which the causal law holds. But it seems to us that such speculations, we emphasize ex-plicitly, are fruitless and meaningless. Physics should only describe formally the relations of perceptions....

Clearly the transition from determinism to indeterminism is linked with the positivistic-pragmatic-operationalist-instrumentalist-phenome-nalist attitude that many physicists adopted in the early 20th century, partly influenced by Ernst Mach and other critics of 19th-century mech-anism, partly as a result of the difficulty of fitting the new phenomena of atomic physics into any consistent theoretical scheme. Positivism (as I will call this complex of views) is a retreat from the aspiration to know and understand everything, an admonition to be content with the partial knowledge that can be attained at a particular stage in the development of theory and experiment. A positivist may call himself an indeterminist, meaning that his science cannot determine that which lies beyond present observation; indeterminism is then the same as "uncertainty." (This was the position of Bohr, Heisenberg, and Eddington.) Or he may call himself a determinist, meaning that his theory correlates all known or knowable facts about the observable world, and that anything beyond the obser-vable world is not his concern anyway. (It is in this sense that Planck and some modern physicists and philosophers say that quantum mech-anics is a deterministic theory.) Conversely an antipositivist or realist may be a determinist in the Laplacian sense, or an indeterminist who insists that there is a fundamental randomness in nature. Since this last position is rather hard to defend, the acceptance of indeterminism was greatly facilitated by the (at least temporary) acceptance of positivism.

Yet the acceptance of indeterminism could not have occurred unless determinism had first been shown to be either untenable or inconvenient in scientific work, since (I claim) positivism is not ordinarily a very attractive stance for physicists. The contribution of 19th-century statistical physics was to show that in many cases the use of deterministic laws is not convenient in dealing with systems of many particles even though the motion of one or two such particles can be completely determined. Developments in the first two decades of the 20th century showed that statistical physics can also deal with systems whose behavior does *not* seem to be determined by known causes. The positivistic attitude allowed the question of ultimate determinism to be set aside even as it encouraged the pragmatic use of statistical methods that carried the flavor of indeterminism.

A lecture by Paul Langevin in 1913 shows how statistical thermodynamics provided a link between the old and the new science. He pointed out that the physics of discontinuity must use the calculus of probabilities, which is the only possible link between the world of atoms and our observations. Thanks to Boltzmann, statistical methods permeate physics.

As I indicated earlier, the problem of molecular randomness was to some extent separate from the problem of the statistical nature of the second law of thermodynamics (irreversibility principle). For those physicists who had not studied in detail the writings of Burbury, Boltzmann, Planck, and Ehrenfest on molecular disorder, the more important point was that the Second Law is not absolutely true but is still good enough for all practical purposes. It was no longer certain that entropy will not decrease, only highly probable.

If the second law of thermodynamics—regarded by many physicists as one of the most important laws of nature to have been discovered since the time of Newton—was "only" statistical, how could one be certain that all other laws are absolutely valid? As physicist J. M. Jauch puts it,

> If statistical laws can behave nearly deterministically for systems with large degrees of freedom, could it not be true that *all* physical process laws, even the most fundamental ones, are perhaps only statistical laws and that they appear deterministic only because we cannot, with the usual observations, discern the fluctuations? (Jauch 1973:19)

Or perhaps the quest for total knowledge of the world was a symptom of the arrogance of scientists, who ought to be satisfied with limited statistical knowledge. There were strong social pressures, especially in

Weimar Germany, urging scientists to abandon the claim that science and only science could lead to certain knowledge. These pressures have been analyzed by the American historian of science Paul Forman (1971), who presents a persuasive argument that several statements by physicists renouncing causality in the early 1920s were made in response to external criticisms of science. To his account I would add the remark that the well-known statistical interpretation of irreversibility suggested a defensible fall-back position for any physicist who felt compelled to retreat from absolute causality. And in fact that was exactly the suggestion of three of the scientists discussed by Forman: Franz Exner (1919), Walther Nernst (1922), and Erwin Schrödinger (1922). Perhaps all laws are only statistical; in particular, the first law of thermodynamics (energy conservation).

The influence of these anticausal statements on the later Born-Heisenberg pronouncements of indeterminism is difficult to estimate. Schrödinger's lecture was not published until 1929, and in 1931 he stated that his and Exner's ideas attracted little attention when first voiced. Perhaps one can say no more than that they reflected as well as promoted a general feeling among physicists in the 1920s that classical determinism had become untenable, and that the only hope of progress was in the exploitation of statistical ideas. In any case, it would not have been enough to say, "make the first law statistical, like the second," without a much more definite formulation of a method for calculating probabilities of atomic processes; and that was attained only by the Heisenberg-Schrödinger quantum mechanics in 1926.[4]

2.10 HEISENBERG'S PRINCIPLE

The final stage in the fall of determinism can be described rather briefly here, since it has ostensibly only an indirect connection with the history of statistical thermodynamics. (Its relation to the development of quantum mechanics will be discussed in the following chapter.) Niels Bohr, noting the success of Eintein's 1916 theory mentioned in §2.8, suggested in 1923 that rather than "seek a *cause* for the occurrence of radiative processes ... we simply assume that they are governed by the laws of probability" (1924:21). The next year, with H. A. Kramers and J. C. Slater, he published a new radiation theory in which *strict* energy conservation

---

[4] Alfred Lande, one of the founders of quantum theory, made probably the strongest possible statement on this point (1953:53): "Starting from the Second Law as fundamental, the necessity of abandoning strict determinism in the molecular distribution in favor of irreducible laws of probability appears as a strange but necessary consequence. The peculiar indeterminacy of quantum physics, however, cannot be based on the Second Law alone, although it, too, is of a thermodynamics origin in spite of its apparently pure mechanical character."

was abandoned in favor of conservation "on the average." Though the Bohr-Kramers-Slater theory was immediately overturned by experiments of Bothe and Geiger and of Compton and Simon, the desirability of introducing explicit randomness into quantum theory began to be widely discussed.

When Schrödinger gave de Broglie's proposal for a wave theory of matter its comprehensive mathematical formulation, the particle nature of the electron seemed to be in doubt. Born, on the basis of recent experiments on interatomic collisions by his colleague James Franck, felt it necessary to maintain the particle conception of the electron, and therefore was led to abandon his earlier belief that "the statement that a system is at a given time and place in a certain state probably has a meaning "(1926a:129). His solution (1926b) was to propose that the wave function in Schrödinger's theory simply determines the probability that the electron is in a certain region of space. The wave function itself and its temporal evolution are determined by the force laws and the initial state of the system, in somewhat the same way that states are determined in Newtonian mechanics, but there seems to be no causal determinacy at the level of individual events.

Werner Heisenberg, who founded modern "quantum mechanics" in 1926, recalled in 1973 that "When I entered Bohr's institute in Copenhagen in 1924, the first thing Bohr demanded was that I should read the book of Gibbs on thermodynamics. And he added that Gibbs had been the only physicist who really understood statistical thermodynamics." (1973:9) Such was the environment created by Bohr, who had been fascinated by the problems of statistical physics since the beginning of his own scientific career. Heisenberg has retrospectively identified the statistical interpretation of thermodynamics as part of the breakdown of classical physics, though he gives a somewhat exaggerated role to Gibbs:

The second place at which the insufficiency of the old Newtonian scheme of concepts became apparent was in the theory of heat; though here the difficulties were much more subtle and less easy to perceive than in the theory of electricity. At first everything seemed to be going smoothly. Statistics could be applied to the motions of large numbers of molecules, thereby rendering the laws of the phenomenological theory of heat intelligible. Only when there was also an effort to justify the hypothesis of randomness appropriate to such a statistics was it observed that the bounds of Newtonian physics would have to be left behind. Probably the first man to see this with full clarity was J. Willard Gibbs, in the 1807s. But it took decades before the Gibbsian view of heat theory gained acceptance. (1974:156)

But Heisenberg's own formulation of the indeterminacy principle in 1927 showed no trace of any indebtedness to Gibbs, Boltzmann, or Maxwell. Like the original phenomenal statements of the second law of thermodynamics by Clausius and Thomson, it was phrased negatively: it is impossible by any measurement to determine both the position and momentum of a particle with such an accuracy that the product of the errors of observation of these two quantities is less than an irreducible minimum value, proportional to Planck's quantum constant.

"Impossible to determine." What does that mean? That whatever values the position and momentum may have, the act of our measurement perturbs them unpredictably? That position and momentum are inherently random quantities? Or that it is meaningless to say they have any particular values in themselves, apart from our observation of them? To answer this question we must now make a brief excursion into the history of quantum theory and its philosophical interpretations.

III. THE QUANTUM THEORY

3.1 The Planck and Einstein Hypotheses

The quantum hypothesis originated in Max Planck's search for a mathematical formula to describe the distribution of energy over frequencies in black-body radiation. This is the continuous spectrum of radiation emitted by any substance in thermal equilibrium, apart from the "line spectrum" composed of a discrete set of frequencies characteristic of a particular substance. The idea that there should be a single universal function, depending only on frequency and temperature, goes back to Gustav Kirchhoff's radiation theory of 1859. The temperature dependence of this function (integrated over some range of frequencies) had been extensively investigated throughout the 19th century; in a sense it is an outgrowth of Newton's "law of cooling" (1701), which states that the net loss of heat from a body is proportional to the difference between its temperature and that of its surroundings. The law suggests that the total radiation is proportional to temperature, although one must first have a valid concept of absolute temperature, and a reliable way to separate radiation from other forms of heat transfer, in order to implement the suggestion.[1]

By 1879, the Austrian physicist Josef Stefan was able to show that, according to available experimental data, the rate of energy emission of a hot body is proportional to the *fourth* power of its absolute temperature T. In 1884, Ludwig Boltzmann derived the T^4 law by a combination of thermodynamic and electromagnetic reasoning, showing that there must be a mechanical pressure as well as an energy density associated with electromagnetic radiation in order to satisfy the second law of thermodynamics, and using the relation between pressure and energy given by Maxwell's electromagnetic theory. According to Boltzmann's theory, one has to imagine that space, at a certain temperature, is filled with electromagnetic waves moving in all directions, with energy density E proportional to T^4. The ideal "black body" is a cavity within which thermal equilibrium is maintained, equipped with a pinhole through which the radiation emerges to be sampled by an outside observer.

---

[1] For further details see Brush (1976a, chap. 13) and Kangro (1970).

In 1893, the German physicist Wilhelm Wien extended Boltzmann's reasoning to derive a "displacement law" for the black-body radiation law. He used the Doppler principle to find a relation between values of the distribution function for different combinations of frequency and temperature, the result being that if one compares the function for two different temperatures one finds that the curve is simply displaced by a certain amount; for each wavelength the displacement is such that the product of wavelength and temperature remains constant. In terms of frequency, this means that the distribution function must have the form

$$\rho(v, T) = v^3 f(v/T). \tag{1}$$

When integrated over all frequencies this gives the Stefan-Boltzmann law regardless of the form of the function f.[2]

Wien also proposed an explicit form for the distribution law in 1896:

$$\rho(v, T) = \alpha v^3 \exp(-\beta v/T) \tag{2}$$

and showed that this function fitted the experimental data available at that time.

During the years 1897–99, Max Planck tried to construct a thermo-dynamic/electromagnetic theory of black-body radiation. His background was in thermodynamics and its applications to physical chemistry, and he had previously been quite skeptical of the utility of the atomic-kinetic theory. Presumably his interest in black-body radiation was due to his presence at the University of Berlin, where there was a very active experimental group working on this subject. He also conceived the idea that one might deduce the general principle of irreversibility (second law of thermodynamics) from electromagnetic theory by considering the interaction of radiation with a system of oscillators which absorb and emit electromagnetic waves. Whereas Boltzmann had tried to derive irreversibility from special molecular models in his kinetic theory of gases (§2.5). Planck argued that general principles of physics should not depend on such models. He intended to prove that if his results held for an idealized system of harmonic oscillators, they must be valid for any system.

In spite of his antiatomistic attitude before 1900, Planck had become somewhat familiar with the mathematical techniques of kinetic theory, probably because he took on the job of editing the lecture notes of

---

[2] $\int_0^\infty \rho\,dv = T^4 \int_0^\infty x^3 f(x)\,dx = aT^4$, where a is a numerical constant independent of T, depending only on the form of f.

Kirchhoff, his predecessor at Berlin, for publication, and Kirchhoff (though himself not very enthusiastic about atomic models) had treated the kinetic theory in his lectures. Planck got involved in a minor polemic with Boltzmann about the derivation of the H-theorm in Kirchhoff's lectures, so it appears that he had to become an expert on the subject in self-defense.

Planck first showed that the distribution law is related to the energy-distribution law for the oscillators by the equation

$$\rho(v,\, T) = \frac{8\pi v^2}{c^3}\, U(v,\, T). \tag{3}$$

Now at this point we have to point out what Planck did *not* do. He did not use the equipartition theorm of statistical mechanics, according to which each mechanical degree of freedom of a system has the same average energy, which means that each oscillator (regardless of its frequency) has energy $U = kT$ (where k is Boltzmann's constant). That would lead to the "Rayleigh-Jeans" law, first mentioned (without this particular value of the constant factor) by the British physicist Lord Rayleigh in 1900:

$$\rho(v,\, T) = \frac{8\pi v^2}{c^3}\, kT. \tag{4}$$

Integration of Eq. (4) over all frequencies to get the total energy would then give the "ultraviolet catastrophe":

$$\int \rho(v,\, T) = \frac{8\pi kT}{c^3} \int_0^\infty v^2\, dv = \infty. \tag{5}$$

Contrary to the textbook accounts of the origin of quantum theory, Planck did not worry about the ultraviolet catastrophe, because he did not believe that the equipartition theorem applied to black-body radiation or that statistical mechanics was the appropriate method for finding fundamental laws of nature. Hence there was no "crisis" at this point. The crisis was invented after 1900 (by Paul Ehrenfest) to justify the abandonment of Newtonian physics, even though Planck himself had no such intentions in 1900.

Now let's return to what Planck *did* do. He analyzed the relation between energy and entropy for various possible distribution laws, and came to the conclusion that Wien's law (2), which seemed to be most satisfactory on experimental grounds (up to 1899), corresponded to a very simple and natural formula for entropy,

$$S = -\frac{U}{av} \ln \frac{U}{ea'v}. \tag{6}$$

This formula has the property that the second derivative of entropy with respect to energy is

$$\frac{d^2S}{dU^2} = -\frac{1}{av}\frac{1}{U}, \tag{7}$$

which is always negative, since a, v, and U are always positive. This means that if the energy of an individual oscillator happens to deviate from its equilibrium value, it will return to equilibrium with an increase in entropy; equilibrium corresponds to an absolute maximum of entropy. Planck thought that he could make his theory perfectly consistent with the second law of thermodynamics by adopting (7) or (6) as the general formula for black-body radiation.

Further experiments by Lummer and Pringsheim at Berlin then showed that Wien's law breaks down at high temperatures and low frequencies; in that case it appears that

$$\rho \propto Tv^2, \qquad \text{hence} \quad U \propto T.$$

The corresponding second derivative of the entropy (computed from the thermodynamic formula $ds/dU = 1/T$) is

$$\frac{d^2S}{dU^2} = -\frac{C}{vU^2}. \tag{8}$$

Planck therefore proposed an empirical interpolation formula which would reduce to (7) for small U and to (8) for large U,[3]:

$$\frac{d^2S}{dU^2} = \frac{\alpha}{U(\beta + U)}, \tag{9}$$

where α is inversely proportional to v. One can then integrate twice with

---

[3] In the light of later developments one could say that he was thereby combining the particle and wave aspects of radiation, which taken separately would lead to Eq. (7) and Eq. (8) respectively (Jammer 1966:44–45).

respect to U and get a formula for the entropy,

$$S = k \left\{ \left(1 + \frac{U}{hv} \right) \ln \left(1 + \frac{U}{hv} \right) - \frac{U}{hv} \ln \frac{U}{hv} \right\}. \qquad (10)$$

From this one can calculate the distribute law,

$$\rho(v, T) = \frac{8\pi v^2}{c^3} \frac{hv}{e^{hv/kT} - 1}. \qquad (11)$$

This is Planck's distribution law for black-body radiation.

Planck announced this law to the Berlin Physical Society in October 1900. There was no reference to Rayleigh's paper (1900a) published a few months earlier, though Planck was probably familiar with it. Rayleigh had not actually proposed Eq. (4) as a law valid for all frequencies, since he did not expect the equipartition law to apply to black-body radiation in general except possibly for low frequencies; instead he proposed a modification of Wien's law having the form $\rho \sim v^2 T e^{-\beta v/T}$. From Rayleigh's viewpoint the main defect of Wien's original law, Eq. (2), was that it implied that the amount of energy at each frequency would go to a constant in the high-temperature limit, which seemed to him physically implausible. The point is that Rayleigh himself did not see any crisis in the discrepancy between the predictions of the equipartition theorem and the formulae needed to fit the experimental data; and Planck was even less willing than Rayleigh to take seriously the predictions from equipartition. So far, then, we have nothing more than a search for a formula that will fit the data but also be consistent with principles such as the Wien displacement law and the second law of thermodynamics.

The invention of quantum theory occurred during the weeks after Planck presented his paper to the Berlin Physical Society (October 19, 1900) and learned that it provided a very precise fit to the most recent experimental data. He decided that thermodynamics and electromagnetic theory by themselves were not sufficient to determine the distribution law (contrary to his earlier expectations), since he had just shown that either (8) or (9) satisfied all the theoretical requirements. Hence there must be another "physical reason" for the validity of his new law.

At this point Planck's familiarity with the *mathematical* aspects of statistical mechanics provided the essential clue. He already knew that he could derive something like an H-theorem for radiation by using a formula similar to Eq. (6)—this had been done in his earlier paper on irreversibility—and he knew that according to Boltzmann's statistical interpretation of entropy, which was connected with the original *H*-theorem in kinetic theory, there is a formula

$$S = k \ln W, \tag{12}$$

where W is the "probability" of a state.[4] W is usually computed by combinatorial analysis, counting all the possible arrangements of atoms subject to certain conditions; the results are usually quotients of factorials, and when one takes the logarithm of an expression like $N!$, where N is a very large number, one uses the Stirling approximation,

$$\ln N! \approx N \ln N - N.$$

Boltzmann himself had done a number of such combinatorial calculations in his earlier papers; a peculiar feature was that it was necessary in his approach to start by considering a discrete set of arrangements of the atoms, in which each atom can have only those energies corresponding to multiples of a basic quantum, ϵ. Boltzmann would then take the limit as $\epsilon \to 0$ at the end of the calculation.

Planck discovered that if one considers a collection of N oscillators among which is distributed a total amount of energy $E = P\epsilon$, in such a way that each oscillator has an integral number of quanta, then the number of arrangements is

$$W = \frac{(N + P - 1)!}{P!(N - 1)!}. \tag{13}$$

One way to prove this is to represent each arrangement by a sequence of P dots with $N - 1$ lines for the boundaries between the oscillators: $(\ldots / \cdot // \ldots / \ldots / \cdot \cdot / \cdot)$. The numerator gives the total number of ways of arranging the dots and lines, regarded simply as $N + P - 1$ permutable elements. The denominator represents permutations of the dots among themselves (which should not count as different arrangements, since each quantum of energy is identical and not "labeled"); it also represents permutations of the lines among themselves (implying that the oscillators are identical and unlabeled).

Eq. (13) could have been obtained from a book on probability theory, or Planck could have derived it himself. When one substitutes this W into

---

[4] Planck himself was the first to write it in this form, which has since become standard. Boltzman used more complicated expressions, giving the difference in entropy of two states in terms of the logarithm of the ratio of the numbers of configurations corresponding to those states. Thus he avoided the problem of giving a precise definition of W for a single state.

Eq. (12) and uses Stirling's approximation, the result can be written in the form

$$\frac{S_N}{kN} = \left(1 + \frac{P}{N}\right) \ln \left(1 + \frac{P}{N}\right) - \frac{P}{N} \ln \frac{P}{N}. \tag{14}$$

for the entropy per oscillator. This has the same form as Eq. (10) if one identifies the energy quantum $\epsilon = h\nu$ (since the average energy per oscillator is $U = P\epsilon/N$). Somehow Planck was able to travel this path backward, starting from Eq. (10) and recognizing that it could have been derived from (13).

Planck's 1900 papers did not produce much excitement in the world of physics, although those who were especially concerned with the blackbody radiation problem recognized that his formula provided a satisfactory fit to the experimental data. Planck himself did not elaborate on the significance of his quantum hypothesis or develop its applications to other situations. Indeed, it is not even clear whether he intended to propose a *physical* quantization of energy at this point, or whether he merely introduced "energy elements" for mathematical convenience in doing combinatorial calculations. Thomas Kuhn (1978) has argued that the "quantum discontinuity" was first seriously proposed not by Planck but by Einstein and Ehrenfest, and only retrospectively associated with Planck's 1900 hypothesis. Although Kuhn's interpretation is not generally accepted by historicans of physics, there is some evidence to support it and none (as far as I know) that decisively refutes it.

Einstein's paper (1905a), often referred to as his "photoelectric effect" paper, has the title "On a heuristic point of view about the creation and conversion of light." He begins by suggesting that despite the enormous success of the wave theory of light, it may still be fruitful to consider the hypothesis that light energy is distributed discontinuously in space; that a light ray consists of "a finite number of energy quanta, localized in space, which move without being divided and which can be absorbed or emitted only as a whole" (quoted from translation by ter Haar 1967:92).

Einstein shows that if one were to assume a thermal equilibrium of energy between ether and matter, then following the equipartition theorem one would arrive at a divergence for the radiation distribution integrated over all frequencies, as in Eq. (5). Nevertheless this assumption does give the correct distribution for low frequencies and high radiation densities, where Planck's distribution law coincides with it ($\rho \propto \nu^2 T$). At high frequencies and low densities, Wien's law is valid; if one calculates the entropy as a function of volume in this case, one gets exactly the same formula as for the entropy of a perfect gas or dilute solution ($S - S_0 =$ $(E/\beta v) \ln (v/v_0)$). Hence radiation at low densities and high frequencies behaves like a gas of mutually independent particles. So Einstein's conclusion seems to be that if radiation in free space consists of quanta, it should obey Wien's rather than Planck's distribution law. He offers no explanation for this odd conclusion at this point.

Assuming that light consists of quanta with energy $R\beta v/N$ (he does not use Planck's symbol h for the combination $R\beta/N$), Einstein proposed the following interpretation of the photoelectric effect:[5]

Energy quanta penetrate into a surface layer of the body, and their energy is at least partly transformed into electron kinetic energy. The simplest picture is that a light quantum transfers all of its energy to a single electron; we shall assume that that happens. . . . An electron obtaining kinetic energy inside the body will have lost part of its kinetic energy when it has reached the surface. Moreover, we must assume that each electron on leaving the body must produce work P, which is characteristic for the body. Electrons which are excited at the surface and at right angles to it will leave the body with the greatest normal velocity. The kinetic energy of such electrons is

$$\frac{R}{N} \beta v - P.$$

If the body is charged to a positive potential Π and surrounded by zero potential conductors, and if Π is just able to prevent the loss of electricity by the body, we must have

$$\Pi \epsilon = \frac{R}{N} \beta v - P,$$

where ϵ is the charge of the electron. . . .

If the formula derived here is correct, Π must be, if drawn in Cartesian coordinates as a function of the frequency of the incident light, a straight line, the slope of which is independent of the nature of the substance studied.

The energy of the photo-electrons, according to the above, must increase linearly with the frequency, but no electrons are emitted at all if the frequency is less than a threshold value (when $hv = P$). The energy of each individual electron is independent of the intensity of radiation, but the number of electrons emitted will be proportional to the intensity, if

---

[5] The effect was discovered by Hertz in 1887.

one assumes that the ejection of each electron is the direct result of the absorption of a single quantum.

While these conclusions agreed qualitatively with those recently found experimentally by Lenard, quantitative verification was difficult. It was finally accomplished by the American physicist Robert A. Millikan in 1916, after several years of work. (According to his Nobel lecture in 1924, Millikan had originally expected that his experiments would disprove Einstein's theory.) Millikan showed that the numerical value of the slope of the line (energy vs. frequency) is 6.57×10^{-27} erg sec, in close agreement with Planck's value determined from the black-body radiation distribution.

Einstein received the Nobel Prize for this theory of the photoelectric effect in 1921; Millikan won it in 1923 for his experimental verification of the theory.

Einstein soon saw another opportunity to apply the quantum hypothesis; he published a paper on the specific heats of solids in 1907, in which he showed how the deviations from the Dulong-Petit law could be explained. He assumed that a vibrating atom in a solid, instead of having its average energy proportional to kT as required by the classical equipartition theorem, would behave like one of Planck's oscillators with energy given by Eq. (3) combined with (11):

$$U(v, T) = \frac{c^3}{8\pi v^2}\, \rho(v, T) = \frac{hv}{e^{hv/kT} - 1}. \tag{15}$$

But instead of assuming some kind of frequency distribution as in black-body radiation, Einstein proposed as a first approximation that all the atoms have the same frequency of vibration in each of three spatial directions; thus the total energy of a solid of N atoms would be

$$E = \frac{3Nhv}{e^{hv/kT} - 1}. \tag{16}$$

The specific heat at constant volume is then

$$c_v = \frac{dE}{dT} = \frac{3R(hv/kT)^2 e^{hv/kT}}{\left[e^{hv/kT} - 1\right]^2}. \tag{17}$$

At high temperatures this reduces to $c_v = 3R$, in agreement with the Dulong-Petit law, but at low temperatures it goes to zero, as $e^{-hv/kT} \to e^{-\infty}$.

Einstein's conclusion that the specific heat of a solid goes to zero at absolute zero temperature harmonized well with Walther Nernst's new heat theorem or "third law of thermodynamics" published in 1906. After extensive experimental work, Nernst announced in 1911 that the specific heats of all substances, together with all other thermodynamic functions, go to zero at absolute zero temperature, and that this provides strong confirmation of the quantum theory. Nernst thus became one of the most enthusiastic supporters of quantum theory among physical chemists.

(Additional references for this section: Dorling 1971; Galison 1981; Garber 1976; Hermann 1969; Hiebert 1978; Hoyer 1980; Klein 1962, 1963a, 1963b, 1965, 1966, 1967, 1974; Miller 1976. An excellent overview of the development of the quantum theory of matter, by a participant, is Slater 1967.)

3.2 LINE SPECTRA

In the previous section we saw that the inventor of quantum theory combined skepticism about late-19th-century atomic theory with mathematical facility and an interest in accurate experimental data. That sounds like the positivist/instrumentalist attitude which some philosophers of science have claimed to be responsible for the revolution in 20th-century physics. But Max Planck had one important characteristic which was definitely antipositivist: the conviction that there are absolute mathematical laws of nature, expressed in terms of fundamental constants; in other words, that reality is more than just a complex of human sensations. Planck's antipositivist attitude showed up very clearly in his 1908 lecture on the "unity of the physical world-picture," in which he attacked Mach's positivism as essentially sterile (Planck 1909; see Toulmin 1970 for the ensuing debate between Mach and Planck). In this lecture he proclaimed the final triumph of Boltzmann's approach to physics, even though that triumph could only have been achieved by someone less committed than Boltzmann himself to the use of detailed mechanical models.

The further development of quantum theory by Niels Bohr and others required all the qualities possessed by Max Planck—mathematical ability, belief in the existence of fundamental laws, and familiarity (but not undue enchantment) with classical mechanistic theories—plus one: the willingness to consider abandoning certain established physical theories. Thus Einstein was willing to set aside the wave theory of light in order to develop the quantum theory of radiation; Planck was not. Later, Heisenberg, Born, Bohr, and others were willing to set aside Laplacian determinism in order to develop the quantum theory of atomic structure; Einstein was not.

In the development of atomic theory from Bohr through Schrödinger and Heisenberg, spectroscopic data and their representation by accurate numerical formulae played an important role. The discovery of these formulae illustrates on a lower level some of the same philosophical attitudes about science that I have just noted about the development of quantum theory. Within a decade after Kirchhoff and Bunsen had established the principles of spectrum analysis, and in particular the idea that each element has a characteristic set of wavelengths at which it may absorb or emit radiation (1859–60), atomic theorists were attempting to relate those wavelengths to properties of the molecule.

Maxwell (1875b) described molecules as systems of atoms held together by forces, capable of vibrating at various frequencies. When molecules collide, they are set into vibration, the amplitudes depending on the force of the collision but the periods depending on the constitution of the molecule. As the molecule moves freely through space it transfers this vibration to the ether; if its mean free path is long enough it may eventually lose all its vibrational energy until it undergoes another collision. Thus the intensity of spectral lines increases with temperature (since higher temperature means higher molecular speeds and more violent collisions). Higher density means more frequent collisions and thus more frequent disturbances of the regularity of molecular vibrations, hence deviations from the normal period and a "broadening" of the spectral line.

Ferdinand Lippich, professor of mechanics at the Technische Hochschule in Graz, pointed out in 1870 that even in the absence of collisions the motions of molecules would give rise to a finite width of spectral lines because of the Doppler effect (some molecules would be moving toward the observer, others away). Lord Rayleigh and others worked out the kinetic theory of spectral-line widths in some detail, taking into account both collisions and the effects of Doppler shifts.

G. J. Stoney, the Irish physicist who was one of the first to estimate atomic sizes from kinetic theory and who later introduced the word "electron," proposed a detailed theory of spectral lines in the 1870s. He assumed that the vibrations of a molecule can be decomposed into a series of fundamentals plus overtones (for example through Fourier analysis). Thus the wavelengths ought to be given by the formula λ/n, where λ is the fundamental wavelength and n is an integer. By taking values of n as high as 30, Stoney was able to fit three of four spectral lines of hydrogen measured by Ångstrom[6] (McGucken 1969:110–16).

---

[6] Taking $\lambda = 131{,}277.14 \times 10^{-10}$ m and $n = 20$, 27, and 32, one gets wavelengths of 6563.93, 4862.11, and 4102.37×10^{-10} m respectively.

In order to apply Stoney's theory it was necessary to assume (1) that many of the overtones were "suppressed" so that only a few of the higher ones are actually observed; (2) that some molecules have more than one fundamental tone. How does one test such a theory? Stoney thought that his theory was confirmed by the fact that several of the observed wavelengths of a particular substance have ratios (taken two at a time) which are very close to ratios of integers. But Arthur Schuster, in 1881, showed that if one allows fairly large integers (up to 100) and examines a large body of experimental data, one will find a number of such coincidences even if the wavelengths are chosen at random. Schuster argued that for several metals whose line spectra had been investigated, the number of coincidences with integer ratios (within the error of measurement) was not much greater than what one would expect to occur by chance.

The positivist response to the apparent failure to find a molecular theory of spectra is illustrated by a statement of Alfred Cornu, a French physicist, in 1885: "the hope of finding a simple law, as that of musical harmonics, is the sign of a preconceived idea which it is important to discard immediately; this law of whole numbers applies only to a very particular kind of sonorous body of which the type is the cylindrical column of great length in relation to section: if the form of the vibrating body departs from this special type, the relation between the frequencies of successive sounds becomes very complex." One should seek regularities in the observational data, but not think in terms of analogies with particular mechanical systems (McGucken 1969:128).

The breakthrough was nevertheless made in the same year by Johann Jakob Balmer (1825–98), a geometry teacher at a girls' secondary school in Basel, Switzerland. He showed that all four of Ångstrom's lines for hydrogen could be accurately represented by the formula

$$\lambda = \frac{m^2 h}{(m^2 - n^2)} \qquad \text{where } h = 3645.6 \times 10^{-10} \text{ m and } n = 2.$$

Thus:

m =		Formula gives	Ångstrom's value	Difference
3	= H_α (C-line)	$9h/5 = 6562.08$	6562.10	+0.02
4	= H_α (F-line)	$4h/3 = 4860.8$	4860.74	−0.06
5	= H_α (before G)	$25h/21 = 4340$	4340.1	+0.1
6	= H_α (h-line)	$9h/8 = 4101.3$	4101.2	−0.1

Balmer says in his 1885 paper that when he first worked out his formula he knew only the values for these four lines, but his formula predicted a fifth line ($m = 7$, $\lambda = 3969.65$), which should lie within the visible spectrum. His friend Jakob Edward Hagenbach at the University of Basel then informed him that the astronomers Vogel and Huggins had measured a number of additional lines in the hydrogen spectrum. Vogel had found a line at 3969, and both Vogel and Huggins had found lines in the ultraviolet spectrum which appeared to correspond to $m = 8, 9, 10$, and 11, though the agreement with the formula was not quite as good as for smaller values of m.

Balmer's formula had no obvious physical meaning in terms of 19th-century mechanical atomic models; in particular, it implied that the observed wavelengths are always *greater* than the "fundamental" wavelength h, rather than smaller as in the acoustic analogy. Yet the quantitative success of the formula forced physicists to take it as a model for further investigation, and to search for wavelengths that corresponded to other integer values of n.

During the decade following the publication of Balmer's paper there were two alternative approaches to generalizing his formula. The first was followed by Heinrich Kayser and Carl Runge at Hannover; the second by Janne Rydberg at the University of Lund (Sweden). Both agreed that instead of using wavelength one should take its reciprocal, for example the "wave number" n or the frequency. Kayser and Runge advocated the formula

$$\frac{1}{\lambda} = A + Bm^{-2} + Cm^{-4}$$

so that Balmer's formula would correspond to the special case $C = 0$. Rydberg proposed

$$n = \frac{1}{\lambda} = n_o - N_O/(m + \mu)^2$$

so that Balmer's formula would correspond to $\mu = 0$.

Rydberg's investigation was published in 1890, though he claimed to have obtained his basic formula even before the appearance of Balmer's paper in 1885. In 1896 Rydberg and Schuster independently discovered that there is a connection between three different series of spectral lines, known as "principal," "sharp," and "diffuse" series, which enables all of them to be written in a simple form. Rydberg expressed the formula as

$$n = \frac{N_O}{(m_1 + \mu_1)^2} - \frac{N_O}{(m_2 + \mu_2)^2}.$$

In the notation which later became standard, the frequency of a line is given by

$$v = Rc\left(\frac{1}{\tau_2^2} - \frac{1}{\tau_1^2}\right),$$

where R is called the "Rydberg constant," c is the speed of light, and τ_1 and τ_2 are integers.

Walther Ritz in 1908 explicitly proposed that every spectral line can be written as a difference of two terms of this form; this is known as the "Ritz combination principle."

With the help of some theoretical guidance from Ritz, Friedrich Paschen was able to discover the series of hydrogen lines corresponding to $n = 3$ in Balmer's original formula in 1908. The series for $n = 1$ was found by Theodore Lyman in 1914, and other series were later discovered by Brackett, Pfund, etc.

By 1913 Rydberg's method of representing spectral lines seems to have definitely won out over the Kayser-Runge approach. It was thus part of the "background knowledge" available to theoreticians when they tried to construct models of the atom. (For further details on this topic see Hindmarsh 1967 and McGucken 1969.)

3.3 THE BOHR MODEL OF THE ATOM

After Einstein's 1905 paper on photons, the next major step in the development of quantum theory was Niels Bohr's atomic model, published in 1913. Using Rutherford's discovery (1911) that most of the mass of the atom is concentrated in a tiny nucleus at the center, Bohr postulated that bound electrons occupy definite orbits, and that transitions from one orbit to another involve the absorption or emission of radiation in accordance with Planck's quantum hypothesis. The model succeeded in accounting for the Balmer-Rydberg formula for the spectral lines of hydrogen (§3.2) and promised to give a method for describing the structure of more complex atoms. The promise was only partly fulfilled, and it was eventually necessary to make a more complete break from Newtonian ideas in order to obtain a satisfactory theory of atomic structure. After this theory was developed by Heisenberg and Schrödinger in 1926, it was again Niels Bohr who took the leadership in proposing what became for two decades the dominant philosophical viewpoint of the new physics, the so-called Copenhagen interpretation of quantum mechanics. Bohr attempted to extend this interpretation to other areas of science and thus to develop a comprehensive view of the nature of human knowledge. In the late 1930s he investigated nuclear structure and proposed the "liquid drop" model, which provided an explanation of fission. He was one of the first to foresee the dangerous consequences of the nuclear arms race, and attempted unsuccessfully to persuade the wartime leaders Roosevelt and Churchill to place the atomic bomb under international control.

As part of his examination for the master's degree, Bohr was assigned by his adviser, Professor C. Christiansen, to write a paper on "the application of the electron theory to explain the physical properties of metals." Bohr became very enthusiastic about this subject, and expanded his master's essay into a doctoral dissertation. He followed the approach of the Dutch physicist H. A. Lorentz, who had used the kinetic theory of gases to treat the motions of electrons in a metal. Where Lorentz, in his 1905 papers, had assumed that the electrons collide like billiard balls with fixed metal molecules, Bohr applied Maxwell's more general kinetic theory with forces varying inversely as the nth power of the distance.

According to historians John Heilbron and Thomas Kuhn (1969), it was his calculations on the diamagnetism and paramagnetism of electrons in metals that led Bohr to consider the electronic structure of atoms.[7] J. J. Thomson (1900) had suggested that free electrons might account for diamagnetism, but his argument was fallacious. The French physicist Paul Langevin had developed a more satisfactory theory of both paramagnetism and diamagnetism in 1905; he had succeeded in explaining the inverse temperature dependence of the paramagnetic susceptibility (Pierre Curie's law, 1895) by attributing a fixed dipole moment to the molecules. If there was no permanent molecular moment, Langevin assumed that a magnetic field would affect the motions of electrons inside the atoms in such a way as to produce a temperature-independent diamagnetic susceptibility, again in qualitative agreement with Curie's experiments.

Bohr showed that Langevin's theory was inconsistent with statistical mechanics, at least if the electrons were bound within atoms by forces obeying Newtonian mechanics. In particular, he proved that a classical system must have zero diamagnetic susceptibility. (This result is known as "Miss van Leeuwen's theorem" because Johanna van Leeuwen published it in 1919, not having seen Bohr's dissertation.) Bohr concluded, more generally, that "it is not possible to explain the magnetic properties of bodies on the basis of the electron theory if such effects as, e.g., the emission of energy are neglected, which might have the effect of preventing a statistical mechanical equilibrium among the bound electrons as that

---

[7] Magnetic susceptibility is defined as $\chi_m = M/H$, which M is the magnetic moment induced by an external magnetic field H. If $M > 0$ it is *paramagnetic*; if $M < 0$ it is *diamagnetic*. Whereas paramagnetism is a manifestation of electron spin, diamagnetism reflects electron angular momentum. For the later history of this subject see §4.7.

assumed for the free electrons. . . . An emission of energy, however, might perhaps explain the paramagnetic but not the diamagnetic phenomena" (1972:383). In Bohr's mind the problem of developing a satisfactory electron theory of metals was thus connected with the problem of radiation emission, and he discussed Jeans' assumption that complete thermal equilibrium is not actually attained in experiments on black-body radiation. (If it were, one would encounter the divergence problem, later known as the "ultraviolet catastrophe.") That assumption did not seem satisfactory to Bohr; citing recent papers of Einstein and Planck, he wrote that "it seems impossible to explain the law of heat radiation if one insists upon retaining the fundamental assumptions underlying the electromagnetic theory. This is presumably due to the circumstance that electromagnetic theory is not in accordance with the real conditions and can only give correct results when applied to a large number of electrons (as are present in ordinary bodies) or to determine the average motion of a single electron over comparatively long intervals of time (such as in the calculation of the motion of cathode rays) but cannot be used to examine the motion of a single electron within short intervals of time. Since it would take us quite outside the scope of the problems treated in the present work, we shall not enter upon a discussion of the attempts made to introduce fundamental changes in the electromagnetic theory, attempts that seem to be leading to interesting results" (1972:378–79).

One of Bohr's main positive results in his 1911 dissertation was that good agreement with the ratio of thermal to electrical conductivity is obtained by taking $n = 3$ in the force law, that is, by assuming that the metal molecules are electric dipoles. Precisely this model had been suggested a year earlier by J. J. Thomson in an attempt to explain the photoelectric effect. This may have been one reason why Bohr decided to go to Cambridge to work in Thomson's laboratory after he received his Ph.D., although as it turned out he did not communicate very well with Thomson and eventually moved on to Manchester to work with Rutherford.

While in Cambridge during the fall of 1911, Bohr attempted without success to get an English translation of his dissertation published. (The fact that it was not available except in Danish until 1972 not only deprived Bohr of some of the credit for his contributions to the electron theory of metals but also obscured the historical relation between his work on that subject and on his model of the atom.) He continued to read the literature on theories of magnetism, in particular a 1911 paper by the French physicist Pierre Weiss which introduced a quantized atomic magnetic moment or "magneton" as a basis for improving Langevin's theory. Heilbron and Kuhn (1969) argue that the idea that the magnetic moment of an atom is restricted to certain definite values implies that the motions of bound electrons are also restricted, and this connection probably influenced Bohr's introduction of quantized orbits.

Bohr moved to Manchester in March 1912, and soon became involved in the radioactivity problems being studied by Rutherford's group there. It will be recalled that Rutherford had published his nuclear model of the atom, based on the alpha-particle scattering experiments of Geiger and Marsden, the previous year. This model provided the starting point for Bohr's theory of electron orbits. In addition to concentrating all the positive charge and almost all of the mass of the atom in a small nucleus, Rutherford's model implies that hydrogen has only one electron, unlike other models such as J. J. Thomson's.

As Bohr worked out his ideas, he became aware of two pieces of experimental evidence and one pertinent theory. (1) R. Whiddington, an English physicist, found in 1911 that when an anticathode is hit by cathode rays of increasing velocity, there is a sudden change in the nature of the emitted radiation at certain critical velocities. This seems to suggest the idea of "energy levels"—a certain threshold energy is needed to knock an electron out of a particular orbit. (2) A Danish spectroscopist, H. M. Hansen, told Bohr (after his return to Copenhagen) about Balmer's and Rydberg's formulas for spectral lines. The Ritz combination principle, together with Planck's quantum hypothesis, suggests that each frequency depends on the *difference* in energies of orbits or energy levels, before and after the radiation is emitted or absorbed. (3) The British astrophysicist J. W. Nicholson published a series of papers on the solar corona (1912) in which he succeeded in accounting for many of the spectral lines with an atomic model involving Planck's constant. While Nicholson's model was different from Bohr's in essential details, Bohr's work was clearly influenced by it, especially in the treatment of excited states.

Bohr's paper "On the Constitution of Atoms and Molecules" was published in the *Philosophical Magazine* in July 1913. To understand his theory we must first recall the relevant formulas for a classical orbit with an electrostatic force between charges e and e'. For simplicity we assume the orbit to be circular. Then the centripetal force mv^2/r must be equated to the electrical force ee'/r^2:

$$\frac{mv^2}{r} = \frac{ee'}{r^2}. \tag{1}$$

The linear speed v is related to the angular frequency of revolution,

$$v = 2\pi\omega r. \tag{2}$$

The total energy of the revolving electron is the sum of kinetic and potential energies,

$$U = \frac{1}{2} mv^2 - \frac{ee'}{r}.$$

Substituting from Eq. (1), we get a negative magnitude for the energy, so we define a new quantity, the "binding energy" W:

$$U = -\frac{ee'}{2r} \equiv -W. \tag{3}$$

There is nothing new so far, but now Bohr looks for a relation between frequency and energy, because he wants to use Planck's relation between frequency and energy to determine the possible energy levels. The postulate that energy of radiation can only come in "quanta"—integral multiples of a basic energy unit hv—is going to be used to determine certain "allowed" energies of an electron in an atom.

To get a relation between frequency and energy, first use Eq. (1) to express v in terms of ee'/r, and then use (3) to express this in terms of W:

$$v = \sqrt{ee'/rm} = \sqrt{2W/m};$$

hence, from (2),

$$\omega = v/2\pi r = \frac{\sqrt{2}\sqrt{W}}{\sqrt{m}\sqrt{2r}} = \frac{\sqrt{2}W^{3/2}}{\sqrt{m}(ee'/2r)(2\pi r)}$$

$$= \frac{\sqrt{2}W^{3/2}}{\pi ee'\sqrt{m}}. \tag{4}$$

Now comes the leap in the dark: Bohr assumes that if an electron at infinite distance from the nucleus is at rest (so its frequency of revolution is $\omega = 0$) and is then captured into an orbit with frequency ω, it will emit radiation whose frequency is just the average of the frequencies of the initial and final orbits, namely $\frac{1}{2}\omega$. Bohr does not try to give any real justification for this assumption, and indeed it would have been impossible to do so, since his theory is incompatible with classical physics. Classically, a particle revolving with frequency ω would emit radiation of the same frequency ω; but now Bohr is assuming there is radiation only when the electron jumps from one orbit to another, so presumably the frequency of the emitted radiation is going to depend on two classical frequencies. In the case when *both* initial and final states have very small frequencies and the electron is far away from the nucleus, Bohr could rationalize this assumption by using what later became known as the "correspondence principle"—for large quantum numbers, the behavior of a quantum system must approach that of a classical system. But this argument could only be applied *after* the scheme of electronic orbital quantum numbers had been worked out by the line of reasoning we shall now trace; hence it does not have much to do with how Bohr made his original discovery.

Having assumed that the electron can radiate at a frequency of $\frac{1}{2}\omega$ when being captured from infinity into an orbit with classical frequency ω, Bohr now applies Planck's postulate in the following form: the energy W may be any integer multiple of the quantum $\frac{1}{2}h\omega$:

$$W = \tfrac{1}{2}\tau h\omega, \tag{5}$$

where τ is the number of quanta emitted. Of course with the advantage of hindsight we can point out that more than one quantum can be emitted only if more than one atom is being considered; Bohr does not seem to realize that a single electron can only emit energy $\frac{1}{2}h\omega$ in the process he is describing, even if we grant his unfounded assumption that the frequency of the emitted radiation is $\frac{1}{2}\omega$. But let us not be too critical, for Bohr is not proposing a logical derivation but rather a train of ideas. Indeed, if we now substitute (5) into (4) and solve for ω, we get

$$\omega = \frac{\sqrt{2}\left(\frac{\tau}{2}h\omega\right)^{3/2}}{\pi ee'\sqrt{m}} \qquad 1 = \frac{\sqrt{2}\tau^{3/2}h^{3/2}\omega^{1/2}}{2^{3/2}\pi ee'\sqrt{m}} \qquad \omega = \frac{4\pi^2 m e^2 e'^2}{\tau^3 h^3}. \tag{6}$$

Hence from (5),

$$W = \frac{2\pi^2 m e^2 e'^2}{\tau^2 h^2} \tag{7}$$

and from (2).

$$2r = \frac{\tau^2 h^2}{2\pi^2 m e e'}. \tag{8}$$

Now let's put in some numerical values and see what happens. Using $e = 4.7 \times 10^{-10} = e'$, $e/m = 5/31 \times 10^{17}$, and $h = 6.5 \times 10^{-27}$, Bohr finds, for $\tau = 1$,

$$2r = 1.1 \times 10^{-8} \text{ cm}, \quad \omega = 6.2 \times 10^{15} \text{ sec}^{-1}, \quad \text{and} \quad W/e = 13 \text{ V}.$$

He remarks that "these values are of the same order of magnitude as the linear dimensions of the atoms, the optical frequencies, and the ionization-potentials." So far so good. But even more interesting is the fact that W is inversely proportional to the square of an integer! "Forgetting" the original definition of τ as the number of quanta emitted when an electron dropped from infinity to a state with given W (that is, different values of τ might correspond to the same W), Bohr redefines τ so as to determine "a series of configurations of the system," which is to say a series of possible values of W. Thus for each τ there is a unique set of W, ω, and r to be calculated from Eqs. (6), (7), and (8). And the frequency of radiation emitted when the electron goes from the set determined by τ_1 to the set determined by τ_2 will be determined by the energy difference, applying Planck's formula a second time:

$$h\nu = W_2 - W_1 = \frac{2\pi^2 m e^2 m'^2}{h^2} \left(\frac{1}{\tau_2{}^2} - \frac{1}{\tau_1{}^2} \right).$$

If we compare this with the empirical spectral formula given in §3.2, we see that they are identical provided that the Rydberg constant R is equal to

$$R = \frac{2\pi^2 m e^2 e'^2}{ch^3}.$$

When Bohr made the comparison with current values of the constants, he found that the discrepancy was "inside the uncertainty due to experimental errors in the constants," that is, about 7%. Subsequently it was found that with better values for the constants the agreement is nearly exact.

As a general rule, Bohr proposed that the angular momentum of the electron in its orbit is equal to an integral multiple of $h/2\pi$:

$$m\nu r = nh/2\pi \tag{11}$$

(as can be found by substituting the values of m, ν and r from the above equations).

Bohr's formula failed to account for one series of lines observed in the spectrum of the star ζ Puppis by the American astronomer E. C. Pickering in 1896, and in vacuum tubes containing a mixture of hydrogen in helium observed by A. Fowler in 1912 (Jammer 1966:82–84). These lines were generally ascribed to hydrogen, but Bohr proposed that they should be assigned instead to once-ionized helium. If Eq. (9) is applied to a system consisting of a nucleus with charge $e' = 2e$, one gets a similar formula as for hydrogen but with an additional factor of 4. The new formula did fit the Pickering-Fowler lines. Bohr suggested that his interpretation could be tested by measuring the spectrum of a mixture of helium and chlorine (or some other electronegative element that would facilitate the ionization of helium). The experiment was done by E. J. Evans (1913) and confirmed Bohr's interpretation. Fowler objected that the observed wavelengths still differed by a small amount from Bohr's formula. Bohr then showed that the discrepancies are eliminated if one uses the "reduced mass" of the electron,

$$m' = \frac{m}{1 + \dfrac{m}{m_{\text{nucleus}}}}.$$

This takes account of the fact that the electron does not really revolve around a fixed nucleus, but rather both revolve around their common center of gravity, which coincides with the nucleus only if the latter has infinite mass. (The same correction is well known in the case of the earth's motion around the sun.)

Further confirmation of Bohr's theory came from the Franck-Hertz experiment in 1914. They bombarded a gas (Hg) with electrons of increasing energy; there are only elastic collisions until one gets to a certain critical energy, then one suddenly observes inelastic collisions with radiation at a particular frequency. The energy loss of the electron corresponds to the energy difference between the ground state and an excited state of the atom. (This experiment was *not* originally done for the purpose of testing Bohr's theory, as is usually stated; it was only interpreted this way afterward.)

Despite these early triumphs, there were certain mysterious aspects of Bohr's theory which made it unsatisfactory, especially if one tried to think of the electron actually moving around in space in one orbit and then jumping to another. Rutherford wrote to Bohr in 1913, on receiving the manuscript of Bohr's paper, as follows:

> There appears to be one grave difficulty in your hypothesis, which I have no doubt you fully realize, namely, how does an electron decide what frequency it is going to vibrate at when it passes from one

stationary state to another? It seems to me that you would have to assume that the electron knows beforehand where it is going to stop. (Moore 1966:59)

(Additional references for this section: Heilbron 1977, 1981; Hirosige and Nisio 1970; Hoyer 1973, 1974; Nisio 1967.)

3.4 WAVE MECHANICS

We now come to the end of Newtonian mechanics as a fundamental basis for physical theory (though almost all of its practical applications remain valid at present). It had already been replaced by Einstein's theory of relativity in one domain (high velocities and/or strong gravitational fields), and the beginnings of the quantum theory needed for another domain (the atomic world). The complete amalgamation of these two theories into a consistent general system, which could play the same role as Newton's laws or Aristotle's natural philosophy, still lies largely in the future.

To set the stage for the introduction of quantum mechanics, let us recall the situation in physics around 1920. First, Planck's quantum hypothesis had been generally accepted as a basis for atomic theory, and the photon concept had been given strong support by Robert A. Millikan's experimental confirmation of Einstein's theory of the photoelectric effect. Yet, despite the early success of Bohr's theory of the hydrogen atom, no one had found a satisfactory way to extend the theory to multielectron atoms and molecules without introducing arbitrary additional hypotheses tailored to fit each particular case. Furthermore, it seemed necessary to use principles of Newtonian mechanics and Maxwellian electromagnetism in some parts of the theory while rejecting them in others.

Although the photoelectric effect, as interpreted by Einstein, seemed to offer definite proof that light energy comes in discrete units, this did not by any means persuade physicists that the wave theory of light should be abandoned in favor of a particle theory. Experiments on the interference and diffraction of light, and indeed all the properties of electromagnetic radiation discovered in the nineteenth century, seemed to require wave properties. Even the quantum theory, which assigned each "particle" of light an energy proportional to its "frequency" ($E = hv$), depended on a wave property, periodicity. Thus one was led to the paradoxical conclusion that light behaves in some respects like particles and in other respects like waves.

Further evidence for the particle nature of light came in 1923 with the discovery of the "Compton effect" by the American physicist Arthur Holly

Compton.[8] As Einstein had already pointed out in 1916 (though Compton may not have been aware of this particular paper at the time), photons should have a definite momentum equal to hv/c, if one applies both quantum theory and special relativity theory to them. Compton reasoned that the law of conservation of momentum should apply to collisions between photons and electrons, so that in the scattering of photons by electrons there should be a reduction in frequency as some of the photons' momentum is transferred to the electrons. Before 1923, experiments on the scattering of electromagnetic radiation by matter had not revealed this effect, since in most cases the scattering was by the massive nucleus, and in this case the momentum transfer is expected to be negligible. Compton showed that when x-rays are scattered, the predicted change in frequency is observed; moreover, electrons knocked out of the target could be detected and were found to have the momentum as a function of direction predicted by the theory.

Another puzzling aspect of the physical theories developed in the first two decades of the 20th century was their failure to prescribe definite mechanisms on the atomic level. Thus Bohr's theory specified the initial and final states of an atomic radiation process, but did not describe just how the electron got from one orbit to another. The theory of radioactive decay was even less definite: it simply predicted the probability that a nucleus would emit radiation, but did not state when or how any particular nucleus would decay. It appeared that the determinism or causality characteristic of the Newtonian world-machine—the predictability, at least in principle, of the detailed course of events of any atomic process, given the initial masses, velocities, and forces—was slipping away. Just as Maxwell and Boltzmann had shown that the second law of thermodynamics was only statistically valid as a description of observable processes, so some physicists began to suspect that Newton's laws of motion and the laws of conservation of energy and momentum might, in the atomic world, be true only "on the average," but not necessarily at every instant of time.

But if neither the wave theory nor the particle theory of light could be proved exclusively correct, and if the classical laws of mechanics and electromagnetism could no longer be relied on to describe atomic processes, clearly something new was needed, some fundamental law from which all these partially or statistically valid theories could be derived. And perhaps even the basic Newtonian persuppositions about the nature and purpose of theories would have to be modified. As they gradually

[8] The Dutch physicist/chemist Peter Debye suggested the effect at about the same time, but his work did not go far enough for him to claim a share in the discovery. For a comprehensive account see Stuewer (1975).

recognized the seriousness of these problems, physicists early in the 1920s began to talk (much more urgently than they usually do) about a "crisis in the foundations of physics."

It is a curious fact that radical conceptual changes in a science are often initiated by men who did not receive their initial professional training in that science. The person who first approaches the problems of a discipline with the mature perspective of an outsider, rather than being indoctrinated previously with the established methods and attitudes of the discipline, can sometimes point to unorthodox though (when seen in retrospect) remarkably simple solutions to those problems. Thus the generalized law of conservation of energy was introduced into physics from engineering, physiology, and philosophy, though its value was accepted readily enough by physicists.

Louis de Broglie received his first degree in history at the University of Paris in 1910, intending to go into the civil service. But he became intrigued by scientific problems as a result of discussions with his brother, the physicist Maurice de Broglie, and by reading the popular scientific expositions of the mathematician Henri Poincaré. Although he managed to acquire sufficient competence in using the standard tools of theoretical physics to satisfy his professors, his real interest was in the more fundamental problems of the nature of space, time, matter, and energy.

In his doctoral dissertation of 1924 (based in part on papers published a year earlier), de Broglie proposed a sweeping symmetry for physics: just as photons behave like particles as well as like waves, so electrons should behave like waves as well as like particles. In particular, an electron of mass m moving with velocity v will have a "wavelength" λ given by the simple formula[9]

$$\lambda = \frac{h}{mv}. \tag{1}$$

where h is Planck's constant.

Although de Broglie's hypothesis was based originally on theoretical arguments, it was soon realized that it could be checked rather directly by experiments of a kind which had already been performed by C. J. Davisson in the United States. In these experiments, electrons emitted from an electrode in a vacuum tube were allowed to strike a metal surface, and the scattering of the electron beam at various angles was measured.

[9] For photons, equating Planck's quantum formula for energy to Einstein's relation between energy and mass gives $hv = mc^2$. Since $v\lambda = c$, we have $m = hv/c^2 = h/c\lambda$, or $\lambda = h/mc$. De Broglie postulates that a similar formula holds for all particles, replacing c by v. But the "phase wave" travels at a speed c^2/v, which is greater than c if $v < c$. Hence it can't carry energy (according to relativity theory) but only "guides" the particle.

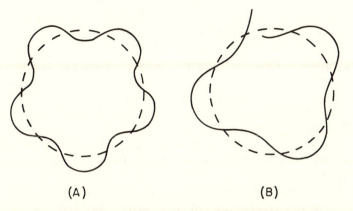

(A)   (B)

FIG. 3.4.1. De Broglie waves must fit around a circle to make a Bohr orbit.

Walter Elsasser in Germany pointed out that Davisson's results could be interpreted as showing a diffraction of the electrons as they passed through rows of atoms in the metal crystal, and that other experimental measurements of electron scattering by C. Ramsauer could be attributed to the interference of waves that somehow determined the path of the electrons. In both cases the apparent wavelength associated with the electrons agreed with the values calculated from de Broglie's equation (1). In 1926, Davisson together with L. H. Germer carried out further experiments which confirmed de Broglie's hypothesis, and in 1927 G. P. Thomson[10] demonstrated the phenomenon of electron diffraction by still another method. Later experiments showed conclusively that not only electrons but protons, neutrons, heavier nuclei, and indeed all "material particles" share this wavelike behavior.

In his 1924 memoir, de Broglie pointed out that the quantization of allowed orbits in the Bohr model of the hydrogen atom can also be deduced from his formula for the wavelength of an electron, if one makes the natural assumption that the circumference of the orbit must be large enough to contain an integral number of wavelengths, so that a continuous waveform results, as in Fig. 3.4.1. The assumption, applied to a circular orbit of radius r, would mean that

$$2\pi r = n\lambda = \frac{nh}{mv}.$$

[10] The son of J. J. Thomson, who had received the Nobel Prize in physics (1906) for his discovery (1897) of the electron and "proof" that it is a particle. In 1937, Davisson and G. P. Thomson were jointly awarded the Nobel Prize in physics for demonstrating the wave nature of the electron.

where n is an integer. If we multiply through by mv, we see that this is just the same as Bohr's hypothesis that the angular momentum mvr is a multiple of $h/2$ (Eq. 3.3.11).

De Broglie's hypothesis suggests that the reason why atomic energy levels are quantized is the same as the reason why the frequencies of overtones of a vibrating string are "quantized": the waves must exactly fill a certain space. Hence a theory of atomic properties might be constructed by analogy with the theory of vibrations of mechanical systems—reminiscent of the old program of Pythagoras and Kepler, who based many of the laws of nature on the laws that govern the operation of musical instruments

Before showing how this clue was followed up, I must remark that the solution to the problem of atomic structure was first discovered by Werner Heisenberg; his theory, subsequently known as "matrix mechanics," was published in 1925, a few months before Schrödinger's "wave mechanics." Although the two theories are generally regarded as being mathematically equivalent, Schrödinger's is much more widely used and seems to convey a more definite (though perhaps misleading) physical picture than Heisenberg's, and therefore will be given a more prominent place here.

Erwin Schrödinger, the founder of wave mechanics, studied at the University of Vienna under Fritz Hasenöhrl (Boltzmann's successor), Franz Exner, and Egon von Schweidler (see §2.9 for the contributions of Exner and Schweidler to the rise of indeterminism). His early publications dealt with the acoustics of the atmosphere, the thermodynamics of lattice vibrations, and statistical mechanics. His discovery of the fundamental differential equation in quantum mechanics was directly related to his interest in the statistical problems of gas theory, especially Einstein's attempt in 1925 to apply the Bose counting procedure to a system of particles with finite mass. As a result of this work on gas theory he became acquainted with Louis de Broglie's dissertation on the wave nature of matter; wave mechanics is the outcome of his attempt to reconcile the apparently inconsistent ideas of Boltzmann, Bose, Einstein, and de Broglie. The role of statistical physics in Schrödinger's discovery has recently been explored in detail by Martin J. Klein (1964), Paul A. Hanle (1977a, 1977b, 1979), and Linda Wessels (1979), and their papers should be consulted for a full account of this fascinating story.

Roughly speaking, the "wave nature of matter" was a way of interpreting the apparent lack of statistical independence of the entities postulated to comprise a physical system. The probability that an entity is in a particular state depends on the presence of other similar entities in the system, even though there is no force (in the classical sense) between the entities. The first indication of this phenomenon appeared in Einstein's 1905 theory of photons, in which it was shown explicitly that a gas of *independent* quanta would behave in accordance with Wien's radiation law rather than Planck's. At that time it was not clear in what sense quanta should be regarded as particles or what kind of interdependence led them to obey Planck's law.

Later, in the 1920s, the problem was formulated in terms of the "indistinguishability" of the entities. In classical physics it is implicitly assumed that a particle preserves its identity even though it may have exactly the same mass, size, charge, etc. as other particles. Thus two microstates of a system of N particles which differ only by the interchange of the labels of identical particles would be considered different microstates. As long as N is fixed, it makes no difference to the thermodynamic properties whether this assumption is made or not, but when one considers a more general system in which N may vary—for example, the grand ensemble of Gibbs (§1.13)—then it does make a difference.

In a system of waves in a mechanical medium, or black-body radiation, N is not fixed and the quanta have no inherent distinct identities; they are properties of the entire system, little more than abstract ways of representing complicated motions through a Fourier harmonic analysis. If one computes the number of microstates (W) on the assumption that the quanta are distinguishable, then one should divide by $N!$ so as not to count again microstates that differ only by the interchange of identical entities. But to treat particles in this way would imply that they are linked together as part of a system—that the behavior of each particle is somehow dependent on all the others even in an ideal gas.

Several physicists, including Ehrenfest, Planck, and Schrödinger, argued about the problem of correcting for indistinguishability in the quantum statistical mechanics of ideal gases. Finally in 1925 Einstein, following an important suggestion by the Indian physicist S. N. Bose, proposed a theory which exploited the analogy between particles and waves. This theory will be discussed in more detail in §4.4; here we simply note that in the Bose-Einstein procedure, the particles have no labels to begin with. A configuration is characterized by stating the number of cells which have 0, 1, 2, . . . particles in them.[11] Einstein justified this procedure by citing de Broglie's suggestion that particles are analogous to crests in a system of waves; in such a system there would be no reason to attribute a separate identity or label to each crest beyond its position and velocity.

[11] The reader is urged to consult Delbrück's (1980) discussion of this point. The problem can be swept under the rug by an operationalist approach (Hestenes 1970).

It was this formulation that appealed to Schrödinger and led him to take de Broglie's work seriously.

"Why was it Schrödinger who developed de Broglie's ideas?" is the title of an article by V. V. Raman and Paul Forman (1969). In other words why didn't the physicists at Göttingen (Heisenberg and Born), Munich (Sommerfeld), or Copenhagen (Bohr), who were more deeply involved than Schrödinger in the theory of atomic structure and spectra, see the significance of de Broglie's matter-waves? Raman and Forman answer that de Broglie probably had a rather low reputation in those places by 1924, because of a number of disagreements and animosities concerning technical points in atomic physics and the discovery of element 72 (claimed by both the French and the Danes). On most points de Broglie turned out to be wrong, so there was no particular reason to pay much attention to his "phantasies about phase waves associated with material particles" (Raman & Forman 1969:296). In fact it was Einstein who first saw the merits of de Broglie's thesis (having received a copy from Langevin), and he pointed out its significance in his 1925 paper on Bose-Einstein statistics.

But Einstein was not concerned with the technical problems of spectroscopy; Raman and Forman argue that de Broglie's ideas could best be appreciated by a person who had one foot in each camp—the Einstein "statistical mechanics" tradition, and the Bohr-Sommerfeld-Heisenberg "spectroscopy" school—without being too strongly committed to either one. That person was Schrödinger. He was impressed by Einstein's reference to de Broglie, and not put off by de Broglie's presumed low reputation in spectroscopy. One expects that he would have been attracted by three aspects of de Broglie's work: (1) the attempt to combine atom mechanics with statistical mechanics; (2) the use of relativity theory as a guide; (3) the stress on Hamilton's analogy between optics and dynamics (see below). A fourth reason for Schrödinger's interest was his own 1922 paper pointing out a connection between the quantum conditions on electron orbits and certain mathematical expressions in Hermann Weyl's formulation of general relativity theory; de Broglie's paper suggested a similar connection between quantum conditions and relativity. Finally, long before Raman and Forman's study, it was generally known that Peter Debye had asked Schrödinger to report on de Broglie's thesis at the physics colloquium in Zurich; this request may have provided the immediate stimulus for Schrödinger to start thinking about wave mechanics (Jammer 1966:257).

Schrödinger explained his theory by using the analogy which the Irish mathematician William Rowan Hamilton, at the beginning of the 19th century, had pointed out between mechanics and geometrical optics. Both mechanics and geometrical optics can be formulated in terms of "variational principles," that is, the postulate that the path followed by a particle or ray during a definite time interval 0 to t is that which makes a certain function, integrated over the path, a minimum. In mechanics, "Hamilton's principle" states that the "action" S, defined as

$$S = \int_0^t L\, dt,$$

where L is the "Lagrangian function" $= T - V$ ($T =$ kinetic energy, $V =$ potential energy), is a minimum. Analytically this means $\delta S = 0$, where δ is an infinitesimal variation of the path. In geometrical optics the corresponding postulate is known as "Fermat's principle," or the principle of least time, $\delta \int ds/u = 0$. Here u is the instantaneous speed of light, which varies with the index of refraction in the medium: $u = c/n$.

Hamilton's theory had been further developed by the German mathematician K. G. J. Jacobi, who showed in 1837 that Hamilton's two partial differential equations for S could be reduced to a single equation having the form

$$\frac{\partial S}{\partial t} + H\left(q_1 \ldots q_n; \frac{\partial S}{\partial q_1} \cdots \frac{\partial S}{\partial q_n}; t\right) = 0, \qquad (2)$$

where H is the "Hamiltonian function," which, for a conservative system, is just equal to the total energy $T + V$. The q_i are generalized coordinates of the system. Eq. (2) is known as the Hamilton-Jacobi equation.

The basic equation for wave propagation, derived by the French mathematician Jean le Rond d'Alembert for the simplest case of a vibrating string in 1746, may be written

$$\nabla^2 \phi - \frac{1}{u^2}\frac{d^2\phi}{dt^2} = 0, \qquad (3)$$

where ∇^2 is the differential operator

$$\nabla^2 = \frac{\partial^2}{\partial x^2} + \frac{\partial^2}{\partial y^2} + \frac{\partial^2}{\partial z^2}.$$

Schrödinger now asked the question: If Eq. (3) represents the generalization to wave optics of geometrical optics (the latter being valid in the limiting case of small wavelengths) and if geometrical optics is analogous to Newtonian particle mechanics because of the mathematical similarity of Hamilton's principle and Fermat's principle, what is the generalization of Newtonian particle mechanics analogous to Eq. (3)?

The answer is based on de Broglie's postulate, Eq. (1), written in the particular form appropriate to Hamiltonian mechanics. The energy of a particle of mass m moving at speed v subject to a force with potential energy V is

$$E = \tfrac{1}{2}mv^2 + V;$$

hence its momentum $p = mv$ is $\sqrt{2m(E - V)}$. Substitute this in Eq. (1); the particle is now characterized by a wavelength

$$\lambda = \frac{h}{\sqrt{2m(E - V)}}.$$

It is therefore represented by a wave, which can be written mathematically in the form

$$\phi = e^{-i\omega t}\psi(x, y, z), \tag{4}$$

where the circular frequency ω is related to the wavelength and the wave velocity in the medium by the equation

$$\frac{2\pi}{\lambda} = \frac{\omega}{u}.$$

Substituting Eq. (4) into Eq. (3), we get

$$\nabla^2\psi - \frac{1}{u^2}(-\omega^2)\psi = 0,$$

which reduces (using de Broglie's postulate) to

$$\nabla^2\psi + \frac{8\pi^2 m}{h^2}(E - V)\psi = 0. \tag{5}$$

This is the *Schrödinger equation,* the most important equation in 20th-century physical science, since it is now believed to determine (apart from relativistic corrections) the behavior of all material systems.

Schrödinger then applied his equation to the hydrogen atom, where the potential is $V = -e^2/r$. Since the problem has spherical symmetry, it is convenient to use polar coordinates r, θ, ϕ; using the standard rules for transforming coordinates, one has

$$\nabla^2 = \frac{\partial^2}{\partial r^2} + \frac{2}{r}\frac{\partial}{\partial r} + \frac{1}{r^2}\frac{\partial^2}{\partial \theta^2} + \frac{\cot\theta}{r^2}\frac{\partial}{\partial \theta} + \frac{1}{r^2\sin^2\theta}\frac{\partial^2}{\partial \phi^2}.$$

The technique of "separating the variables" is now used to solve the differential equation in three variables by reducing it to three equations each in one variable:

$$\text{let} \quad \psi = R(r)\Theta(\theta)\Phi(\phi);$$

then we have

$$\left\{\frac{d^2}{dr^2} + \frac{2}{r}\frac{d}{dr} + \frac{8\pi^2\mu}{h^2}\left(E + \frac{e^2}{r}\right) - \frac{\lambda}{r^2}\right\}R = 0, \tag{6}$$

$$\left\{\frac{d^2}{d\theta^2} + \cot\theta\frac{d}{d\theta} + \lambda - \frac{m^2}{\sin^2\theta}\right\}\Theta = 0, \tag{7}$$

$$\left\{\frac{d^2}{d\phi^2} + m^2\right\}\Phi = 0. \tag{8}$$

Here, to agree with conventional notation, I have used the "reduced mass" $\mu = mM/(m + M)$, where M is the mass of the proton (nucleus of the hydrogen atom) and m is the electron mass. In the case $M/m \to \infty$, $\mu \to m$. The letter m in Eqs. (7) and (8) no longer means mass but a numerical constant which, along with λ, has to be introduced in the process of separating the variables.

Since the constants m and λ turn out to play a significant role in wave mechanics, it may be helpful to recall why they are needed in solving partial differential equations. The reason is that if you start from an equation of the form

$$f(x, y, z) = 0$$

(where f includes various derivatives as well as ordinary functions) and try to find a solution of the form

$$f = X(x)Y(y)Z(z),$$

you will usually get, on substitution and carrying out the differential operations, an equation that looks like

$$\{X\} + \{Y\} + \{Z\} = 0,$$

where $\{X\}$ stands for some combination of derivatives of X and other functions of x, etc. If this equation is to be true when x, y, z vary independently, each expression in braces must be a constant, so that the sum of the three constants is zero:

$$\{X\} = a, \{Y\} = b, \{Z\} = c, a + b + c = 0.$$

This is how you get your three separate equations in one variable each—but at the price of introducing some arbitrary constants not present in the original equation.

Up to this point in Schrödinger's theory *no quantization has been introduced.* Planck's constant h is merely a numerical parameter relating the wavelength of a particle to its momentum. But now quantization suddenly appears *as a mathematical consequence of the natural requirement that the wave function satisfying the Schrödinger equation be single-valued.* And the "quantum numbers" emerge from the constants needed to separate the variables in the differential equation. This can be seen most simply in Eq. (8), which describes the dependence of the wave function on the angle ϕ. The solution is

$$\Phi = \sin m\phi \quad \text{or} \quad \cos m\phi \quad \text{or} \quad e^{im\phi}.$$

If you add 2π to the angle you will be back in the same place, and Φ must have the same value; but this can be true only if *m is an integer.*

The solution of Eq. (7) was well known; it is derived from the so-called Legendre polynomials introduced by the French mathematician Adrian-Marie Legendre in problems on the gravitational attraction of spheroids, and extensively developed by Laplace. The main point here is that the solution depends on both parameters m and λ, and that λ must have the value $l(l + 1)$, where l is an integer; also, $|m|$ must be less than or equal to l.

Similarly the first equation (6) can be solved, using "Laguerre polynomials" (discovered in 1879 by the French mathematician Edmond Nicolas Laguerre, in connection with his investigation of properties of the integral $\int_x^\infty e^{-x}/x \, dx$). The result is that two more integers must be introduced, n and n', called the total quantum number and radial quantum number respectively; they are related by

$$n = n' + l + 1.$$

It is then found that the energy of the system depends only on n, through the equation

$$E = -|E| = -\frac{\mu e^4}{2\hbar^2 n^2} \tag{9}$$

where $\hbar = h/2\pi$ in the usual notation. *This is just the equation for the energy levels in Bohr's model of the hydrogen atom* (§3.3, Eq. 7).

Thus Schrödinger succeeded immediately in reproducing all the *observable* properties of the Bohr atom, and at the same time provided a general equation that could be applied to more complicated systems.

The solution of the Schrödinger equation in general yields a set of *eigenfunctions* (German for "proper" or "characteristic" functions), each of which has a definite numerical parameter known as the *eigenvalue* (half-translated from the German *eigenwert*, proper value), corresponding to the energy. Depending on V, this set of eigenvalues (known as the "spectrum") may be finite or infinite, and may include both discrete and continuous values. Thus in the case of the hydrogen atom there are an infinite number of *negative* energy levels given by Eq. (9), corresponding to *bound states* of the electron, plus a continuous set of positive energies, corresponding to *free states*.

In the simple cases in which Schrödinger's equation can be solved exactly or almost exactly, it was found to predict the correct values of spectral frequencies and other observable properties. As electronic computers have been applied to more complicated systems, the scope of applicability of the equation has been extended to cover all atoms and many of the smaller molecules. But it should be noted that the Schrödinger equation for a system of N particles is a partial differential equation in $3N$ variables, so the difficulty of solving it increases rapidly as one considers larger molecules.

The major defect (if it can be called that) of Schrödinger's theory, like de Broglie's theory from which it evolved, is that it fails to explain just what is vibrating or "waving"; it is definitely *not* the electron, conceived as a point-particle analogous to a billiard ball, because that would still imply the classical conception of a particle localized at a particular spatial position and having a particular velocity, both of which change in a deterministic way from one instant to the next. Instead, the electron itself *is* in some (not visualizable) sense the wave. If the electron is in a definite *state* or energy level in the atom (corresponding to one of the allowed orbits in Bohr's theory), then it would seem to occupy simultaneously the entire orbit and to have simultaneously all values of the velocity vector corresponding to the different possible directions of motion in a circular orbit at constant speed. Even this statement is unduly restrictive, for the theory can be interpreted to mean that the electron is "spread out" over a region of space around the Bohr orbit and has a large range of differential speeds. This is a natural description of a wave disturbance in a continuous medium, but it seems difficult to reconcile it with our earlier conception of the electron as a particle whose mass, charge, and velocity can be experimentally determined. This is just as difficult, indeed, as it is to think of an electromagnetic wave as a vibration of some ethereal medium and at the same time as a particle having a definite momentum and energy.

In the case of electromagnetic waves, one could at least say that the vibrating quantities were electric and magnetic field intensities. For Schrödinger's electron waves, we can say that the vibrating quantity is a *wave function* ψ, determined by Eq. (5) or its generalization to time-dependent systems. The word "function" is used because, like the electric and magnetic fields E and H, ψ is a mathematical function of space and time variables; that is, its value depends on x, y, z, and t. (For an electron in a stationary state corresponding to a Bohr orbit, ψ does not depend on t.) However, unlike E and H, the values of ψ may be complex numbers (numbers which depend on the imaginary unit $\sqrt{-1}$), and therefore the function defies any simple physical interpretation.

According to Schrödinger's theory, one must first construct the wave equation for a physical system by converting the expression for the kinetic energy of each particle into a mathematical operation on the wave function ψ, that is, the sum of its second partial derivatives with respect to the space coordinates. Thus for a single particle, we see from Eq. (5) that the "kinetic energy operator" KE_{op} is

$$KE_{op} = \frac{-h^2}{8\pi^2 m} \nabla^2. \qquad (10)$$

In order to determine the stationary states of an atomic or molecular system, Schrödinger's theory requires us to add this kinetic energy operation to a potential energy operation PE_{op}, where PE_{op} is defined simply by multiplying ψ by the algebraic expression for the total potential energy of the system, V. (V usually consists simply of a sum of terms of the form $e_i e_j / r_{ij}$, where e_i and e_j are the charges of two interacting electrons or nuclei and r_{ij} is the distance between them.) The total operation is called the Hamiltonian[12] operator, H:

$$H = KE_{op} + PE_{op}.\qquad(11)$$

The fundamental equation of Schrödinger is then simply

$$H\psi = E\psi,$$

where E is a numerical constant which is equal to the energy of the system.

The Schrödinger equation (12) must now be solved for the unknown wave function ψ and the unknown energy E. In general there are several possible ψ functions which are solutions of this equation; each has its own eigenvalue E. Thus the fact that an atom has a discrete set of possible states and a discrete set of corresponding energy values emerges automatically as a mathematical property of the equation; the set of rules for constructing the Schrödinger equation replaces the previous postulates about allowed orbits in the Bohr theory. These rules apply to all conceivable systems, whereas in Bohr's theory the postulates had to be constructed by trial and error for each case, with only the physicist's intuition and experience to guide him.

(Additional references for this section: Gerber 1969, Kragh 1979, Kubli 1970, Ludwig 1968, Miller 1978, MacKinnon 1976, 1982.)

3.5 The Philosophy of Quantum Mechanics

The early success of quantum mechanics led physicists to try to interpret its concepts in terms of the classical categories applicable to particles. The first such attempt was made by Max Born (1926b). He proposed that the electron is really a particle, and that the wave function simply represents the *probability* that it follows a particular path through space. The use of the word "probability" here is not the same as in classical physics, for example in the kinetic theory of gases, in which it was assumed that the molecules have definite positions and velocities which are determined at each instant of time by Newton's laws and can in principle be calculated exactly. (As we saw in §§2.4 and 2.5, however, this assumption was not consistently maintained by kinetic theorists even before 1900.)

In Born's view, the positions and velocities of each subatomic particle are basically random; they cannot be said to have any definite values, even in principle, but have only certain probabilities of exhibiting

---

[12] Since it is analogous to Hamilton's function in Newtonian mechanics, see Eq. (2).

particular values. Born argued that the contrary assumption (adopted at first by Schrödinger), namely, that the electron is really a wave spread out in space with a charge density given by the wave function, is inadmissible. For if the electron is not bound in an atom, its wave function eventually will spread out over an infinite space; yet experiment indicates that all of the electron is actually present at some particular place. This result is still consistent with the idea that the electron has a certain probability of being at any given place; performing the experiment to find its position converts this probability into a certainty—it either *is* or *is not* there, but it cannot be "partly there." It is already evident that any attempt to pin down a definite property of the electron involves an act of measurement, on this view.

Max Born was awarded the Nobel Prize for physics in 1954, primarily for his statistical interpretation of quantum mechanics. The fact that the award came so long after the original publication indicates that this interpretation was not at once regarded as a "discovery" but that it gradually came to be accepted as a fruitful viewpoint in physical theory. Insofar as many scientists do accept the notion that natural processes are fundamentally random rather than deterministic, we have to admit that the Newtonian world-machine, or the mechanistic viewpoint advocated by Descartes and Boyle in the 17th century, has been abandoned.

Another way of characterizing the consequences of quantum mechanics is through Heisenberg's *indeterminacy principle*, often called the uncertainty principle. In 1927, Heisenberg proposed as a postulate that there are certain pairs of physical properties of a particle which cannot simultaneously be measured to an arbitrarily high degree of accuracy. The more accurately we try to measure one property in this pair, the less accurate will be our measurement of the other one. For example, if we try to determine both the position (x) and momentum (p) of an electron, then the product of the average errors or uncertainties of measurement (symbolized by δ) must be at least as great as $h/2\pi$:

$$(\delta x) \cdot (\delta p) \geq \frac{h}{2\pi}.$$

Although the principle applies to the position and velocity of any object, it is a significant limitation only for atomic or subatomic particles, since Planck's constant h is such a small number.

The indeterminacy principle can be deduced from quantum mechanics, but its validity was also established by considering various possible experiments which might be designed to measure both x and p. Such experiments were analyzed in some detail in a famous discussion between

Niels Bohr and Albert Einstein, beginning in 1927. The conclusion was that any measurement involves an *interaction* between the observer (or his apparatus) and the object being observed. For example one could measure the position of an electron by bouncing photons off it, in the same way that we "see" ordinary objects; but in order to do this very accurately it would be necessary to use very short-wavelength photons— otherwise they would simply be diffracted around the electron rather than reflected. But a short-wavelength photon has a large momentum and would therefore give the electron a strong kick when it hit it, changing the electron's momentum in an unpredictable way, thereby making it impossible for us to determine p accurately at the same time.

Heisenberg's principle might be interpreted as merely a restriction on how much we can *know* about the electron, taking into account the limitations of existing experimental methods, without thereby rejecting the belief that the electron really does have a definite position and momentum. The term "uncertainty principle" would then be appropriate, with its implication that the principle applies to the observer's knowledge rather than to nature itself.

There are two objections to that view, however. The first is the feeling of many physicists that science should be concerned only with concepts that have "operational" definitions, following the doctrine of the American physicist P. W. Bridgman. The 19th-century lumeniferous ether was frequently brought forward as a horrible example of a hypothetical entity on which considerable time and effort had been wasted, because,it could not be operationally defined. A second reason for being skeptical about the assumption that an electron really has a definite position and momentum is that no one has yet succeeded in constructing a satisfactory theory based on this assumption, despite numerous attempts.

Heisenberg did not suggest that the terms "position" and "momentum" be banished from physics, for they are needed to describe the results of experiments. Even the term "path" or "orbit" of an electron is still admissible as long as we take the attitude that "the path comes into existence only when we observe it" (Heisenberg 1927, Jammer 1966:329). This remark appears to some philosophers to align quantum mechanics with the traditional philosophy of subjective idealism, according to which the real world consists only of the perceptions of an observer, and physical objects have no objective existence or properties apart from such human observations. Taken together with Born's assertion that atomic quantities are inherently random, Heisenberg's viewpoint would seem to deny the possibility of finding out anything certain about a physical system taken by itself, since the observer or the measuring apparatus is an essential part of any physical system we may choose to study. (For example, the photon which was used to "see" the electron in the earlier experiment becomes part of the system being studied.)

A somewhat similar situation is quite familiar in the behavioral sciences. In almost any psychological experiment, the subject knows that his behavior is being observed, and may not behave in the same way that he would if no psychologist were present. But at least in that case we feel confident that there *is* such a thing as "behavior when the psychologist is not present," even if we do not observe it, and we probably believe that if the psychologist were sufficiently clever he could carry out his experiment without the knowledge of the subject—if his scientific and moral code of ethics did not forbid it.

A better analogy might be the intelligence test, which purports to measure an inherent property of the individual, but which is now recognized to be so strongly dependent on the circumstances of testing and on cultural presuppositions that it is doubtful whether "intelligence" really has an objective existence, except in the operational sense of ability to perform *on a particular test* and others like it ("intelligence is what the intelligence test tests").

Quantum mechanics is rescued from the quicksands of subjectivism by the fact that it makes perfectly definite predictions about the properties of physical systems, and these properties can be measured with the same results (including the measurement uncertainty) by all observers. If one prefers, one does not *have* to abandon the idea that there is a real world "out there," and even the fact that the measuring apparatus is, strictly speaking, inseparable from the things we are measuring does not in practice make very much difference.

Moreover, in spite of the fact that our knowledge of the position and momentum of electrons, photons, and all other particles is restricted to probabilities, the wave function which determines those probabilities is itself completely determined by the theory, and changes with time in a very definite, "causal" way. If ψ is given at a particular time and the Hamiltonian operator of the entire system is known, then ψ can be computed at any other time. (The "entire system" includes of course any measuring device that may interact with the system.) Thus, as Heisenberg himself pointed out, "probability" itself acquires a certain "intermediate reality," not unlike that of the *potentia* (possibility or tendency for an event to take place) in Aristotle's philosophy. An isolated system which is evolving in this way but not being observed is, according to Heisenberg, "potential but not actual in character"; its states are "objective, but not real" (Pauli 1955:27).

This viewpoint can only be applied to atomic systems; otherwise it is likely to lead to absurdities, as Schrödinger showed with his famous "cat paradox" (1935). Suppose a cat is placed in a chamber with a radioactive sample and a Geiger counter connected to a device that will kill the cat if a radioactive particle is detected, and suppose the probability that this happens in a definite time interval (say one hour) is arranged to be $\frac{1}{2}$. At the end of the experiment the cat (according to a literal interpretation of quantum mechanics in Heisenberg's sense) is represented by a wave function which can be written[13]

$$\psi = \psi_A + \psi_D,$$

where ψ_A means the cat is alive and ψ_D means it is dead. According to Heisenberg's viewpoint the cat is not "really" alive or dead but in an intermediate "potential" state—until we open the chamber and look at it. At the instant when we *observe* the cat, its wave function "collapses" to either ψ_A or ψ_D, and it becomes either alive or dead. (The paradox is that this interpretation violates "common sense" somewhat more blatantly than do other aspects of modern physical theories.)

Although quantum mechanics has stimulated much discussion by philosophers, it would be inaccurate to claim that the adoption of this theory in physics requires the acceptance of either of the traditional doctrines in philosophy, idealism or realism. The significance of quantum mechanics is rather that it forces us to take an attitude toward the descriptive categories when doing atomic or subatomic physics that is different from the attitude we employ in talking about the physical world in other cases. (This attitude can itself be considered a philosophical doctrine, often called "instrumentalism.")

One expression of the new attitude is the *complementarity principle*, proposed by Niels Bohr in 1927. According to this principle, two modes of description of a given system may appear to be mutually exclusive, yet both may be necessary for complete understanding of the system. Thus, on the one hand, we may wish to emphasize *causality* by pointing to the fact that there is a well-determined evolution of the state of the system (defined by its wave function), as mentioned above. Yet this description is meaningful only if we refrain from making any observations of space and time variables, since the process of making such observations would disturb the state in an unpredictable way and destroy the causality. On the other hand, we might prefer to emphasize the spatiotemporal description and sacrifice the causality. Bohr held that each description gives a spatial glimpse of the total "truth" about the system taken as a whole.

Bohr accepted the indeterminacy principle as a useful quantitative limitation on the extent to which complementary (and therefore partial)

---

[13] Strictly speaking it should be "normalized," $\psi = (1/\sqrt{2})(\psi_A + \psi_D)$.

descriptions may overlap—that is, an estimate of the price that must be paid for trying to use both descriptions at once. But he did not follow Born and Heisenberg in their insistence that the "wave nature of matter" meant nothing more than the inherently random behavior of entities that are really particles. Instead, Bohr argued that the wave and particle theories are examples of complementary modes of description, each valid by itself though (in terms of Newtonian physics) incompatible with each other.

The complementarity principle is a major component of what is now called the *Copenhagen interpretation* of quantum mechanics. This interpretation assumes that the physical world has just those properties which are revealed by experiments, including the wave and particle aspects, and that the theory can deal only with the results of observations, not with a hypothetical "underlying reality" that may or may not exist. Any attempt to go further than this, to specify more precisely the microscopic details of the structure and evolution of an atomic system, will inevitably encounter nothing but randomness and indeterminacy.

The Copenhagen interpretation met, and still meets, with strong objections from a minority of leading physicists, including Einstein, de Broglie, and Schrödinger. There is understandably a deep-seated resistance to abandoning the traditional conception of physical reality, and much appeal in Einstein's famous critique of inherent randomness—"God does not play dice." But so far no satisfactory alternative interpretation of quantum mechanics (nor any substantial experimental refutation of the theory itself) has been found. (See Brush 1980 for further details.)

IV. QUANTUM MECHANICAL PROPERTIES OF MATTER

4.1 SPECIFIC HEATS OF SOLIDS AND GASES

The first application of statistical mechanics to quantum theory was that of Planck (1900b) and Einstein (1905), that is, the invention of quantum theory itself. I have also mentioned Einstein's theory of the specific heats of solids (1907). In this section I will summarize some additional applications of the quantum theory to thermal properties of matter in the second decade of the 20th century; we are still dealing here with the "old" quantum theory, before the discovery of wave and matrix mechanics, and before the concepts of Bose-Einstein and Fermi-Dirac statistics were developed. The successes of these applications, at a time when a comprehensive theory of atomic structure and spectra had not yet been established, encouraged theorists to continue their search for such a theory based on the quantum hypothesis.

In 1912 Peter Debye, a Dutch physicist/chemist who was temporarily at the University of Zurich, improved Einstein's theory of the specific heats of solids, making it theoretically more satisfactory and experimentally more accurate. Whereas Einstein had assumed that all atoms vibrate with the same frequency v, so that the energy of N atoms is given by

$$\frac{3Nhv}{e^{hv/kT} - 1},$$

Debye assumed that the solid is like a continuous solid to the extent that it has a continuous spectrum of possible frequencies starting with zero, but going up to a maximum frequency v_m determined by the condition that the total number of modes of vibration is $3N$. The average energy of the system is then given by

$$U = \frac{9N}{v_m^3} \int_0^{v_m} \frac{hv(v^2 \, dv)}{e^{hv/kT} - 1} = 9NkT \left(\frac{T}{\theta}\right)^3 \int_0^{\theta/T} \frac{\xi^3 d\xi}{e^\xi - 1},$$

where $\theta = hv_m/k$ is now called the "Debye theta." (Recall that, just as in the classical theory of black-body radiation, the number of modes of vibration between v and $v + dv$ is proportional to v^2.) The specific heat is then

$$c = (3Nk)\left\{\frac{12}{x^3}\int_0^x \frac{\xi^3 d\xi}{e^\xi - 1} - \frac{3x}{e^x - 1}\right\} \qquad \text{where} \quad x = \theta/T.$$

In the limit $T \to \infty$ the expression in braces goes to 1, leaving just the Dulong-Petit law $c = 3Nk$. In the limit $T \to 0$, the expression in braces goes to $(4/5)\pi^4 x^{-3}$, so the specific heat is proportional to the third power of the absolute temperature at very low temperatures. (This is what one would expect from the analogy with black-body radiation, since the Stefan-Boltzmann law, $E \propto T^4$, implies $C \propto T^3$.)

Debye's formula was shown to agree better with experimental data than Einstein's, which vanishes more rapidly at low temperature, as $e^{-hv/kT}$ (Schrödinger 1919, Moelwyn-Hughes 1961:103–4). Further refinements were introduced in another theory published at about the same time (1912) by Max Born and Theodore von Kármán; they attempted to take account of the actual frequency spectrum for an atomic lattice rather than estimating the spectrum by that of a continuous solid. Their theory is the basis for more recent research, but Debye's established the fruitfulness of the quantum theory in a very simple, direct way.

As mentioned earlier, one of the scientists most active in developing the applications of quantum theory to thermal properties of matter was the German chemist Walther Nernst (Hiebert 1978). Nernst and his collaborators provided much of the experimental specific-heat data at low temperatures, both to test his "new heat theorem" (forerunner of the third law of thermodynamics) and to test the predictions of quantum theory. In addition to solids, they studied several gases.

According to Lord Kelvin (1901), one of the two "clouds over the 19th century theories of heat and light" was the apparent failure of the equipartition theorem to apply to the internal motions of atoms and molecules. It it were correct, the ratio of specific heats should be

$$\frac{c_p}{c_v} = \gamma = \frac{n + 2}{n},$$

where n = number of degrees of freedom of the molecule. (The specific heat at constant pressure is greater than that at constant volume because if the substance is allowed to expand as it is heated one has to provide additional energy to do work against the external pressure.) If atoms were point-masses, then one should have $\gamma = 5/3$ for monatomic gases, $\gamma = 4/3$ for diatomic, etc., counting just three degrees of freedom per atom.

If the atom has internal structure, for example is composed of electrons and a nucleus, then according to classical statistical mechanics each of its parts should share the total energy, n would be much larger, and γ should be close to one. But it was found by experiment that, to a good approximation,

(1) $\gamma = 5/3$ for monatomic gases (mercury, argon, helium, etc.);
(2) $\gamma = 1.4$ for diatomic gases at room temperature (H_2, O_2, N_2, etc.).

From (1) it was concluded that the equipartition theorem does apply to atoms taken as point-masses but not to their internal parts. From (2), it was concluded that $n = 5$ for diatomic molecules. This could be explained, according to Boltzmann (1876) and Bosanquet (1877), by assuming that the diatomic molecule is a rigid solid of revolution, perhaps like a dumbbell. Theoretically such an object has six degrees of freedom, since it can rotate around any of three perpendicular axes as well as moving in the three spatial directions. However, Boltzmann argued that rotation around the axis of symmetry (the line joining the two atoms) would not be changed by collisions between molecules, and therefore would not contribute to the specific heat of the gas; one may subtract one degree of freedom to get $n = 5$, $\gamma = 7/5 = 1.4$.

Maxwell (1877) protested that such models are unsatisfactory; they assume that the molecule is so rigid that its parts cannot be set into vibration by collisions, except perhaps at very high temperatures. But this assumption is inconsistent with the results of spectroscopy. The bright lines in the spectrum of a gas are fixed at the same position over a range of temperatures, which shows (Maxwell argues) that they result from harmonic vibrations; that is, the restoring force is proportional to the displacement, which means that "all impacts, however slight, will produce vibrations." Moreover, many gases have absorption bands even at room temperature, "which indicate that the molecules are set into internal vibration by the incident light." Since internal motions of molecules do exchange energy with the ether, the equipartition theorem must apply to them (Maxwell 1877:246). Like Newton, who refused to feign hypotheses, Maxwell concluded that the discrepancy between theory and experiment might "drive us out of all the hypotheses in which we have hitherto found refuge into that state of thoroughly conscious ignorance which is the prelude to every real advance in knowledge" (1877:245).

Despite Kelvin's "two clouds" lecture, the specific-heats anomaly did not play a prominent role in the early development of quantum theory. It was not seriously discussed until 1911, when Nernst proposed that the Planck-Einstein quantum formula should be applied to the energy of vibration of atoms in a diatomic molecule. By this time it was known that specific heats of gases are not constant but vary with temperature; above 2000°C, diatomic molecules do appear to have six degrees of freedom. The Danish chemist Niels Bjerrum showed that Nernst's formula does account for the specific-heat data on hydrogen, nitrogen, and carbon dioxide fairly well at temperatures from 18° up to more than 2,200°C. This explanation of the specific-heat anomaly had the further advantage that the vibration frequency assumed in the formula $E = hv/(e^{hv/kT} - 1)$ was not merely an arbitrary parameter but could be related to spectroscopic determinations of molecular vibrations.

The reason why the vibrations of the atoms do not contribute to the specific heat of a diatomic or polyatomic molecule at room temperature is that the fundamental vibration frequency is so high that $hv \gg kT$; hence the average energy in this degree of freedom is very small according to quantum theory. Similarly, the motions of electrons do not contribute unless radiation is emitted or absorbed, which will not happen unless the temperature is high enough to excite electrons from their ground states. Finally, the atom may be regarded as a point-mass for the purpose of computing its specific heat, and rotation of a diatomic molecule around the interatomic axis may be ignored, because the energy of rotation is inversely proportional to the moment of inertia. If the moment of inertia is very small (as it is when almost all of the mass is concentrated in the nucleus), the quantum of rotational energy is much larger than kT at room temperature. (The absence of this rotational specific-heat contribution is considered obvious by modern writers; I have not been able to find a detailed quantitative justification for it in the early literature.)

Nernst, in 1916, extended his thermodynamic theory to predict that all gases must reach a state of "degeneration" at very low temperatures. According to his heat theorem, the specific heat of all substances must tend to zero at absolute zero temperature; in this limit, the pressure becomes independent of temperature and depends only on volume. He suggested that at low temperatures the translational motion of gas molecules is converted into circular motions around equilibrium positions. He thereby arrived at the following pressure-volume relation ("equation of state") in the limit $T \to 0$:

$$p = \frac{h^2}{4\pi m} (N/V)^{5/3}.$$

It is remarkable that this is just the same, aside from a numerical factor, as the equation of state of an ideal "Fermi-Dirac" gas at zero temperature, although Fermi-Dirac statistics had not yet been invented (see §4.5).

4.2 ABSOLUTE ENTROPY AND CHEMICAL REACTIONS: IONIZATION AND STELLAR ATMOSPHERES

Another application of quantum theory was made by Otto Sackur in 1911. He remarked that if the full potentialities of quantum statistical mechanics are to be realized, we must find out how to apply it to chemically reacting systems. For this purpose we need an *absolute* definition of entropy, or at least one that will allow us to compare the entropies of different chemical species; classical thermodynamics defines only entropy *changes*. Thus, for an ideal gas (for which $\partial U/\partial v = 0$ and $p = RT/v$), we have from the second law:

$$dS = \frac{dQ}{T} = \frac{dU}{T} + \frac{p\,dv}{T} = \frac{1}{T}\frac{\partial U}{\partial T}dT + \frac{1}{T}\frac{\partial U}{\partial v}dv + \frac{p\,dv}{T}$$

$$= c_v\frac{dT}{T} + \frac{R\,dv}{v} \tag{1}$$

If we integrate over T and v to find S, we have to introduce an integration constant characteristic of the particular gas, S_i:

$$S = c_v \ln T + R \ln v + S_i.$$

In a chemical reaction of the type

$$A + B \rightleftharpoons AB,$$

it can be shown that the equilibrium constant is given by

$$\ln \frac{c_{AB}}{c_A c_B} = \ln K = \frac{Q}{RT} + \sum \frac{c_v}{R} \ln T + \sum \frac{S_i}{R}, \tag{2}$$

where c_A, c_B, c_{AB} are the concentrations; Q is the heat of reaction.

If one adopts the Nernst heat theorem, which makes $S = 0$ at $T = 0$ for all substances, then the entropy constants S_i can be computed from vapor-pressure data. Or, if one tries to compute them from Boltzmann's relation between entropy and probability, one can count the number of ways of distributing N atoms into "elementary regions" of phase-space defined by differentials of position and velocity:

$$d\sigma = dx\,dy\,dz\,d\xi\,d\eta\,d\xi.$$

This gives

$$S_i = R\left(\frac{3}{2} + \ln\left[(2\pi R/M_i)^{3/2}\frac{1}{d\sigma}\right]\right). \tag{3}$$

If one lets $d\sigma \to 0$, as one would expect classically, this expression becomes infinite. This is only to be expected from "Boltzmann's formula" $S = k \ln W$, where W is infinite if one counts all possible configurations in a continuous space. In fact Boltzmann did not write the formula this way but only used it in the form of *differences* between entropies of a substance at different temperatures or pressures; the constant S_i then cancels out.

Arnold Sommerfeld (1911) suggested that the elementary region $d\sigma$ should be related to Planck's constant. Since it has the dimensions of distance × momentum, one may put.

$$h = m\,dx\,d\xi \tag{4}$$

and thus

$$d\sigma = h^3.$$

Sackur and H. Tetrode used this assumption in 1912 to compute the entropy constants (known as "chemical constants") for many gases, and found that the values agreed with those determined experimentally by Nernst's method.

In another of his amazingly fertile remarks, Nernst (1918:154) pointed out that the formula for the chemical constant could just as well be applied to the electron, and that the theory of chemical reactions could thus be applied to the thermal ionization of a gas, that is, the dissociation of a neutral atom into a positive ion and a negative electron. One needs to know the heat of dissociation (the ionization potential), which might be calculated from Bohr's atomic model, though this is still "somewhat hypothetical."

As we noted in §3.3, one of the first victories of the Bohr model was the explanation of the anomalous lines in the spectrum of the star ζ-Puppis, which had previously been attributed to hydrogen but which Bohr ascribed to ionized helium. Thus by 1918 there was not only good evidence that atoms in stars are ionized, but also that the energy levels of the remaining outer electrons can be described in terms of the quantum theory.

In a paper on radiative equilibrium published in 1917, Arthur Eddington, the British astrophysicist, introduced the hypothesis that atoms in the interior of stars are completely ionized. The average molecular weight of the stellar material played a major role in his theory, and in an earlier paper he had set this parameter equal to 54. He noted that Newall, Jeans, and Lindemann had independently suggested to him that the high temperatures involved—estimated then at 10^6 to 10^7 degrees—would probably disintegrate the atoms into nuclei and electrons, so that the average atomic weight would be 2. The result of this modification was to improve the agreement between theory and observation in some minor details, and more significantly to provide "an explanation of why stellar masses are of the order of magnitude actually found" (1917:597).

John Eggert, who was apparently working in Nernst's laboratory at the time, was the first to apply Nernst's chemical equilibrium formula (including the chemical constants) to ionization in stars, but he considered only the rather artificial case in which a ring of eight electrons is ejected all at once from an iron atom, and did not attempt to give an interpretation of spectral-line intensities (Eggert 1919).

Eggert's paper did stimulate the initial interest of the person who accomplished the major breakthrough in the application of statistical ionization theory to astrophysics. Meghnad Saha, an Indian mathematician who went into physics because of a personality conflict with the chairman of the mathematics department at the University College of Science (Calcutta), went to Alfred Fowler's laboratory in London in 1919, where he became acquainted with current experimental work on stellar spectra. In 1920 he published a paper on "Ionization in the solar chromosphere" in which he discussed the spectral lines "which are relatively more strengthened in the spark than in the arc, and which Lockyer originally styled as enhanced lines." Lockyer (with whom Alfred Fowler had worked for several years) argued that enhanced lines were produced by the products of atomic dissociation at high temperatures. But this explanation was not very satisfactory, since one would have to assume that the chromosphere is hotter than the photosphere, which is to say that the temperature of the sun increases outward from the center (Saha 1920a:472).

When Saha applied the equilibrium equation to the ionization of calcium atoms, using the Sackur-Tetrode formula for the chemical constant, he found that the fraction x of calcium atoms once ionized is given by the equation.

$$\log \frac{x^2}{1 - x^2} P = -\frac{U}{4{,}571T} + 2.5 \log T - 6.5, \qquad (5)$$

where U is the heat of dissociation, estimated (from the experimentally determined ionization potential) to be 1.409×10^4 cal. The equation immediately revealed the hitherto unsuspected importance of *pressure*: if P is increased, x must decrease. This is of course nothing more than a direct consequence of the principle of Henry Louis Le Chatelier, developed during the 1880s in connection with the equilibrium of gas reactions in the blast furnace: any change in conditions causes the system to respond in such a way as to produce an opposite effect. Since the ionization reaction, $Ca \rightarrow Ca^+ + e^-$, increases the number of particles, one can predict that an increase in the external pressure will push the reaction back toward the left.

Saha could now explain why the enhanced spectral lines, assumed to be due to ionized atoms, are stronger in the chromosphere than in the photosphere even though the temperature is probably lower: the reduction in pressure promotes ionization more than the reduction in temperature discourages it.

The new ionization theory also explained why the spectral lines of alkali metals seem to be completely absent from sunlight: the metals would be completely ionized at the pressure and temperature estimated for the sun's atmosphere, and the lines of the ionized atoms lie in the ultraviolet, where they are hard to detect. However, Saha predicted in a second paper (1920b:809) that the lines of rubidium and cesium might be found in the spectrum of sunspots, which are cooler than the rest of the sun's atmosphere. This prediction was confirmed in 1922 by H. N. Russell, who found that lines of rubidium were present in the spectra of sunspots recorded by Meggers and Brackett, and that the lines of potassium were very much strengthened.

Saha recognized that his theory might be used to assign temperatures to stars of various spectral classes, by obtaining theoretical expressions for the temperatures at which lines of the various ionized species appear. His first attempt to do this, in 1921, introduced an erroneous formula for second-stage ionization; he failed to realize that the electron concentration that occurs in the equation for each ionization reaction should be the total concentration of electrons from all ionization reactions going on the system, not just that one. The error was immediately corrected by E. A. Milne (1921), who noted that the degree of second-stage ionization is substantially reduced from Saha's value when the electrons from first-stage ionization are taken into account. But, apart from this correction, Milne was enthusiastic about the potential value of Saha's theory and forecast its wide application to many problems in astrophysics, including an elucidation of the differences in spectra between giants and dwarfs.

The next major advance was made by the British physicist Ralph Howard Fowler, at that time a fellow of Trinity College in Cambridge

University. Together with Charles Galton Darwin (grandson of the evolutionist Charles Robert Darwin), he developed a new approach to statistical mechanical calculations, using the theory of complex variables and the method of steepest descents to avoid some of the usual approximations. The Darwin-Fowler method incorporated Planck's "Zustandssumme" (§1.13),

$$Z = \sum_i e^{-E_i/kT},$$

where the E_i are quantized energy levels of the system or the individual atom. Clearly a detailed theory of spectral-line intensities would have to be based on formulae that specified not only the concentrations of various kinds of ions but also the fraction of those ions that are in the particular electronic state appropriate for the absorption or emission of a given quantum of energy.

But a difficulty immediately arises if one simply uses the energy levels E_i given by Bohr's theory of the hydrogen atom. According to this theory (§3.3), E_i is inversely proportional to i^2 for an infinite series of bound states, $i = 1$ to ∞, so the series for Z would diverge. This problem had already been noted by von Kármán (Herzfeld 1916). Fowler argued that the series must be cut off at a finite value of i on physical grounds, since the radius of the ith orbit is proportional to i^2 and such orbits could not, for large i, remain undisturbed by the presence of other atoms in the gas. Yet Fowler could see no satisfactory way of deciding how many terms in the series to include: "pure statistical theory knows no half-way house between an impossible state which can be ignored and a state which is to receive its full share of energy" (1923:20). An approximate solution of the problem was suggested independently by Enrico Fermi and Harold Urey a few months later: one can use the "excluded volume" b in the van der Waals equation of state (§1.12) to compute a correction to the entropy and free energy, which leads to an exponential damping factor in the higher terms of the series and forces it to converge (Fermi 1923, 1924; Urey 1924).

Fowler (1923) noted that even if one retains only the first term in the series, it must depend on the quantum "statistical weight" of the ground state of the atom or ion. Saha's thermodynamic treatment implicitly assumes that this weight is 1, but according to Bohr's theory it should be 2 for the hydrogen atom.

Fowler and Milne then published (1923) an improved version of Saha's method for determining the temperatures and pressures in stellar atmospheres from observations of the changes in intensities of absorption lines in the sequence of spectral types. They pointed out that Saha's method depended on the "marginal appearance" of a line being due to a small number of atoms, but gave no definite way of estimating how many atoms would be needed. The point of marginal appearance would be affected by variations in the abundance of the element from one star to another. A more reliable method, they suggested, would be to look for the place in the stellar sequence at which a given line attains its *maximum* intensity. The conditions for this maximum depend theoretically only on temperature and pressure, not on the relative abundances of the elements. For lines due to the neutral atom in the ground state there would not be any maximum, since the fraction of such atoms can only decrease as the temperature increases. But for all other series of lines (due to neutral excited atoms, or ionized atoms) there must be such a maximum, since the number of such atoms starts out at zero for low temperatures and must again become zero at extremely high temperatures, when all electrons have been lost.

Fowler and Milne found that their statistical formulas for maximum intensities could be fitted to observed data, provided that pressures in the reversing layers of stars are of the order of 10^{-4} atm, much smaller than those assumed by Saha (1 or 10^{-1} atm). This result was in agreement with other recent astrophysical evidence, though earlier it had been assumed that pressures of several atmospheres were found in stellar atmospheres. Since the Saha theory of ionization equilibrium was based on the assumption that the ideal gas laws are valid, it appeared that his theory would be *more* accurate in this case than had previously seemed likely (Payne 1925b:111).

As Milne had hinted in 1921, ionization theory also provided an explanation for the differences between spectra of giants and dwarfs of the same class; class is determined by temperature, but pressure comes in as an additional variable. The atmospheres of giant stars are at much lower pressures than those of dwarfs; hence there will be much great ionization and the enhanced lines will be more intense (Milne 1924:105).

Cecilia Payne, a British-born astronomer working at the Harvard Observatory, showed that ionization theory could also be used to estimate the abundance of elements in stellar atmospheres from their spectra. Assuming that all such atmospheres have the same composition and that at marginal appearance of a line the same number of atoms is involved for all elements, she used the Fowler-Milne method of maxima to obtain the fraction of the total number of atoms of a given kind active in absorption at a given temperature; the relative abundances of the corresponding atoms are then given directly by the reciprocals of the fractional concentrations at marginal appearance (Payne 1925a:193; 1925b:183).

Payne's calculations based on the Saha-Fowler-Milne ionization theory showed that the abundances of hydrogen and helium in stellar atmospheres are much greater than those of any other elements. But she refused to believe this result, since at that time it was generally assumed that the composition of stars is approximately the same as that of the earth's crust. She argued that "The stellar abundance deduced for these elements is improbably high, and is almost certainly not real" (1925a:197; cf. 1925:186–88). A year earlier K. T. Compton and H. N. Russell at Princeton had found that hydrogen lines are very intense in giant stars, suggesting "an absurdly great abundance of hydrogen" unless one assumed nonequilibrium effects, possibly due to metastable quantum states of the hydrogen atom (Compton & Russell 1924).

Abandoning the assumption of thermodynamic equilibrium would have vitiated much of the work based on Saha's theory. Yet current theories of the origin of the solar system made it difficult to accept a composition for the sun radically different from that of the earth; as Payne pointed out, "If . . . the earth originated from the surface layers of the sun [as supposed by Harold Jeffreys and others], the percentage composition of the whole earth should resemble the composition of the solar (and therefore of a typical stellar) atmosphere" (1925:185).

The problem of the anomalous astrophysical behavior of hydrogen and helium worried astronomers for several years. S. Rosseland (1925) suggested that electrical forces would tend to expel hydrogen from the core of a star so that it would accumulate at the surface, even though its total abundance might be small, and this explanation was tentatively accepted by Unsöld (1928) and McCrea (1929). A comprehensive analysis of all available data on the sun's atmosphere by H. N. Russell (1929) led him, and soon afterward most other astronomers, to accept the conclusion that stellar atmospheres do in fact contain a very large proportion of hydrogen and helium. In particular, the solar atmosphere consists of 60 parts of hydrogen (by volume), 2 of helium, 2 of oxygen, 1 of metallic vapors, and 0.8 of free vapors; this became known as the "Russell mixture." Rather than trying to devise mechanisms for separating small amounts of hydrogen from the core of a star, astronomers in the 1930s decided that they could explain a numbers of puzzling facts by assuming that hydrogen is the most abundant element practically everywhere (Strömgren 1932). The evolution of stars from one class to another (the path in the "Hertzsprung-Russell diagram) was then seen to be associated with changes in hydrogen content, and ultimately with a sequence of nuclear fusion reactions in which the heavier elements were built up from hydrogen (Strömgren 1933, 1938; Sitterly 1970; for further information on the astronomical problems mentioned in this section see Struve & Zebergs 1962, chap. XI).

4.3 ZERO-POINT ENERGY

In 1910 Max Planck published an article reviewing developments in radiation theory since 1900. Surprisingly (from the viewpoint of a modern reader) he refused to accept Einstein's hypothesis that electromagnetic radiation is quantized. Planck warned that one should not be so hasty in throwing out the wave theory of light, after all the struggles to establish it and after all its successes in explaining so many phenomena. He still believed in the strict validity of Maxwell's equations for empty space, thus excluding the possibility of discrete energy quanta in a vacuum.

The following year Planck published his "new radiation hypothesis." He now argued that his original assumption—that an oscillator can absorb energy only in discrete amounts—is really untenable; for suppose the oscillator is irradiated with radiation of very low intensity. In this case it will take a long time for the incoming energy to add up to the required quantum value $h\nu$; this is especially true for high-frequency radiation, which is present only in small amounts according to the Planck distribution. But one can simply remove the source of radiation after a short time, before a single quantum has been absorbed. The oscillator will then have absorbed only part of a quantum and cannot get any more, in contradiction to the hypothesis that absorption can occur only in integer multiples of a quantum. Because of this difficulty, Planck assumed instead that *absorption* is *continuous*, while *emission* is discrete.

An oscillator may now have, at any given time, an energy U which includes n quanta of amount $\epsilon = h\nu$, plus a fractional remainder ρ:

$$U = n\epsilon + \rho \qquad (n = \text{integer}, \ 0 \leq \rho \leq \epsilon).$$

If $U < \epsilon$, then $n = 0$ and the oscillator cannot radiate. It is clear that on the average the remainder will be equal to $h\nu/2$, and the average energy of the oscillator is thus

$$\bar{U} = \frac{h\nu}{e^{h\nu/kT} - 1} + \frac{h\nu}{2}.$$

The first term is the same as Planck's original formula, but the second represents an average residual energy that persists even at zero temperature.

Einstein and Stern (1913) suggested that in order to test the validity of the new radiation formula, one should look at a system in which the frequency changes with temperature. Such a system, they claimed, is formed by a gas of rotating dipoles in equilibrium with radiation. They assumed that the average energy of the dipoles,

$$E = \frac{J}{2}(2\pi v)^2 \qquad (J = \text{moment of inertia}),$$

should be equated to the average radiation energy at the same frequency.

$$
E = \begin{cases}
N\,\dfrac{hv}{e^{hv/kT} - 1} & \text{(a)} \\[3ex]
N\left[\dfrac{hv}{e^{hv/kT} - 1} + \dfrac{hv}{2}\right] & \text{(b)}
\end{cases}
$$

where (a) or (b) applies according to Planck's first or second theory, respectively. This assumption determines v as a function of T by means of a transcendental equation which can be solved numerically, and one can then compute the specific heat as a function of T. They found that (b) fits the experimental data on hydrogen better, especially at low temperatures. However, they were not convinced that the zero-point energy should really be exactly $\frac{1}{2}hv$ rather than some other value.

Later it was found that zero-point energy is a direct consequence of quantum mechanics. Indeed, it follows from the indeterminacy principle that even at the temperature of absolute zero a particle cannot be at rest; moreover, the pressure of a gas at zero temperature does not vanish, as it does in classical physics, but has a finite value depending on its density (see §4.5 and the first footnote in §4.6).

4.4 BOSE-EINSTEIN STATISTICS

After the establishment of quantum mechanics it was found that a system of identical particles will have markedly different statistical properties, depending on whether the wave function remains the same or changes its sign when the coordinates of two particles are interchanged. These two possibilities are now called "Bose-Einstein statistics" and "Fermi-Dirac statistics" respectively. The first kind was actually developed before quantum mechanics, though its full significance was not appreciated until later (Delbrück 1980).

In June 1924 the Indian physicist Satyendranath Bose sent Einstein a short paper in English on the derivation of Planck's radiation law, asking him to arrange for its publication in the *Zeitschrift für Physik*.[1] Einstein translated it into German himself and added a note stating that he considered the new derivation an important contribution; it was published a few weeks later.

Bose derived Planck's distribution law without making any use of classical mechanics and electrodynamics or of Bohr's correspondence principle. Instead he used a purely combinatorial method, assuming that radiation consists of quanta of various frequencies, there being N_s quanta of frequency v_s. According to Einstein's 1916 paper, each quantum of energy hv_s has momentum hv_s/c. If volume in phase-space is divided into cells each of volume h^3 (see §4.2) and one takes account of the two possible directions of polarization,[2] then the number of cells corresponding to a frequency interval $(v, v + dv)$ will be $8\pi Vv^2\, dv/c^3$. Bose assumed, in effect, that the quanta themselves are indistinguishable, so that permutations of quanta do not count as separate arrangements. This is the key to the new derivation, though Bose did not discuss it explicitly. He simply stated that the number of ways of distributing the N_s quanta in these cells such that there are $p_0{}^s$ vacant cells, $p_1{}^s$ with one quantum, $p_2{}^s$ with two quanta, etc. is just

$$\frac{(A_s)!}{p_0{}^s!\, p_1{}^s! \dots} \qquad \text{where} \quad A_s = \frac{8\pi v_s{}^2}{c^3}\, dv_s,$$

provided that the p's satisfy the condition

$$\sum_i p_i{}^s = N_s.$$

[1] Bose was at that time a lecturer at Dacca University in East Bengal; he had previously collaborated with Meghnad Saha on the theory of the equation of state of gases, and on an English-language anthology of Einstein's relativity papers. His famous quantum statistics paper was rejected by the *Philosophical Magazine*, where he first tried to publish it in 1923, and the original English version does not seem to have survived. In 1971 William Blanpied interviewed Bose and attempted to find out the origin of his ideas on quantum statistics. Bose said "that he had taught radiation theory for years and had worried excessively about the problem of entropy in black body radiation." Asked "if his studies in Indian philosophy had provided him with an insight denied the typical Western scientist," he found this suggestion absurd and asserted that Indian philosophy had no influence on his physics. After sending his paper to Einstein, Bose spent two years in Europe, but did not succeed in his hopes of developing a collaboration with Einstein, apart from some brief meetings. He returned to a professorship at Dacca in 1926, and spent most of his time there teaching and advising graduate students, publishing little of significance. For biographical details see Singh et al. (1974).

[2] Bose told Blanpied in 1971 that he deleted a section discussing polarization, on Einstein's recommendation.

Applying this argument to all frequencies, one finds that the number of microstates corresponding to a particular set of values of the N's is

$$W = \prod_s \frac{(A_s)!}{p_0{}^{s}! p_0{}^{s}!} \cdots$$

If W is maximized with respect to variations of the N's, taking account of the condition that the total energy $E = \sum_s N_s h\nu_s$ is fixed, one arrives by a simple calculation at Planck's formula.

Einstein immediately followed up Bose's approach by applying it to a monatomic ideal gas. In this case the maximization of W has to be subject to the additional condition that the total number of particles $N = \sum_s N_s$ is fixed. He found that if the sum over energy levels is replaced by an integral, the equation for the total number of particles has a maximum possible value; for specified volume V and temperature T, this value is

$$N = \frac{(2\pi m k T)^{3/2} V}{h^3} \sum_{n=1}^{\infty} \frac{1}{n^{3/2}}.$$

Einstein suggested that if the number is increased beyond this value, some atoms will have to go into the lowest quantum state with zero kinetic energy; the phenomenon would be analogous in some ways to the condensation of a saturated vapor. Yet there are no attractive forces to hold the atoms together in any particular region of space.

Einstein's prediction of a quantum condensation of gases, published in 1925, was not taken seriously by other physicists for more than a decade. One reason for this was that G. E. Uhlenbeck, in his dissertation in 1927, criticized Einstein's mathematical treatment of the problem; he argued that the result was due only to replacing the sum over energy levels by an integral, and that if the sum was computed accurately no condensation would occur. Although Uhlenbeck's dissertation was published only in Dutch, his research director was Ehrenfest, and presumably his criticism became known to the physics community through informal communications.[3] Uhlenbeck retracted his criticism in 1938 (Kahn & Uhlenbeck 1938:410).

4.5 FERMI-DIRAC STATISTICS

Before discussing the later history of the Bose-Einstein condensation, we must first introduce the alternative Fermi-Dirac statistics. This is based on the "exclusion principle," proposed in 1925 by Wolfgang Pauli

---

[3] See for example Ehrenfest's letter to Einstein, quoted by Uhlenbeck (1980). For the views of Schrödinger and others on this point see Hanle (1975:232–50).

in order to explain the anomalous Zeeman effect in atomic spectra, and more generally to provide a systematic method of accounting for the filling of allowed orbitals by electrons. By introducing a new two-valued quantum number for the electron (which was subsequently identified as "spin" by Uhlenbeck and Goudsmit), it was possible to lay down the general principle that no more than one electron in an atom may have the same set of four quantum numbers. In other words, each set of quantum numbers defines a state or cell which may be "occupied" by either zero or one electron. (The chemical aspects of this theory will be taken up in §5.8.)

The Italian physicist Enrico Fermi presented his quantum theory of the ideal gas, based on the Pauli exclusion principle, to the Accademia dei Lincei in February 1926, and submitted a comprehensive paper on this theory to the *Zeitschrift für Physik* on March 26; Dirac's paper was presented to the Royal Society of London on August 26, 1926.

Both Fermi and Dirac presented essentially the same derivation for their distribution law, though Fermi worked out its consequences in more detail. Let A_s, as before, represent the number of cells or quantum states with energy $E_s = hv_s$. The number of atoms (N_s) that can occupy these cells is now limited to the set of values $0, 1, \ldots, A_s$. Hence the number of arrangements is

$$W = \prod_s \frac{A_s!}{N_s!(A_s - N_s)}.$$

When this expression is maximized, keeping fixed the total number of particles and the total energy, the result is that the equilibrium value of N_s is

$$N_s = \frac{A_s}{e^{\alpha + E_s/kT} + 1}. \qquad \text{(Fermi-Dirac)}$$

The corresponding result for Bose-Einstein statistics is

$$N_s = \frac{A_s}{e^{\alpha + E_s/kT}}. \qquad \text{(Bose-Einstein)}$$

In each case the parameter α is to be determined from the conditions on the total number of particles and total energy.

Dirac pointed out that the difference between Bose-Einstein and Fermi-Dirac statistics corresponds to the difference between wave functions that are symmetric and antisymmetric with respect to interchanges of particles.

"Symmetric" in this case means that the function remains the same, whereas antisymmetric means that it changes sign. In both cases the square of the wave function (or the wave function multiplied by its complex conjugate, if complex) remains the same, that being the quantity that has a direct physical significance. Moreover, if two particles are in the same quantum state and the wave function is antisymmetric, it must be zero. (For example, if the wave function is expressed in terms of products of single-particle wave functions, where the label of the particle is in parentheses and the quantum number is written as a subscript, $\psi_{n_1}(1)\psi_{n_1}(2) = -\psi_{n_1}(2)\psi_{n_1}(1)$ can be true only if the product is zero.) Thus the Pauli principle is automatically satisfied for antisymmetric wave functions.

Fermi in his 1926 paper showed that at low temperatures the equation of state of the Fermi-Dirac gas has the form

$$p = \frac{ah^2 n^{5/3}}{m} + b\frac{mn^{1/3}k^2 T^2}{h^2} + \ldots,$$

where a and b are numerical constants and n = number of particles in unit volume. This is almost identical to Nernst's result for "degenerate" gases in the limit $T \to 0$ (see above, §4.1). The fact that the gas has a finite pressure at zero temperature is a consequence of its quantum "zero-point energy," which in turn may be regarded as a consequence of the indeterminacy principle—if the molecules were *not* moving at absolute zero, one would know their momentum with zero uncertainty, while the uncertainty in position would still be finite if the same gas could be confined in a container.

Pauli (1927) and Sommerfeld (1927, 1928) used Fermi-Dirac statistics to construct a theory of electrons in metals. It had already been realized by others that many properties of metals could be explained by assuming that the electrons are free to move throughout the system (Riecke 1898, Drude 1900, Lorentz 1909), but this assumption seemed inconsistent with the fact that electrons made hardly any contribution to the specific heat of the metal.

According to Fermi's theory, the distribution law for the particles has the form

$$f(\epsilon) \sim \frac{1}{e^{(\epsilon - \epsilon_0)/kT} + 1},$$

where ϵ_0 is a constant determined by the total number of electrons, called the Fermi energy. It is the maximum energy one gets by filling all the possible quantum states starting with the one of lowest energy. This is in fact the way the states will be filled at zero temperature,

$$\lim_{T \to 0} f(\epsilon) = \begin{cases} 1, & \epsilon < \epsilon_0 \\ 0, & \epsilon > \epsilon_0. \end{cases}$$

As the temperature is increased, the electrons start jumping to "excited" levels above the Fermi energy, but at ordinary temperatures most of the electrons remain in states below that energy, and make hardly any contribution to the specific heat.

If we are to apply the theory of the *ideal* Fermi-Dirac gas to electrons, we must first recognize that there will be deviations from ideal behavior produced by the electrical forces between the charges. Nevertheless it is remarkable that at sufficiently high densities the interactions have relatively little effect. The Fermi energy increases as the 2/3 power of the density, as may easily be verified from the quantum formula for the volume occupied by a quantum state in phase-space. If one treats the interactions by quantum-mechanical perturbation theory, using the wave functions for the noninteracting particles as a basis set, then one sees that interactions can be effective only if two interacting particles can be scattered into unoccupied states. But at low temperatures all the lowest states are occupied, so the particles must go into states above the Fermi energy. The probability for such a scattering process, involving a rather large energy change (especially at high densities), will therefore be small; in effect, only the electrons with energy close to the Fermi energy can interact with each other and produce deviations from the ideal (Fermi) gas law. In particular, it turns out that the ratio of the potential to the kinetic energy goes to zero as the density goes to infinity. (From dimensional considerations alone one might expect that the potential energy of a system of charged particles should vary as the 1/3 power of the density.)

One interesting consequence of this argument is that the ineffectiveness of interparticle forces, due to the lack of available states in phase-space, makes it possible to construct a fairly accurate "shell model" of the nucleus, in which the quantum states can be determined even though the nature of the forces between nucleons is not very well known. This was the achievement of Maria Goeppert Mayer and J.H.D. Jensen (1955).

4.6 QUANTUM DEGENERACY IN THE STARS

According to quantum mechanics, all substances should become metallic at sufficiently high pressures, because the bound states of electrons in atoms and molecules will disappear. Eventually this "pressure ionization" will liberate all the electrons, not just the ones in the outer valence shell. Astrophysicists realized almost immediately that the Fermi-Dirac theory of the electron gas could be used to describe the state of matter in white dwarf stars, which were known to have very high density; and the detailed development of quantum statistical mechanics, especially its relativistic version, was strongly stimulated by such applications.

A peculiar feature of the Fermi-Dirac gas at low temperatures is that when the electrons simply fill up the lowest energy states, they are unable to respond to small changes in energy. This is because an electron cannot absorb energy except by jumping to a higher-energy quantum state, and all the states just above it are already filled. Only the electrons in the highest filled states—at the "Fermi level"—can absorb small amounts of energy by going into empty states just above them. Conversely, even when the gas is cooled to the temperature of absolute zero, the electrons still have finite energy and exert a pressure; this pressure depends mainly on the density and hardly at all on the temperature, unless the temperature is high enough to put a significant number of electrons into the upper quantum states.

R. H. Fowler showed in 1926 that some of the peculiar properties of white dwarf stars could be attributed to these "quantum degeneracy" effects. Even though their temperatures are not low by terrestrial laboratory standards, the density is so high that most of the electrons are in the lowest possible energy states. Fowler's paper was the first demonstration that the new quantum statistics could explain an important property of gross matter, and it constituted a major breakthrough in astrophysical theory (Chandrasekhar 1944, 1964).

In late 1929 Edmund C. Stoner at the University of Leeds, following up an earlier suggestion by Wilhelm Anderson in Dorpat, showed that when the density is very high, relativistic effects will become important and will make the kinetic energy of the electrons vary as the 1/3 rather than the 2/3 power of the density.[4] He suggested that the pressure of the degenerate gas would not be able to balance its gravitational attractive forces if the mass were greater than a limiting value, which he estimated to be about 2.19×10^{33} grams (slightly more than the sun's mass, 1.98×10^{33} grams). He did not speculate about what might happen to a more

---

[4] These expressions can be most easily derived from the postulate that the element of volume in phase-space is $\Delta p_x \Delta q_x = h$. Thus if we compress the system so that the coordinate q_x is decreased, momentum p_x will increase as the 1/3 power of volume. According to the relativistic relation between energy and momentum, $E^2 = p^2 c^2 + m^2 c^4$, where m_o = rest mass and c = speed of light. If $p \ll mc$ this reduces to $E = m_o c^2 + (p^2/2m_o)$, in which the first term (rest-mass energy) is just an additive constant; thus $E \propto p^2 \propto V^{-2/3}$. In the extreme relativistic case, $p \gg mc$, we have $E \sim pc \sim V^{-1/3}$.

massive star but simply noted that "gravitational kinetic equilibrium will not occur" (1930:963).

S. Chandrasekhar, an Indian-born astrophysicist who was at that time working with R. H. Fowler and E. A. Milne, undertook a systematic development of the theory of Fermi-Dirac gases in conjunction with models for stellar structure, attempting to solve the equations for hydrostatic equilibrium for a fluid whose pressure is proportional to $\rho^{5/3}$ for low densities and to $\rho^{4/3}$ for high densities. (This would correspond to energies proportional to $\rho^{2/3}$ and $\rho^{1/3}$ in the nonrelativistic and relativistic limits, respectively.) In 1931 he pointed out that in the extreme relativistic limit the equation of state $P \propto \rho^{4/3}$ can be used in the Emden equation for polytropic gas spheres and gives a maximum mass of 1.822×10^{33} grams. This estimate avoided Stoner's assumption that the density is uniform throughout the star. But both Stoner and Chandrasekhar assumed that the mean molecular weight of the gas is 2.5, a value that would soon have to be substantially reduced in the light of the evidence presented by H. N. Russell and B. Strömgren that stars contain large amounts of hydrogen (§4.2).

L. D. Landau (1932) independently made the same estimate from the Emden equation but used a mean molecular weight of 2 and thus obtained a maximum mass of 2.8×10^{33} grams, or about 1.5 solar masses. For a larger mass there would be nothing in the known laws of physics to prevent the star from collapsing to a point, so Landau concluded that existing stars with masses greater than $1.5 M_{\odot}$ must have regions in which the laws of quantum mechanics are violated, perhaps due to a breakdown of energy conservation when nuclei come in close contact.

In 1935 Chandrasekhar succeeded in deriving the maximum mass formula for the degenerate Fermi-Dirac gas, using the complete and rather complicated formula for the pressure-density relation valid at all densities.[5] In the case where radiation pressure is absent the maximum mass is

$$M_3 = \frac{5.728}{\mu^2} M_{\odot},$$

where $M_{\odot}$ = mass of sun and μ = mean molecular weight. For $\mu = 2.5$ this gives essentially the same result which he had obtained in 1931 ($.91 M_{\odot} \approx 1.8 \times 10^{33}$ grams). When radiation pressure is included the limit is somewhat increased (Chandrasekhar 1939:423–38).[6]

---

[5] It was necessary at the same time to defend the validity of this formula against the attack of Sir A. S. Eddington (1935); see Møller & Chandrasekhar (1935). For an account of this episode see Wali (1982).

[6] Various additional physical effects may change the numerical coefficient; for example rotation may allow systems of much higher mass to be stable (Ostriker et al. 1966; Ostriker & Bodenheimer 1968).

It was still not clear what would happen to a star with mass greater than the maximum, if it continued to contract under its own gravitational forces. Chandrasekhar speculated that "at some stage in the process of contraction the flux of radiation in the outer layers of the stars will become so large that a profuse ejection of matter will begin to take place. This ejection of atoms will continue till the mass of the star becomes small enough for central degeneracy to be possible" (Chandrasekhar 1935b:257). Or it might continue to contract until it reaches a density so high that the nuclei crumble and a "neutron star" forms by the "inverse beta decay" reaction ($p + e^- \rightarrow n + v$). This possibility, which had been suggested by Baade and Zwicky in 1934, was investigated theoretically by L. D. Landau in Russia and by J. Robert Oppenheimer and his students in the United States. In either case—explosion or continued contraction—one could imagine that heavy elements would be synthesized inside the star and eventually scattered into space, so they would be available for the formation of a second generation of stars.

The general idea that heavier elements could be formed in stars by the fusion of hydrogen atoms was discussed by several astrophysicists in the 1930s, but was first worked out quantitatively by Hans Bethe, a German physicist who had emigrated to the United States in 1938. Bethe analyzed a sequence of reactions by which protons could be put together to form helium nuclei, and showed that an equivalent process could occur if carbon nuclei were present to act as a catalyst (the "carbon cycle"). This was the work for which Bethe received the Nobel Prize in physics in 1967. The problem of how elements heavier than helium could first be produced was left open until 1952, when the Australian-born physicist E. E. Salpeter at Cornell University found a possible solution. He proposed that two helium nuclei could stick together as a beryllium nucleus which, though unstable, would last long enough to capture another helium nucleus and form a carbon nucleus (see also Öpik 1951:71).

Salpeter's reaction was used in a detailed theoretical study of stellar evolution in 1955 by the British astronomer Fred Hoyle and the American astronomer Martin Schwarzschild. They found that as a star with mass slightly greater than the sun continues to generate heat by the fusion of hydrogen nuclei into helium, it gradually gets hotter; the proportion of helium reaches about 50% before the internal temperature becomes high enough (about 200 million degrees) to start the Salpeter helium-fusion reaction. Even though the temperature is extremely high by terrestrial standards, the density is so great that the electrons in the core of the star form a degenerate quantum gas. This means that when energy generation from helium fusion does begin, the gas cannot respond in the normal way by increasing the pressure, which would force the star to expand and thereby cool itself. Instead there is an unstable "runaway" process in which the star gets even hotter, which in turn boosts the rate of energy production. There is a sudden increase of luminosity called the "helium flash," which is terminated only when the temperature gets so high that the electrons are kicked up to high energy levels, removing the degeneracy and converting the interior of the star momentarily back to a normal (nondegenerate) state. In the meantime the outer parts of the star have expanded enormously, producing what we call a "red giant" star.

In 1962 the Japanese astrophysicist Chushiro Hayashi and his colleagues put together a comprehensive account of stellar evolution which illustrates how quantum statistics has changed our ideas about the possible states of matter. They pointed out that the last stage of evolution, in which a red giant star burns up its nuclear fuel and collapses into a degenerate white dwarf, is strangely reminiscent of the evolutionary theory adopted at the beginning of the 20th century by H. N. Russell, James Jeans, and others. In that theory the star begins as a gaseous red giant, collapses to a hot white star, and then changes to a liquid star, which cools down because a liquid, being relatively incompressible, cannot generate much heat by gravitational contraction. (An ideal gas has the curious property that it can radiate energy while getting hotter as it contracts, but liquids and solid cannot do this.) In the new theory there is a comparable change in the direction of evolution at the point when the star no longer generates enough heat to compensate that which it loses by radiation. The phase change from gas to liquid which was responsible for stopping the contraction and changing the nature of the evolution in the old theory has been replaced by a change from "normal" to "degenerate" matter in the new theory.

There is a further similarity between the old and the new theories. Before the development of the quantum theory of atomic structure, one would have expected that a cooling star would ultimately freeze, and so a solid would be the final state of stellar evolution. Then Jeans and Eddington pointed out that at the high pressures which must exist in stellar interiors, the atoms would be ionized and behave like a gas. The nuclei would have positive electric charges; hence they would repel each other. According to older ideas about states of matter, a solid is formed as a result of short-range attractive forces between particles, and it would seem that particles which repel each other at all distances could not solidify. However, in 1960 the Russian physicist D. A. Kirzhnits proposed that the charged nuclei embedded in a degenerate electron gas in a white dwarf would condense into a solid- or liquidlike structure. A few months later another Russian physicist, A. A. Abrikosov, and E. E. Salpeter published similar suggestions arguing that a regular solid lattice would be the most likely arrangement until the density became so high that quantum effects become important for the nuclei. (The nuclei themselves would eventually behave like a degenerate gas; at even higher densities they would dissolve into a neutron gas, as mentioned above, and at still higher densities the star would become a "black hole.") This suggestion was confirmed by a computer calculation, based on the principles of statistical mechanics applied to a system of similarly charged particles in a uniform background of the opposite charge, carried out by S. G. Brush, H. L. Sahlin, and E. Teller (1966). As in the case of the hard-sphere gas studied by B. J. Alder and T. E. Wainweight earlier (1957), it was shown that a fluid-solid transition can occur in a system with purely repulsive forces (see §6.6 for further discussion).

(Additional references for this section: Sitterly 1970, Hufbauer 1981.)

4.7 MAGNETIC PROPERTIES OF THE ELECTRON GAS

Wolfgang Pauli, in 1927, was able to explain the relatively small paramagnetic susceptibility of metals with the help of his exclusion principle and the concept of electron spin. One might have expected that if the electrons in metals are free and possess an intrinsic spin and magnetic moment, then they could easily be aligned by an external magnetic field. However, at zero temperature there will be two electrons with opposite spin in each cell in phase-space, and it is not possible for a spin to line up with an external magnetic field, unless (a) the other spin in the same state is aligned opposite to the field, in which case there is no net magnetization at all; or (b) the electron jumps to a higher energy level that is not occupied. But the energy required for the latter process is likely to exceed the potential energy that can be gained by alignment with the field. The result is that the paramagnetism of the system is much less than for classical systems of particles with magnetic moments.

On the other hand, a quantum gas has a diamagnetism which is not present classically. In classical theory, the energy of an electron, regarded as a point charge with no magnetic moment, is unaffected by a magnetic field, since the force is at right angles to the motion and hence does no work. This is not to say the *motion* of the electron is not affected, for of course the electrons will tend to move in circles in a certain direction in a plane normal to the field. It might appear that this circular motion would produce a net current and hence a diamagnetic susceptibility. However, in an ideal gas all points in space are equally likely to be the center of such a circular orbit, which implies that the net current flowing past any point in space will be zero on the average. In particular, this is true even for points very close to the surface of the container, where the electron orbit intersects the surface; in this case, the electron is repeatedly reflected and in effect traverses a very large orbit in a direction opposite to the other orbits. As a result the average magnetic moment must be exactly zero. This result was obtained by Bohr (1911) and other writers, and in its most general form is known as Miss van Leeuwen's theorem (van Leeuwen 1919, 1921). Its validity is not restricted to ideal gases; one assumes only that the system consists of particles exerting any kind of force on each other that can be represented by a term in the Hamiltonian function, excluding magnetic dipole forces. The theorem shows that one cannot construct a classical model of a molecular magnetic dipole by starting with charged particles, unless some kind of artificial constraint is introduced to make the charges move in a certain direction. (For an extensive discussion of this theorem and its consequences, see van Vleck 1932:94–97.)

When the problem is treated by quantum mechanics, it is found that the electron gas may have a finite diamagnetic susceptibility, which depends in a rather complicated way on the magnitude of the field and the shape and size of the container. Landau (1930) showed that if one solves the Schrödinger equation for an electron in a magnetic field, neglecting surface effects, the continuum of free-particle energy levels is partly changed to a discrete set, each level having a degeneracy proportional to the magnitude of the field. Since the lowest energy is slightly higher than the lowest energy level for free particles in zero field, it turns out that the average energy increases with field, thus giving a small diamagnetic susceptibility, which is 1/3 of the Pauli paramagnetic susceptibility for a degenerate electron gas ($T = 0$). Landau's theory of diamagnetism was anticipated to some extent by Francis Bitter, who derived a similar result slightly earlier by an approximate calculation (Brush 1970).

4.8 SUPERCONDUCTIVITY

In addition to solving some of the older problems in the theory of metals, quantum statistics also provided an explanation of two spectacular phenomena discovered in the 20th century, superconductivity and superfluidity. These are generally regarded as "macroscopic" quantum effects—phenomena whose very existence cannot be explained by classical physics. Both were discovered by the Dutch physicist Heike Kamerlingh Onnes at Leiden, in 1911 (though the nature of superfluidity did not become clear until many years later).

It was supposed for some time that a superconductor is simply a perfect conductor, as indicated by the fact that a persistent current can be generated by induction in a superconducting ring; the current keeps flowing for several hours. If one simply sets the conductivity σ equal to

∞ in Ohm's law $\sigma E = J$, then it follows from the Maxwell theory of magnetic induction (curl $E = -\dot{B}/c$) that it is impossible to change the magnetic induction ($\dot{B} = 0$). This would imply that any magnetic flux present in the superconductor when it was cooled into the superconducting state would be frozen in. In this case the actual state of the superconductor would depend on its previous history and there would not be any thermodynamic equilibrium state. However, in 1933 Meissner and Ochsenfeld reported an experiment that indicated that when a metal is cooled below its superconducting transition temperature in a magnetic field, the lines of magnetic induction are completely pushed out of the metal. Thus the superconductor is not only a perfect conductor but also a perfect diamagnetic ($B = 0$), and there *is* an equilibrium state that is independent of the of the past history. (It should be noted that the "ideal Meissner effect" is rather difficult to detect experimentally because it is destroyed by very small impurities.) Thus the superconducting transition is reversible in the sense of the thermodynamics. One may draw a curve in the (H, T) plane representing the variation of critical field (for destruction of superconductivity) H_c with temperature, and this curve plays the same role as the phase-boundary curve in the pressure-temperature plane for the gas-liquid transition. The thermodynamic analysis of the transition was developed by Gorter and Casimir (1934) and may be found also in London (1950:16–26).

After the discovery of the Meissner effect, the brothers Fritz and Heinz London developed a macroscopic theory of superconductivity (1935). They proposed that instead of trying to interpret superconductors within the framework of classical electrodynamics by simply assuming infinite conductivity and zero magnetic permeability, one should change the basic equations relating electric currents and magnetic fields. F. London also suggested that these new phenomenological equations could be explained in terms of long-range ordering of the electrons in momentum-space.

Landau (1938) predicted theoretically that when a strong magnetic field is applied to a superconductor, a domain structure (with alternating normal and superconducting regions) will be produced. This effect was observed experimentally by Shalnikov (1945) and Meshkovsky (1947). According to his original theory, Landau expected that the domains would become very small at the surface. However, he later suggested (Landau 1943) that there might be a domain structure observable also at the surface because of the surface energy associated with the creation of an interface between a normal and a superconducting region in the presence of a magnetic field. The usual thermodynamic theory (for example Schafroth 1960) indicates that the free energy at the surface should be negative, thus favoring the creation of surfaces partly penetrated by the magnetic field. There is not then a "sharp" Meissner effect with complete expulsion of the field at $H = H_c$, but instead a range of fields over which the field is gradually pushed out. Such behavior is exhibited by some ("hard") superconductors but not by others ("soft").

Bloch's theory of metals (1928) was based on the model of an electron in a periodic potential; if the ions that determine this potential are rigidly fixed, then the energy-level spectrum consists of bands, and an electron in an allowed level moves freely (except that its mass must be replaced by an "effective mass"). The electrical resistance of the metal is caused by displacements of the ions from their equilibrium positions; the electrons can then be scattered by the deviations of the potential from periodicity. Indeed one may treat the lattice vibrations as quantized excitations, "phonons," which can interact with the electrons. Just as the Coulomb interaction itself may be regarded as being produced by the exchange of quanta of the electromagnetic field between electrons, an interaction between electrons will also be produced by means of the absorption and emission of phonons.

Fröhlich (1950) suggested that superconductivity might be due to this interaction. His work represents one of the first applications of quantum field theory methods in statistical mechanics. In this connection a slightly earlier paper by Fröhlich, Pelzer, and Zienau (1950) should also be mentioned; these authors had treated the problem of an electron in a polar crystal. The displacement of the ions from their equilibrium positions by the Coulomb field of the electron creates a "polarization field," and the electron's interaction with the quanta of this field (now called "polarons") produces something like a "self-energy" of the electron (since it is the source of the field).

The effective interaction between electrons, taking account of the intermediate interactions with phonons, is attractive instead of repulsive only when the energy difference between the electron states is sufficiently small, and this attractive interaction can give rise to cooperative many-electron states whose energy is lower than the normal state. The first attempts to construct a quantitative theory on this basis were unsuccessful. One reason for the failure, which was made clear only later on, was that perturbation theory was used in a case where it is inapplicable, owing to the high degeneracy of the unperturbed system, which practically has a continuous energy spectrum only (Fröhlich 1953). The successful theory of Bardeen, Cooper, and Schrieffer (1957) leads to a ground-state energy of the form

$$E \sim -Ae^{-B/V},$$

where V is the effective interaction between electrons, and A and B depend on constant parameters and on the density of electron states at the Fermi surface. Although $E \to 0$ as $V \to 0$, no expansion of E in powers of V is possible.

Important support for the Fröhlich theory came from the experimental discovery of the isotope effect (Maxwell 1950, Reynolds et al. 1950). The superconducting transition temperature was found to be proportional to $M^{-1/2}$ (where M = mass of atomic ion) in agreement with Fröhlich's theory. (Later experiments showed that the isotope effect does not occur in all superconductors, however; see Matthias et al. 1963.) The lattice of atomic ions must therefore play a role in superconductivity, even though the quantitative theory was not worked out until 1957.

In view of the theoretical existence of an attractive interaction between electrons, and some earlier speculations about "superelectrons" that do not obey Fermi-Dirac statistics and appear to have double charge, it seems natural (with the advantage of hindsight) to guess that pairs of electrons might act as a bosons and that superconductivity might thus be an example of a Bose-Einstein condensation. It is rather curious that this proposal was actually first made by an experimental chemist, Ogg, in 1946, but was generally ignored by physicists until about 1954. Ogg studied dilute solutions of alkali metals in liquid ammonia and found persistent ring currents, which he attributed to the formation of electron pairs. According to Blatt (1964),

Ogg's experiments proved to be difficult to repeat, his theoretical ideas were discounted by the theorists, and the whole thing was forgotten. In retrospect, I can recall listening to a seminar lecture by Ogg, right after the war, and to that extent the Sydney work may well have been influenced by Ogg. But this influence was entirely subconscious—neither Schafroth, Butler, nor I remembered anything of Ogg's work at the time the Sydney work started, in the latter half of 1954. Only much later, in 1961, was I reminded of Ogg's work in the course of a correspondence with P. T. Landsberg. . . . . There is no doubt about Ogg's priority of the suggestion, and we all owe him apologies for forgetting so completely about it in our publications.

Blatt's account documents what probably happens rather often in science: one person picks up an idea from another in conversation or a seminar, forgets it, then sometime later produces it in the honest belief that he thought of it all by himself. Since informal communication and discussion before publication seem to aid the progress of science in the long run, it seems inevitable that there will be occasions when the wrong person gets credit for an idea.

In spite of arguments in its behalf by Schafroth (1954), the idea of bound electron pairs in metals did not become respectable until it was worked out in detail by Leon Cooper in 1956. It was then used by Bardeen, Cooper, and Schrieffer (1957) to construct a many-electron ground state for a superconductor. The resulting ground state has lower energy than the normal state, and leads to a Meissner effect and other properties in good agreement with experiment. The "BCS" theory is now generally accepted as a major breakthrough toward the correct explanation of superconductivity; later refinements have not changed its essential features, and it has even been shown that their method of calculation gives the thermodynamic properties of their model (in which only pairs of particles whose total momentum is zero interact) exactly in the limit of an infinitely large system (Thouless 1970).

(Additional references for this section: Bardeen 1963, 1973; Cooper 1973; Schrieffer 1973; Mendelssohn 1977; Trigg 1975.)

4.9 Superfluidity of Liquid Helium

Helium, a relatively late addition to the family of known chemical elements, plays a major role in modern physical science, for example as a crucial intermediate in cosmic evolution and as a showpiece for macroscopic quantum effects. The ^{4}He isotope, having the smallest even number of protons, neutrons, and electrons, enjoys unusual stability in both nuclear and chemical environments.

Helium was first identified in solar spectra and proposed as a distinct element in 1868 by Normal Lockyer, British astronomer and founder of *Nature*. (Its name was derived from the Greek *helios* = sun, and first appeared in print in a paper by William Thomson in 1871.) It was not found in a terrestrial laboratory until its discovery by the British chemist William Ramsay in 1895. A few years later Ernest Rutherford, the New Zealand-born founder of nuclear physics, showed that it is a frequent product of radioactive decay. With the development of the nuclear model of the atom it turned out that the "alpha particle," having a double positive charge and a mass about four times that of a hydrogen atom, is simply the nucleus of the helium atom. Before Werner Heisenberg proposed his proton-neutron model of the nucleus in 1932, the helium nucleus was thought to consist of four protons and two electrons; according to Heisenberg's model it consists of two protons and two neutrons. That is the most abundant isotope, now denoted by ^{4}He; the isotopes ^{3}He and ^{6}He, though extremely rare in nature, are of theoretical interest in testing the dependence of helium's properties on its quantum nature (statistics).

At the end of the 19th century scientists believed that every substance ought to exist, or be made to exist, in the solid, liquid, and gaseous states. The race to liquefy the new gas, helium, was won in 1908 by the Dutch physicist Heike Kamerlingh Onnes, who also took the leadership in discovering its properties at low temperatures. (Mendelssohn (1977) gives a lively account of this work.) He had worked with Gustav Kirchhoff as a graduate student, and became interested in molecular theory and the properties of matter as a result of contacts with J. D. van der Waals in the early 1880s. After being appointed professor of physics at Leiden in 1882, he launched a research program to determine the physical properties of many substances over a wide range of temperatures and pressures, partly in order to test van der Waals' "law of corresponding states."[7] Following the successful liquefaction of air by the French physicists Louis Paul Cailletet and Raoul Pictet in 1877, Kamerlingh Onnes procured a similar liquefaction apparatus at Leiden and modified it to reach ever lower temperatures.

"It was a wonderful sight," Kamerlingh Onnes recalled in his Nobel Prize lecture five years later, "when the liquid, which looked almost unreal, was seen for the first time. It was not noticed when it flowed in. Its pressure could not be confirmed until it had already filled up the vessel. Its surface stood sharply against the vessel like the edge of a knife" (1967:325).

Mendelssohn (1977:241) considers it remarkable that Kamerlingh Onnes and other experimenters, who must have seen the dramatic change in the appearance of liquid helium as it was cooled down to 2°K, failed to mention it in their published articles for almost a quarter of a century.[8] But for Kamerlingh Onnes what counted was quantitative results (van der Handel 1973:220). In 1911 he announced that the density of the liquid reaches a maximum at a temperature of about 2.2°K, then decreases as the temperature falls below that point. Apparently he was distracted from following up this finding by his discovery of superconductivity a few weeks later (Mendelssohn 1977:87). In his Nobel Prize lecture he suggested that the density maximum "could possibly be connected with the quantum

[7] The law, proposed in 1881, states that all substances obey the same equation of state when their pressure, temperature, and volume are expressed as multiples of the values of those variables at the gas-liquid critical point.

[8] "It was well known by the technicians in Leiden, even in Onnes's days, that helium underwent a change at 2.2K, because the boiling suddenly calmed down; this was long before the physicists ever talked about the phase transition" (D. ter Haar, private communication, 1980).

theory" just as superconducitivity may be connected with "the energy of
Planck's vibrators" (Kamerlingh Onnes 1967:327, 330), but no one knew
what to do with such hints at the time.

In 1924 Kamerlingh Onnes returned to the "startling result" of liquid
helium's density maximum and published more detailed experiments in
collaboration with J.D.A. Boks. They found a cusplike maximum at
$2.29°K$; at temperatures below that, the density appeared to approach a
constant value about 0.7% below its maximum value. Further experiments
were planned with an American visitor, Leo Dana; they measured the
specific heat and found very high values below $2.5°K$. Not believing these
results to be valid, Kamerlingh Onnes and Dana refrained from publishing them; their paper of 1926 presented only the rather uninteresting
behavior above $2.5°K$. Further investigation of this point was stopped
by Kamerlingh Onnes' death and Dana's return to America.

On the death of Kamerlingh Onnes, leadership in the Leiden program
was taken over by Willem Hendrik Keesom. A student of J. D. van der
Waals in Amsterdam, Keesom had worked with Kamerlingh Onnes between 1904 and 1917, then went to Utrecht, where among other projects
he pioneered the use of x-ray diffraction methods to find intermolecular
distances in liquids. Returning to Leiden in 1923 as professor of experimental physics, he was the first to solidify helium in 1926. This required
an external pressure of about 25 atmospheres, since helium appears to
remain liquid under atmospheric pressure down to the lowest attainable
temperature. (Kurt Bennewitz and Franz Simon 1923 had already suggested that the difficulty in solidifying helium might be due to its quantum
zero-point energy.)

In 1927 Keesom and M. Wolfke noted first a sudden jump and then
also peculiar variations in the specific heat, heat of vaporization, and
surface tension of helium on cooling through the temperature of maximum density. "The thought suggested itself," they wrote, "that at that
temperature the liquid helium transforms into another phase, liquid as
well . . . we call the liquid, stable at the higher temperatures 'liquid
helium I,' the liquid, stable at the lower temperatures 'liquid helium II.'"

The first published description of the change in visual appearance that
accompanies the transformation from helium I to helium II is found in a
paper by J. C. McLennan, H. D. Smith, and J. O. Wilhelm in 1932. They
wrote that "the appearance of the liquid underwent a marked change,
and the rapid ebullition ceased instantly. The liquid became very quiet
and the curvature at the edge of the meniscus appeared to be almost
negligible." From above "the surface of the liquid was indiscernible, while
the lack of impurities and scattered light in the column caused the tube

to appear empty. From the side, however, the surface could be seen distinctly" (1932:165).

With a German visitor Klaus Clusius, Keesom determined the specific-heat curve more accurately and reported that it has an "extremely sharp maximum," although there is no latent heat for the transition from helium I to helium II. They confessed ignorance as to the "inner causes" of the transition but mentioned that similar phenomena had been observed in certain solids, attributed to changes in molecular rotation, and in ferromagnets at the Curie point. "Even though the assumption of rotations has no meaning for helium, there seems to be a close relationship between [rotational transitions and] the effects observed in helium" (Keesom & Clusius 1932, Galasiewicz 1971:86, 88).

With his daughter Anna Petronella Keesom, W. H. Keesom repeated the specific-heat measurments with greater accuracy yet; their paper contains the first public proposal of the now-established name for the He I–He II transition:

> According to a suggestion made by Prof. Ehrenfest we propose to call that point, considering the resemblance of the specific heat curve with the Greek letter λ, the lambda-point. (Keesom & Keesom 1932:742)

Paul Ehrenfest, an Austrian physicist who succeeded H. A. Lorentz as professor of theoretical physics at Leiden in 1912, was the acknowledged expert on quantum theory, statistical mechanics, and thermodynamics (Klein 1970); the last paper he wrote before his suicide in September 1933 dealt with the classification of phase transitions (Ehrenfest 1933).

Several physicists who were familiar with quantum statistics failed to recognize the possible connection between the Bose-Einstein condensation and the lambda transition of liquid helium. One was Ehrenfest, who probably accepted the argument of his student G. E. Uhlenbeck (1927) that Einstein's conclusion about a condensation was only the result of a mathematical error (§4.4). Another was the British theorist John Edward Lennard-Jones, best known today for his interatomic force law (see below, §5.5). At a meeting of the Physical Society of London on February 14, 1930, M. C. Johnson of the University of Birmingham discussed the "degeneracy" of helium gas on the basis of thermodynamic data between 4° and 6°K, using Fermi's correction to the pressure of an ideal gas (§4.5). Lennard-Jones pointed out that he should have used Bose-Einstein statistics, but this suggestion was not followed up (Johnson 1930:179–80).

A detailed discussion of the role of zero-point energy was published in 1934 by F. Simon, who argued that helium below the λ-point is in an ordered state which he himself had previously characterized as "liquid degeneracy" (Simon 1927:809). Keesom (1933) had called it "quasi-crystalline," and Clusius, in an unpublished paper read in Breslau in 1933, spoke of a "crystalline" state "and adds the more specialized assumption of an association starting in the λ-region." Simon argued that since one has to do work against some kind of repulsive forces in condensing the liquid to the solid, and since "at the interatomic distances realized in liquid helium there can be no appreciable repulsion originating in the atomic fields," the zero-point energy must play a large role in counteracting the attractive forces that would ordinarily cause the atoms to crystallize at low temperatures.

Fritz London (1900–1954), a German physicist who had already established his reputation with his work on the quantum theory of inter-atomic forces and the chemical bond (see below, §§5.4, 5.8), came in 1933 to Oxford along with Simon;[9] he later moved to Paris, and in 1939 went to Duke University in Durham, North Carolina. London first considered the possibility of a lattice structure for liquid helium; by October 1934 he had estimated the energy of various possible structures and concluded that a diamond-type lattice would be the most stable. But he did not think very highly of this model, and was continually distracted by the apparently more interesting problem of superconductivity. At Simon's insistence he gave a short manuscript to F. A. Lindemann, head of the Clarendon Laboratory at Oxford, for submission to the Royal Society, but publication was delayed because the referee had never heard of the λ-point and wanted a more detailed description of the properties to be explained (correspondence in London archive at Duke University). The paper finally appeared in 1936; in it, London acknowledged Simon's proposal that the stability of the liquid state at the lowest temperature attainble is due to zero-point energy, but thought that there might be at least a "statistical preference" for some kind of lattice structure even though the atoms could not be definitely localized.

Herbert Fröhlich, then at Leiden, quickly took up London's diamond-lattice model in conjunction with the order-disorder theory recently developed by Bragg and Williams and by Hans Bethe (§6.4). The λ-phenomenon was seen as an order-disorder transition in a system of atoms and "holes" (vacant lattice sites), similar to the "lattice gas" model of Cernuschi and Eyring, and of Lee and Yang (§6.7) (Fröhlich 1937; London-Fröhlich correspondence in the London archive at Duke University).

---

[9] Both emigrated when Hitler came to power in Germany; for further details see the articles on London, Simon, and Lindemann in the *Dictionary of Scientific Biography*.

London's brief note suggesting a connection between "The λ-Phenomenon of Liquid Helium and the Bose-Einstein Degeneracy" was sent from Paris, where he was at the Institut Henri Poincaré, on March 5, 1938, to *Nature*, where it appeared in the issue of April 9. Rejecting Fröhlich's model along with all other static spatial structures, he recalled that Einstein had discussed a "peculiar condensation phenomenon . . . but in the course of time the degeneracy of the Bose-Einstein gas has rather got the reputation of having only a purely imaginary existence." While the λ-point of helium "resembles a phase transition of second order" in Ehrenfest's classification (discontinuity of specific heat) in contrast to the third-order transition of the ideal Bose-Einstein gas (only the derivative of the specific heat is discontinuous), "it seems difficult not to imagine a connexion" between them. The Bose-Einstein condensation for an ideal gas of particles with the mass of helium atoms would occur at 3.09°K, compared with 2.19°K for the λ-transition. London noted that the ideal Bose-Einstein gas model could not account for the specific heat at very low temperatures, where presumably the Debye term would be important (§4.1). If each helium atom could be considered to move in a self-consistent field formed by the other atoms, it might be described by a wave function similar to those appearing in Bloch's theory of metals; but "the quantum states of liquid helium would have to correspond, so to speak, to both the states of the electrons and to the Debye vibrational states of the lattice in the theory of metals. It would, of course, be necessary to incorporate this feature into the theory before it can be expected to furnish quantitative insight into the properties of liquid helium" (London 1938a:643–4).

The initial responses to London's proposal were encouraging. Simon told him that the experimental data might be compatible with a third- rather than a second-order transition, and Uhlenbeck partially retracted his critique of Einstein's condensation. H. Jones at Cambridge University, codiscoverer with J. F. Allen of the "fountain effect," was "very interested in your recent letter to *Nature*" and, with the help of the statistical mechanics expert R. H. Fowler, constructed a more rigorous proof that the ideal Bose-Einstein gas undergoes a condensation.[10]

Keesom, in the first of a series of letters to London, dated 13 May 1938, wanted to discuss the pros and cons of various lattice structures. In his reply London stated (as he had not made clear in *Nature*) that he was now contemplating a single quantum state which has no spatial order at all.

---

[10] Correspondence with Simon, Uhlenbeck, and Allen in the London archive; Fowler & Jones (1938). Their proof was criticized by Gerhard Schubert (1946) and Günther Leibfried (1947), who found it necessary to give alternative derivations of the same result; see also Fraser (1951) and Landsberg (1954). There is a comprehensive account of the mathematical properties of the ideal Bose-Einstein gas by Ziff, Uhlenbeck, and Kac (1977).

Keesom, no mean theoretician himself, pointed out that a system of hard spheres must have some correlation in positions and hence at least partial spatial ordering, while the wave functions are not pure momentum eigenfunctions. London agreed, noting that the quantity with respect to which the system is ordered is not the momentum but the wave number, which is proportional to momentum only for free particles. But Keesom was still skeptical about London's theory when he published his book on helium (1942:347).

In the same period when London was working out his condensation theory, several experiments were revealing further peculiar properties of liquid helium.[11] Perhaps the most important of these was *superfluidity*, discovered by the Russian physicist P. L. Kapitza at the end of 1937 and published in *Nature* on January 8, 1938. Kapitza found that helium II flows easily between two extremely flat glass disks only half a micron apart, and must therefore have a viscosity at least 1,500 times smaller than that of helium I. This "is perhaps sufficient to suggest, by analogy with supraconductors, that the helium below the λ-point enters a special state which might be called a 'superfluid.'" (Kapitza 1938:74). J. F. Allen and A. D. Misener at Cambridge reported at the same time that helium II flows through a thin capillary tube with extremely low viscosity (their letter, dated 19 days after Kapitza's, appeared in the same issue of *Nature*). But they did not use a dramatic term like "superfluidity" to describe the phenomenon, and they have not generally been given as much credit as Kapitza.

Laszlo Tisza, a Hungarian physicist visiting in Paris, discussed London's theory with him and published a note in *Nature* extending it to transport phenomena in liquid helium. Tisza pointed out that in the Bose-Einstein condensation a finite fraction of the atoms will be in the lowest energy level, and he suggested that these atoms do not take part in the dissipation of momentum. "Thus, the viscosity of the system is entirely due to the atoms in excited states." If the viscosity is measured by the damping of an oscillating cylinder, it should be approximately the same as the viscosity of helium gas at the same temperature. But the velocity of flow through a capillary tube, which is ordinarily determined by the viscosity of the fluid, will now be determined in a different way by some complicated combination of the flow properties of the excited atoms and those in the ground state. "In this case, the fraction of substance consisting of atoms in the lowest energy state will perform—like a

---

[11] For accounts of this work see Keesom (1942), H. London (1960), and Trigg (1975: 64–38). A comprehensive bibliography on superfluid helium has just been published by Hallock (1982).

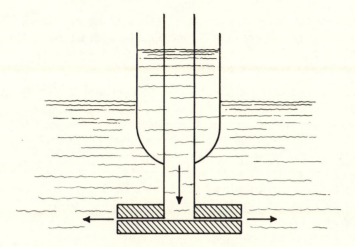

FIG. 4.9.1. Kaptiza's experiment.

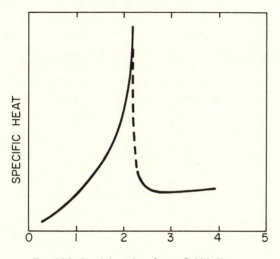

FIG. 4.9.2. Lambda point of superfluid helium.

'superfluid liquid' of viscosity $\mu \sim 0$—some sort of turbulent motion the flow velocity of which does not depend on the bore of the capillary." Since the atoms in excited states behave like a gas at temperature T and pressure $\sim nkT$, and will diffuse like molecules subject to a gradient of osmotic pressure, they will flow at a different rate, the result being to set up a temperature gradient. Conversely, if one maintains a temperature gradient between the ends of a capillary, "a gradient of density of excited atoms, *n*, and thus of pressure is produced" (Tisza 1938). In this way Tisza managed to explain several of the peculiar phenomena associated with helium II.

According to Mendelssohn (1977:258),

> When Tisza's paper was published, London was at first furious because he deplored the rash use of his own cautious suggestion. He also saw more clearly than anybody else the physical impossibility of admitting two fluids made up of the same species of atoms which by definition must be indistinguishable. Beyond this Tisza had postulated properties for the superfluid which did by no means follow from Bose-Einstein condensation. On the other hand, Tisza had been so incredibly successful in clearing up the tangled mess of experimental results by providing an interpretation which was at least logical that even London had to consider it as significant.

In November 1938, Tisza presented the first of two short notes developing his theory to the Académie des Sciences in Paris. He suggested that while the system is homogeneous in ordinary space, it consists of two phases in momentum-space, I and II, which could be expressed by identifying atoms of type I and II. Inhomogeneities of temperature would produce inhomogeneities of the densities and pressures of the two phases. Since the system as a whole is at rest, there is no net current, so the speeds of the two currents must be coupled together and will tend to "*equalize each other according to the propagation equation*

$$\Delta T - \frac{1}{V^2} \frac{\partial^2 T}{\partial t^2} = 0,$$

[where] *V* is the speed of temperature waves,

$$V = \sqrt{\frac{\rho^{II}}{\rho_0} \frac{dp^I}{dp^{II}}} \sim \sqrt{\frac{KT}{m}} \left[1 - (T/T_0)^5 \right] \quad " \quad \text{(Tisza 1938b:1036).}$$

J. F. Allen called this "perhaps the most interesting deduction from the Bose-Einstein theory" but noted that "In spite of its being printed in italics in Tisza's paper, it unfortunately passed largely unnoticed at the time" (Allen 1952:90).[12] The "temperature waves" subsequently became known as "second sound."

---

[12] Also unnoticed was a footnote to the second edition of Nernst's *New Heat Theorem*, suggesting that at very low temperatures, because of high thermal conductivity, "an oscillatory discharge of thermal differences of potential might occur under circumstances" (Nernst 1926:25).

London's emigration to America was accompanied by the publication of two papers in American journals, one in *Physical Review* introducing and discussing the phrase "condensation in momentum space" (1938b: 951) and the other in the *Journal of Physical Chemistry* (1939) on possible modifications of the ideal gas theory which might make it applicable to a liquid. For example, the density of states might be assumed to vary as the 4th power of the wave number k for small k (instead of k^2 as for free particles); this would give a jump in the specific-heat curve. "Such a decrease in the density of levels for the lowest states is quite probable, since the introduction of any interaction between the atoms would produce such an effect" (1939:61–62).

A. Michels, A. Bijl, and J. de Boer, at the Van der Waals Laboratorium in Amsterdam, accepted London's suggestion that below T_λ the system can be divided into atoms condensed in the lowest energy state and others having finite kinetic energy, but proposed a more dramatic way of reducing the number of low-energy states in the liquid. They introduced a finite excitation energy "which must be added to a condensed particle before it can participate in the velocity distribution of the moving atoms. The specific heat increases then exponentially at low temperatures . . . the excitation energy is due to the effect of exchange, which gives rise to a different interaction when molecules are in the same and when they are in different states" (Michels et al. 1939:595; further details in the paper by Bijl, de Boer, & Michels 1941). Bijl (1940) suggested that the specific heat associated with this excitation energy should be supplemented by a T^3 contribution from sound waves (Debye term).[13]

A radically different theory of liquid helium was proposed in 1941 by Lev Davidovich Landau.[14] Landau had started publishing papers on quantum mechanics in 1926, while still a student at Leningrad University; he spent some time in 1929–30 working with Niels Bohr, Rudolph Peierls, and other Western physicists, then returned to the USSR, where he led a theoretical group at Kharkov from 1932 to 1937. In 1937 Landau became director of the section of theoretical physics of the Institute of Physical Problems of the USSR Academy of Sciences in Moscow. In addition to his research in quantum and nuclear physics, he published significant papers on the theory of diamagnetism in metals (1930), superconductivity (1937), and the theory of phase transitions (1935, 1937), before turning his attention to the helium problem in 1941. (In his 1937 paper on phase

---

[13] Bijl was taken prisoner by the Germans during the war because of underground work, and he died in captivity (D. ter Haar, private communication, 1980).

[14] The excellent biography by Anna Livanova (1980) includes a substantial chapter on the superfluidity of liquid helium.

transitions he had suggested that He II is a "liquid crystal" with cubic symmetry, but did not pursue this idea.)

Landau's career was suspended in 1938, when he was arrested, accused of being a German spy, and thrown into prison. Presumably his irreverent attitude toward the scientific and political establishment of the Soviet Union, together with his many international contacts, made him a target of Stalin's purge. After he had spent a year in prison, Kapitza managed to see him, found him in miserable health, and threatened to resign his own position if Landau was not released immediately. Kapitza was considered so valuable a scientist that his demand was met, and Landau was freed (Dorozynski 1965:55–63, 206; Shoenberg 1978:18).

Landau rejected Tisza's explanation of superfluidity; he could not see how atoms in the ground state would move through the liquid without colliding with excited atoms and exchanging energy with them. Instead of treating liquid helium as a Bose-Einstein gas, Landau proposed to develop a "quantized hydrodynamics" in which the macroscopic density and velocity of the fluid would be replaced by noncommuting quantum-mechanical operators. Rather than attempt to derive hydrodynamics from the Schrödinger equation, Landau was following one of the paths by which the Schrödinger equation itself could have been derived in an axiomatic treatment of quantum mechanics. Such a theory called for an experimental test in the classic hypothetico-deductive tradition, even though (here as in other cases) scientists do not let theories stand or fall on the basis of experiment alone.

But further assumptions, largely of an intuitive nature, were needed to derive definite predictions from the theory. Landau focused attention on the energy spectrum of collective excitations in the liquid. He began by postulating two kinds of excitations, one in which the curl of the velocity operator is zero (corresponding to classical irrotational "potential motion" in which the velocity field can be derived from a scalar potential function),[15] the other in which it is not (corresponding to vortex motion). In order to explain superfluidity he assumed that the lowest energy level of vortex motion lies a finite amount Δ above the lowest level of potential motion. Weakly excited states in the potential spectrum would be

---

[15] In the usual notation,

$$\text{curl } \hat{v} = \nabla \times \hat{v} \equiv \left(\frac{\partial v_z}{\partial y} - \frac{\partial v_y}{\partial z}, \frac{\partial v_x}{\partial z} - \frac{\partial v_z}{\partial x}, \frac{\partial v_y}{\partial x} - \frac{\partial v_x}{\partial y} \right).$$

If the velocity $\hat{v}$ can be derived from a potential, that is, $v_x = \partial \phi/\partial y, v_y = \partial \phi/\partial y, v_x = \partial \phi/\partial z$, then each component of curl $\hat{v}$ is identically zero by equality of cross-derivatives.

longitudinal sound waves, quanta called "phonons."[16] Their energy (like that of photons) should be a linear function of their momentum p,

$$\epsilon = cp,$$

where $c =$ the speed of sound. The corresponding elementary excitation in the vortex spectrum he called a "roton," noting that the name was suggested by another Russian physicist, I. E. Tamm. Landau argued that its energy should be approximately

$$\epsilon = \Delta + \frac{m^2}{\mu},$$

where μ is the "effective mass" of the roton.

As far as the specific heat is concerned, Landau's 1941 theory is practically equivalent to that of Michels, Bijl, and de Boer: at low temperatures one has a T^3 contribution from the phonons plus an exponential term for the rotons.[17] But he pushes it further to a quantitative test against the experimental data of B. Bleaney and F. Simon (1939) for temperatures between 0.25 and 0.8°K. The result is disappointing: the experimental specific heat is about 10 times greater than that estimated from phonons, but follows the T^3 law; this seems to refute the prediction that the nonphonon contribution has an exponential temperature dependence.

True to the tradition of theoretical physics (in contrast to the hypothetico-deductive method of the philosopher of science), Landau does not abandon his theory but suggests that something may be wrong with the experimental data (1965:310).

Landau's second prediction from his model is that if the liquid moves at a speed v less than the speed of sound c, phonons cannot be excited; and if $v < \sqrt{2\Delta/\mu}$, rotons cannot be excited; if both conditions are satisfied, one has superfluidity. Landau did not attempt a quantitative comparison with experiment here, but it soon became evident that if one uses reasonable values of his parameters c, Δ, and μ, one gets a "critical velocity" for the breakdown of superfluidity at least an order of magnitude greater than is observed. (This discrepancy could have been discovered

[16] This term had been used earlier by the physicist J. Frenkel (1932:267). The connection between Landau's and Frenkel's works—both the idea of treating liquids like solids, and their general approach to physics—deserves further study; see for example V. J. Frenkel (1974).

[17] Landau does cite Bijl's 1940 paper but states incorrectly that Bijl omitted the T^3 term for sound waves (Landau 1965:308).

in 1941 by analyzing the Kapitza and Allen-Misener experiments on superfluidity, according to Wilks 1957:67.)

Landau is able to recover the qualitative explanations of the Tisza two-fluid model by identifying the normal fluid with his system of phonon and roton excitations, and the superfluid with the ground state. But he does not divide the *atoms* into superfluid and normal components; the two fluids are defined as *motions*, not identifiable portions of matter.[18] This feature is quite consistent with the spirit of quantum mechanics.

Development of the equations for the propagation of sound from Landau's quantum hydrodynamics led him to an equation which was quadratic in u^2 (where u is a quantity having the dimensions of velocity); hence there will be two velocities of sound. One of them is related in the usual way to the compressibility,

$$u_I{}^2 = c^2 = \frac{\partial p}{\partial \rho},$$

while the other one—later called "second sound"—depends on the fraction of superfluid present and depends strongly on temperature; it varies from 0 at T_λ to $u_I/\sqrt{3}$ at $0°K$.

Landau was at this time unaware of Tisza's prediction of temperature waves, and the experimentalists who first tried to detect second sound (Shalnikov and Sokolov in 1940, according to Peshkov 1946) looked only for the propagation of acoustic disturbances. Their failure meant that Landau's theory was "born refuted": it could not account for any properties of helium II that were not already explained by the London-Tisza theory; it offered no explanation of the λ-transition; its predictions for specific heat and critical velocity were in disagreement with experiment; and the expected property of "second sound" propagation had not been observed. Nevertheless Landau's colleague V. L. Ginzburg made the following statement two years later: "The condensation of a Bose gas has nothing to do with the properties of helium. The latter have been explained by Landau's theory only" (1943:305).

In 1944 another of Landau's colleagues, E. M. Lifshitz, made a more detailed theoretical analysis of second-sound waves and showed that they should be detectable as temperature variations. This was quickly confirmed by V. Peshkov, who reported a measurement of second-sound waves generated by alternating current flowing through a heater made of thin constantan wire. The velocity was 19 ± 1 m/sec at $1.4°K$, compared with 26 m/sec estimated by Lifshitz (Peshkov 1944). Peshkov published

---

[18] In popular lectures he compared the fluid of excitations with the old "caloric" fluid used to explain thermal phenomena (Livanova 1980:167–68).

more detailed results in 1946: u_{II} rises rapidly from zero at T_λ to just over 20 m/sec at 1.7°K, then gradually decreases to about 19 m/sec at 1.2°K, the lowest temperature at which he could measure it. Peshkov suggested that second sound might be imagined as "sound" vibrations in the gas of phonons and rotons, just as ordinary sound consists of vibrations of atoms or molecules. But more important was the fact that his results "fully confirm the phenomenological part of the theory of superfluidity and in particular the hydrodynamics of helium II." Landau's phonon-roton model did not come out as well; in order to account for the temperature variation of u_{II} it was necessary to assume that the "constants" Δ and μ vary with temperature.

At this point it would seem that the second-sound data favored Tisza's theory over Landau's, as Tisza himself pointed out on several occasions. But Peshkov rejected the London-Tisza theory as "very artificial and unconvincing" without any detailed discussion of its predictions regarding second sound.

London (1945) also preferred to discuss the two theories on theoretical rather than experimental grounds. He pointed out that Landau's commutatation rule for the velocity operator is inconsistent with the uncertainty principle, and questioned other aspects of his quantum hydrodynamic equations.

Landau ignored London's criticisms and proceeded to modify his roton theory to account for Peshkov's results. He replaced the term $p^2/2\mu$ by $(p - p_0)^2/2\mu$ and proposed to combine both phonon and roton excitations in a single energy-momentum curve which should look like Fig. 4.9.3; "with such a spectrum it is of course impossible to speak strictly of rotons and phonons as of qualitatively different types of elementary excitations" (1947, 1965:465), though Landau's followers continued to do just that for many years.

Another important experiment directly inspired by Landau was E. Andronikashvili's measurement (1946) of the superfluid fraction by observing the period of torsional oscillationn of a pile of disks attached to a wire passing through their centers. Only the normal component is

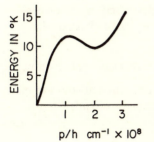

FIG. 4.9.3 Landau's spectrum of excitations in superfluid helium (after Landau 1947:91–92).

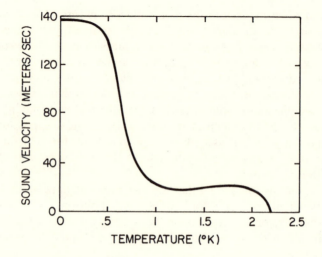

F<small>IG</small>. 4.9.4. Velocity of second sound in superfluid helium (after E. F. Lifshitz, "Superfluidity," *Scientific American*, June 1958).

dragged along by the disks and contributes to the moment of inertia. The results showed that the superfluid fraction varies from about 70% at 1.76°K to 0 at T_λ. While supporting the validity of the phenomenological two-fluid hypothesis, the Andronikashvili experiment did not favor either the London-Tisza or the Landau theory.

In 1947 Tisza, who had moved to the Massachusetts Institute of Technology, reviewed the competing theories of superfluidity and concluded that neither of them could yet give a satisfactory atomistic explanation of the properties of liquid helium; he advocated the development of a more phenomenological or "quasi-thermodynamic" theory. He argued that the existing experimental data on second-sound velocity and specific heat were unfavorable to Landau's theory, and pointed out the need for new experiments to settle the issue. In particular,

> A distinguishing feature of the [Landau] vortices compared with the excited atoms of the Bose-Einstein theory is the existence of an internal rotational energy which should manifest itself in an absorption and dispersion of the relaxation type for first and particularly for second sound.

Also, Tisza suggested that experiments on the flow of the ^{3}He isotope

> should show conclusively whether the Bose-Einstein statistics are of any fundamental importance for the phenomenon of superfluidity. (1947:853)

Tisza criticized Landau's modification of his theory to make it agree with Peshkov's experiment; he argued that the phonons and rotons should be more sharply distinguished, whereas Landau's latest version blurred the distinction between them.

Landau, in response (1948), stated that Tisza's "entire quantitative theory (microscopic as well as thermodynamic-hydrodynamic) is, in my opinion, entirely incorrect." Rejecting the demand for a more precise microscopic characterization of the phonons and rotons, Landau stated his own position quite firmly:

> It follows unambiguously from quantum mechanics that for every slightly excited macroscopic system a conception can be introduced of "elementary excitations," which describe the "collective" motion of the particles and which have certain energies ϵ and momenta p . . . .
> It is this assumption, indisputable in my opinion, which is the basis of the microscopic part of my theory. On the contrary, every consideration of the motion of individual atoms in the system of strongly interacting particles is in contradiction with the first principles of quantum mechanics.

Tisza had violated those principles, he charged, in making statements such as "every vortex element can be associated with a definite mass contained in the volume in which the vorticity is different from zero" (Tisza 1947:852).

Landau pointed out that his theory differed from Tisza's by including phonons in the normal part of the liquid, whereas Tisza's excluded them on the grounds that they are associated with the liquid as a whole. A decision between these two hypotheses could be made by extending the measurement of second-sound velocity down to the temperature region where phonons become important: "I have no doubt whatever that at temperatures of 1.0–1.1°K the second sound velocity will have a minimum and will increase with the further decrease in temperature (Landau 1948, 1965:476).

Tisza, in reply, critized some of Landau's assumptions and noted that the observed critical velocity "may be as much as 100 times smaller than the one computed by Landau" (1948:886). He maintained that his own formula for second-sound velocity should be more accurate at lower temperatures, thus agreeing with Landau's statement that measurements below 1.2°K should provide a crucial test of the competing theories.

Just at the time when the London-Tisza theory based on Bose-Einstein condensation seemed to be most sharply in conflict with the Landau program based on a phenomenological set of equations for the energy spectrum and hydrodynamic behavior, a major step toward unifying these theories was taken by Nikolai Nikolaevich Bogoliubov (see Mitropol'skiy & Tyablikov 1959 for biography). Bogoliubov had already established his reputation in applied mathematics and mechanics before turning his attention to problems of theoretical physics, where his first major contribution was a monograph on "Problems of Dynamical Theory in Statistical Physics" (1946), a formulation of the hierarchy of equations for the distribution functions of classical systems of particles (Brush 1971:65–68). He held chairs at the universities of Kiev and Moscow from 1936 to 1950, then was appointed director of the Laboratory of Theoretical Physics of the Joint Institute for Nuclear Research. Aside from his contributions to kinetic theory he is known for his work on the mathematical proof of "dispersion relations" in quantum field theory and for his formulation of a theory of superconductivity similar to the Bardeen-Cooper-Schrieffer theory.

Bogoliubov began his 1947 paper on superfluidity by asserting that the most natural starting point for a consistent atomic theory of helium is the Bose-Einstein gas with weak interactions between its particles. Attempting to make the theory more realistic by introducing strong interactions would be futile, at least at the beginning, since there is no hope for an accurate theoretical treatment of such a system; the most one can expect at present is a qualitative explanation of superfluidity, in particular a proof that under certain conditions the "degenerate condensate" of a "nearly perfect" Bose-Einstein gas can move without friction at sufficiently small velocity.

Whereas Landau had postulated the existence of elementary excitations (phonons and rotons), Bogoliubov proposed to derive them from the basic equations for a nonideal Bose-Einstein gas. His procedure was to start with those equations in the "second quantization" form developed earlier for the field theory of elementary particles, and to apply certain approximations that seemed to be reasonably accurate for systems of a large number of identical particles obeying Bose-Einstein statistics. (His 1967 monograph is a good introduction to the subject.)

The wave function of an N-particle system can be expressed as a sum of products of eigenfunctions of the Schrödinger equation for a single free particle. (This does not entail any physical assumption about the system but is a mathematical fact, comparable to the fact that any function subject to certain rather general conditions can be expanded in a Fourier series of trigonometric functions.) The sum may be subdivided into terms, each of which represents all the possible permutations of the N particles among the single-particle quantum states such that n_0 are in the state labeled zero (for example the zero-momentum state), n_1 are in the state labeled 1, etc.; if the particles are identical and obey Bose-Einstein statistics, one simply adds these products together in making up the wave function of the system. The N-particle wave function is now represented as a superposition of states, each characterized by a set of "occupation numbers" $\{n_0, n_1, \ldots\}$. One then introduces "creation" and "annihilation" operators, $a_i{}^+$ and a_i, which when operating on a state $\{\ldots n_i \ldots\}$ convert it to a state $\{\ldots n_i + 1 \ldots\}$ or $\{\ldots n_i - 1 \ldots\}$ respectively. The eigenvalue of the operator $a_i{}^+ a_i$ is n_i:

$$a_i a_j{}^+ \{\ldots \ n_i \ldots\} = n_i \{\ldots n_i \ldots\},$$

but operators for the same i do not commute:

$$a_i a_j{}^+ - a_i a_j{}^+ = \begin{cases} 1, & \text{if} \quad i = j \\ 0, & \text{if} \quad i \neq j. \end{cases}$$

Bogoliubov assumed that in a superfluid the number of particles in the zero-momentum state, n_0, is large compared with 1, and therefore the noncommutativity may be ignored and $a_0{}^+$ and a_0 may be treated as ordinary numbers equal to $\sqrt{n_0}$. He was then able to compute the eigenvalues of the N-particle Hamiltonian, assuming weak interactions between the particles. He showed that for small momenta the energy spectrum has the form

$$\epsilon = cp + \sigma p^2,$$

where c is the speed of sound at absolute zero. For large p it can be expanded in powers of the interaction constant, and the first term is just the free-particle energy $p^2/2m$.

While Bogoliubov found that the energy spectrum could be expressed in terms of "quasi-particles"—excitations whose momentum dependence is a combination of linear and quadratic terms—he concluded that "no division of quasi-particles into two different types, phonons and rotons, can even be spoken of" (Bogoliubov 1947, Galasiewicz 1971:260). Thus Bogoliubov's work favored the more recent form of Landau's theory in which there is a single energy-momentum curve for excitations. Bogoliubov also found that even at zero temperature not all molecules have zero momentum. Because of this "depletion" effect, one can no longer simply identify the superfluid fraction with the fraction of particles in the zero-momentum state, as was done in the original London-Tisza theory. Thus in the unification of the two theories each had to give up some of its original features.

On October 13, 1948, S. G. Sydoriak, E. R. Grilly, and E. F. Hammel at Los Alamos succeeded in condensing pure ^{3}He prepared by E. S. Robinson and R. M. Potter; the ^{3}He had been "grown" from pure tritium solutions by β-decay of the tritium (Sydoriak et al. 1949). The fact that ^{3}He could be obtained as a liquid at atmospheric pressure might itself be considered a point against the London-Tisza view, since both London and Tisza had predicted in the same year that ^{3}He could not be liquefied, because of its high zero-point energy (London & Rice 1948:1193; Tisza 1948:26). But that mistake was quickly forgotten when Darrell Osborne, Bernard Weinstock, and Bernard Abraham at Argonne National Laboratory showed that ^{3}He has no superfluid transition between 1.05°K and 3.02°K; they offered it the opportunity to flow through a "superleak" between a platinum wire and a capillary tube, a task which helium II could easily perform; but ^{3}He couldn't do it (Osborne et al. 1949). Their conclusion was: "the experimental results lend support to the hypothesis that the lambda-transition of He^4 is due to Boise[sic]-Einstein statistics."

The absence of superfluidity in ^{3}He was generally taken, by physicists in the West, as a very damaging blow to Landau's theory, since in its original form that theory made no use of the assumption that the atoms must obey Bose-Einstein statistics. But Landau and his followers could now dodge the blow since, as Landau had already noted in his 1948 paper (1965:474), the general form of his energy spectrum could be derived from the Hamiltonian of a weakly interacting Bose-Einstein gas by Bogoliubov's method. Landau could therefore quite reasonably develop a separate theory for Fermi liquids, while maintaining that his theory of superfluidity pertains specifically to Bose-Einstein liquids.[19]

Almost simultaneously with the ^{3}He experiments, new measurements on the speed of second sound at low temperatures provided dramatic confirmation for Landau's theory. Peshkov (1948) found that u_{II} reaches a minimum at 1.12°K, then rises as the temperature falls still lower. This result was confirmed by John Pellam and Russell Scott at the National Bureau of Standards (1949), and by R. D. Maurer and Melvin A. Herlin at the Massachusetts Institute of Technology (1949). It was generally agreed that the marked rise of u_{II} below 1°K refuted Tisza's prediction that $u_{II} \to 0$ as $T \to 0$, and supported Landau's prediction that $u_{II} \to u_I/\sqrt{3}$.

As experimenters pushed their measurements to lower and lower temperatures, it began to appear that u_{II} might not level off at $u_I/\sqrt{3} \approx$ 137 m/sec. Values as high as 190 m/sec were reported, and some physicists

---

[19] When superfluidity was finally discovered in ^{3}He in late 1971, Landau's theory of Fermi liquids was found best suited to describe it. See the lectures in Armitage & Farquhar (1975).

guessed that the limiting value might be $u_{II} = u_I$ (De Klerk, Hudson, & Pellam 1954; Lane 1967:370). But detailed examination of the shape of the heat pulses observed in experiments below $0.5°K$ led to the conclusion that the mean free path of phonons is greater than the size of the container, so the heat is no longer propagated in second-sound waves but converted to a more "normal" mode of heat transfer (Gorter 1952, Kramers 1957, Peshkov 1960). Since u_{II} is no longer an observable quantity below $0.5°K$ (except perhaps in the zero-frequency limit, according to the theory of Humphrey Maris 1974), one might conclude that Landau's prediction cannot be considered completely confirmed; however, some of his followers insist that u_{II} does converge to $u_I/\sqrt{3}$ (Khalatnikov 1976:31; Galasiewicz 1970:123).

The mere fact that u_{II} does not go to zero as $T \to 0$ not only refutes Tisza's hypothesis about the nature of temperature waves but offers severe difficulties to any theory that takes the ideal Bose-Einstein gas as the first approximation. It seems almost essential to have an energy spectrum of the form $\epsilon \approx cp$ for small p rather than the free-particle energy $\epsilon = p^2/2m$, and "so radical a departure from the 'gas-like' condition is a strong argument in favour of Landau's approach from the 'solid-like condition'" (Atkins 1952:203). Bogoliubov's method, or one of its many equivalents, becomes very attractive to followers of London's program for this reason.

As Tisza pointed out (1947:853), Landau's rotons ought to produce absorption and dispersion of first and second sound, and indeed such phenomena soon provided additional support for Landau's theory. Khalatnikov (1950), following Landau's approach, predicted a large peak in the absorption coefficient of first sound at about $1°K$, on the basis of a model in which most of the rotons have energy close to their minimum value while most of the phonons have their maximum energy. When the liquid is compressed by the wave, the phonon level is displaced upward and the roton level downward, so the gap between them decreases. There is then a redistribution of energy between phonons and rotons which causes irreversible dissipation. Subsequently Atkins (1951) and Chase at Toronto found an extremely rapid rise in absorption, in measurements down to $1.2°K$, and Chase (1953) confirmed the predicted maximum with his measurements down to $0.85°K$ (see Wilks 1967:201–6). Similarly Khalatnikov (1952) published a formula for the absorption coefficient of second sound which predicted that it would rise steadily as the temperature dropped below $1.8°K$, and the first such measurements, made after the prediction was published, confirmed it semiquantitatively (Atkins & Hart 1954, Hanson & Pellam 1954, Zinov'eva 1956).

Without going into the many other "kinetic phenomena" of liquid helium, we may fairly say that Landau's program has had an astonishing record of success since 1944. Its major failures, as compared with atomistic theories based directly on Bose-Einstein statistics, have been in its treatment of rotational motion and in not producing an explanation of the λ-transition.

These failures were to a large extent remedied by the work of Richard Phillips Feynman, an American physicist who unified the London and Landau theories by an approach quite different from Bogoliubov's but equally valuable. Feynman, who had recently moved from Cornell to the California Institute of Technology, was already well known for his solution of the renormalization problem in quantum electrodynamics through the use of "Feynman diagrams," a technique widely adopted by physicists. Through a series of brilliant papers and lectures covering a wide range of topics, Feynman became the leader of the post–World War II generation of American physicists, much as Landau did in the USSR.

Feynman's first paper on liquid helium was a short note summarized by its first sentence:

We show why the interatomic potential does not alter the existence of an Einstein-Bose condensation in He^4. (1953a:1116)

The argument was based on an alternative formulation of quantum mechanics which Feynman had proposed five years earlier. The wave function of a particle at time t' can be written in terms of the wave function at an earlier time t'' with the help of a "propagator" or "Green's function"[20] K:

$$\psi(x', t') = \int K(x', t'; x'', t'')\psi(x'', t'')\,dx''.$$

Rather than begin with the assumption that ψ is a solution of the time-dependent Schrödinger equation, which would also provide a method for calculating K, Feynman postulated (1948) that K can be written as a certain integral over all paths $x(t)$ by which the particle can go from the point x'' at time t'' to the point x' at time t':

$$K(x', t'; x'', t'') = A \int \exp\left\{\frac{i}{\hbar} S[x(t)]\right\} d[x(t)],$$

where the "action functional" $S[x(t)]$ is defined, as in Hamiltonian mechanics, in terms of the Lagrangian function $\mathscr{L}$:

$$S[x(t)] \equiv \int_{t''}^{t'} \mathscr{L}\, dt.$$

---

[20] After the British mathematician George Green (1793–1841), who introduced this method for solving differential equations in 1828.

(The normalization constant A depends on how the path integral is defined.)

Feynman's formulation can be shown to be equivalent to the Schrödinger or Heisenberg formulation of quantum mechanics, if one does not worry too much about the mathematical difficulties involved in defining an integral over functions $x(t)$. It has the pleasing feature that it exhibits the correspondence principle quite explicitly: as one takes the limit $\hbar \to 0$, the argument of the exponential function oscillates more and more rapidly, leading to cancellation of the contributions from all paths except those very close to the classical path determined by the extremum of the action, $\delta S = 0$ (cf. §3.4). There is also an interesting analogy between Feynman's path integral and Norbert Wiener's functional-integral theory of Brownian movement (Brush 1961).

Feynman's quantum-mechanical partition function uses a similar representation, but now the "time" variable is replaced by an inverse temperature $\beta \equiv 1/kT$, and the final position $x'(\beta)$ must be identical to the initial position $x''(0)$. The imaginary unit i disappears, so the analogy with Brownian movement is even closer. For a system of N identical particles obeying Bose-Einstein statistics, one has to sum over all paths in which the final positions are any permutation of the initial positions. Thus, as in the "Feynman diagram" approach to quantum field theory, the problem is formulated in such a way that one can visualize various possible motions of the particles, even though these are "virtual" motions, corresponding to terms in a mathematical formula, rather than "real" motions in space and time. The advantage of the Feynman approach is that one has a systematic way of keeping track of terms in a complicated expression, and an opportunity to use physical intuition in devising suitable approximations. (See Wiegel 1975 for a recent review of this method.)

Feynman argued that the nature of the λ-transition is a combinatorial problem involving the number of paths of various lengths corresponding to permutations of identical atoms. To simplify the problem, suppose that the initial positions are points on a simple cubic lattice, and that the largest contribution to the partition function comes from permutation cycles which are polygons with each side equal to the distance d between neighboring lattice sites. Feynman showed that each side contributes a factor

$$y = \exp\left\{-\tfrac{1}{2}md^2/\beta h^2\right\} = e^{-aT}.$$

So a permutation in which the total number of sides of all the polygons is L will give a contribution y^L to the partition function, and if there are $g(L)$ ways of constructing such a permutation, the partition function will have a factor

$$q = \sum_L g(L)y^L.$$

The factor y depends on temperature (e^{-aT}) and at high temperatures will be very small, so permutations with large L will have negligible effect. But as T falls, at some point the contribution from very large polygons becomes important and produces a discontinuous change in the behavior of q.

Thus Feynman reduced the physical problem to a definite though difficult mathematical problem. "The difficulty is that the polygons must not intersect." If one assumed that, "when drawing polygons, if K sides are already drawn, the next atom site has a chance of $(N - K)/N$ of being unoccupied," then one gets a third-order transition like that of the ideal Bose gas. "A solution of the idealized problem which more rigorously describes the geometrical correlations might well give a transition of a different order" (1953:1117).

Ryoichi Kikuchi at the University of Chicago proposed (1954) to solve the combinatorial problem by applying an approximate method he had previously developed in connection with the Lenz-Ising model for order-disorder transitions (§6.4). The result was a second-order transition. Takeo Matsubara and Hirotsugu Matsuda at Kyoto University proposed (1956, 1957) a similar lattice model using second-quantized annihilation and creation operators like those of Bogoliubov, but imposing additional anticommutation rules to prevent two atoms from occupying the same lattice sites, and approximately taking account of zero-point energy. They showed that such a model is equivalent to a ferromagnetic spin system with anisotropic exchange coupling. Thus the techniques developed for dealing with combinatorial lattice problems, including Kikuchi's, could be applied. By allowing vacant lattice sites, Matsubara and Matsuda could explain qualitatively the fact that T_λ decreases with pressure. Lattice models with holes can also account for the gas-liquid transition (Brush 1957, 1958), as can other methods of evaluating Feynman's path-integral partition function (Chester 1955).

Further calculations with Feynman's partition function showed that with reasonable approximations it could yield a gas-liquid transition as well as a second-order λ-transition (Chester 1955, Brush 1958). But this approach did not lead to much progress in understanding superfluidity below T_λ, and theories which predicted only a second-order λ-transition did not seem very promising during the decade after 1958, when the transition was thought to have a logarithmic singularity in the specific heat (see below).

Feynman asserted that "London's view is essentially correct. The inclusion of large interatomic forces will not alter the central features of the Bose condensation" (1953b:1291). But the following paper in the series (Feynman 1953c) showed that Landau's view is also correct: one can justify the phonon-roton model from quantum mechanics applied at the atomic level. The essence of the argument is as follows:

> The central feature which dominates the properties of helium II is the scarcity of available low energy excited states in the Bose liquid. There do exist excited states of compression (i.e.: phonons) but states involving stirring or other internal motions which do not change the density cannot be excited without expenditure of an appreciable excitation energy. This is because, for quantum energies to be low, long wave lengths or long distances are necessary. But the wave function cannot depend on large-scale modifications of the liquid's configuration. For a large-scale motion, or stirring, which does not alter the density, only moves some atoms away to replace them by others. It is essentially equivalent to a permutation of one atom for another, and the wave function must remain unchanged by a permutation of atoms, because ^{4}He obeys the symmetrical statistics. The only wave functions available are those which change when atoms move in a way which is not reproducible by permutation, and therefore either, (1) movements accompanied by change in density (phonons), (2) movements over distances less than an atomic spacing, therefore of short wave length and high energy (rotons and more complex states), or (3) movements resulting in a change in the position of the containing walls (flow).... (Feynman 1955, AAPT: 75)

Feynman thus "atomized" Landau's theory, showing how one can infer from the symmetry properties of the N-atom wave function the scarcity of low-energy states other than phonons. He also attempted, with Michael Cohen, to construct a wave function to describe roton states (Feynman & Cohen 1955, 1956):

> The wave function suggests that the roton is a kind of quantum-mechanical analog of a microscopic vortex ring, of diameter about equal to the atomic spacing. A forward motion of single atoms through the center of the ring is accompanied by a dipole distribution of returning flow far from the ring. (Feynman & Cohen 1956:1189)

Cohen and Feynman (1957) also developed the theory of inelastic scattering of cold neutrons from liquid helium as a means for direct experimental determination of the energy spectrum, analogous to the method proposed for solids by G. Placzek and L. van Hove. The experiment was first done by H. Palevsky (visiting from Brookhaven National Laboratory), K. Otnes, K. E. Larsson, R. Pauli, and R. Stedman using the neutron beam from the Stockholm reactor (1957). The results of this and other neutron-scattering experiments confirmed the energy spectrum suggested by Landau and explained atomistically by Feynman and Cohen.

When is a roton not a roton? Merged with the phonons into a single energy-momentum curve, these theoretical entities gradually lost their identity as rotational motions and became an inappropriate label for whatever excitations might correspond to the region near the minimum at p_0. T. D. Lee and F. Mohling at Columbia University pointed out (1959) that neutron-scattering data showed no evidence of helicity, and Chester (1966:254) reported that even the Feynman-Cohen wave function for rotons could not really be interpreted as a small vortex ring. But if the original Landau roton had died, it had also been reincarnated on a much grander scale, as a "vortex line."

How does one explain the fact that resistance to superfluid flow sets in at a speed several hundred times smaller than that predicted by Landau's theory? Here Feynman revived a suggestion made by Lars Onsager in 1949 but generally overlooked: vortex lines are created by the motion of the fluid, causing turbulence. (On the relations between Feynman and Onsager see Longuet-Higgins & Fisher 1978:458–59). There is a critical velocity because the vortex lines are quantized—the integral of the momentum per atom around a closed circuit must be equal to a multiple of h; hence a certain minimum energy is needed to create the vortex.

Feynman's extensive discussion of quantized vortex lines (1955) is rather similar to his earlier discussion (1953a) of the hypothetical permutation cycles included in his path-integral formulation of the Bose-Einstein partition function. In both cases an additional link in the ring corresponds to a factor sy, where s is the factor by which the number of possible configurations is increased by adding the link. But for vortex lines $y = \exp\{-\epsilon/kT\}$, where ϵ is the energy needed to add a link; so $sy < 1$ at *low* temperatures, whereas for permutation cycles the temperature enters the numerator of the exponent, and $sy < 1$ at *high* temperatures. Thus the λ-transition is now described as a combinatorial effect in which the number of vortex lines suddenly increases on raising the temperature, and they pierce the liquid through and through in the disordered state of helium I (Feynman 1955, AAPT: 108).

The quantized vortex lines predicted by Onsager and Feynman were first observed in 1958 by W. F. Vinen at the Royal Society Mond Laboratory in Cambridge (Vinen 1961). A wire was stretched down the center of a cylindrical vessel containing liquid helium, and circulation was established by rotating the apparatus around the axis of the wire and cooling from above the λ-point. The wire was set into transverse vibration, and the circulation around it could be obtained from the rate of precession of the plane of vibration. Other experiments showing quantized circulation were done by G. W. Rayfield and F. Reif at Berkeley (1963) and by S. C. Whitmore and W. Zimmerman, Jr., at the University of Minnesota (1965). There was also some evidence of vortex lines above T_λ, supporting Feynman's idea that the transition from helium II to helium I is due to sudden formation of long filaments by union and elongation of vortex rings (Andronikashvili & Mamaladze 1966). There were some attempts to develop a quantitative theory of such a transition (Byckling 1965, Wiegel 1975, 1978, and Popov 1973), but so far the mathematical methods used to attack the problem have not been powerful enough to match the intuitive exploits of Landau and Feynman or to keep ahead of experiments.

Theories of the λ-transition seem to have lagged behind the results of laboratory measurements. One exception was the case of the logarithmic singularity in the specific heat, reported by W. M. Fairbank, M. J. Buckingham, and C. F. Kellers (1957) at Duke University, and generally accepted until about 1973. The first logarithmic singularity in the physics of phase transitions was that of the two-dimensional Lenz-Ising model, discovered by Lars Onsager in 1944 (§6.4); this result, though it pertained only to a theoretical system, prompted Tisza (1951) to propose a reformulation of the general thermodynamic theory of phase transitions to allow for such singularities. But the immediate incentive for the Fairbank-Buckingham-Kellers experiment was a conflict between theoretical predictions about liquid helium by a group of Australian physicists—S. T. Butler, J. M. Blatt, and M. R. Schafroth (1956)—and Feynman. The Australians argued that the correlation length for momentum-space ordering should be finite in a real Bose-Einstein liquid, whereas Feynman claimed that ordering would persist to indefinitely large distances. According to the former, the peak in the specific-heat curve should be rounded, whereas according to the latter it should be sharp even at a hundred-thousandth of a degree. Blatt asked Feynman, "How is your intuition so good that you can say that there is a correlation between helium atoms separated by the size of the universe?" (Fairbank and Kellers 1966:72).

The results of the experiment supported Feynman's intuition (Fairbank 1963) but may have discouraged theorists who had succeeded in extracting a second-order transition from approximate treatments of the interacting Bose-Einstein gas and now had to go back and look for a logarithmic singularity. Tisza was surprised by the result, since according to his own general theory of phase transitions such a singularity must be associated with a change of symmetry and it implied as well that the ground-state wave function is doubly degenerate, which "is inconsistent with existing quantum mechanical theories whether based on Bose-Einstein statistics or interacting particles or on elementary excitations" (1958:588; see also Tisza 1966:102; a possible way of satisfying Tisza's requirements was suggested by Girardeau 1978).

Theories of the interacting Bose-Einstein gas were encountering other difficulties in explaining superfluidity. A fundamental problem was whether the interactions between particles substantially change the macroscopic occupation of the zero-momentum state, which was the basis for the concept of Bose-Einstein condensation. One might object that when one turns on the interaction, a wave function made up of single-particle states is no longer an exact solution of the Schrödinger equation of the N-particle system. In fact that statement is not correct; it sounds plausible if one thinks that the wave function for N noninteracting particles has the form

$$\Psi(x_1, \ldots, x_N) = \phi_1(x_1)\phi_2(x_2) \ldots \phi_N(x_N),$$

and this is indeed a solution of the Schrödinger equation. But when one takes account of the Bose-Einstein statistics by constructing a Ψ from a sum of all possible products formed by permuting identical particles and summing over all the one-particle states, then the interaction does not change the set of states but only their occupation numbers in the second quantized representation (Bogoliubov 1947, 1962). One can still define (and, in principle, compute) the number of particles in zero-momentum states, for example by using the "density matrix."[21] This was done by Oliver Penrose and Lars Onsager (1956), who proposed that the criterion for Bose-Einstein condensation should be that the largest eigenvalue of the "reduced" one-particle density matrix is of order 1. They argued that

---

[21] For readers familiar with Dirac's "bra" and "ket" vectors the simplest way to define the density matrix is

$$\rho = \sum_m |m > p_m < m|,$$

where p_m is the probability that the system is in state m. The reduced density matrix for the first particle is computed by taking the trace of ρ with respect to particles $2 \ldots N$, as follows: $\rho_1 = N \, tr_{2 \ldots N}(\rho)$. Most modern textbooks on statistical mechanics include a detailed discussion of the properties of the density matrix and its applications.

by this criterion Bose-Einstein condensation does occur in liquid ^{4}He at 0°K. However, they estimated that only about 8% of the atoms are condensed into the lowest state. Thus n_0/N can no longer be identified with the superfluid fraction p_s/ρ; yet this background of condensed particles may somehow be essential for superfluidity of the entire system (Onsager 1958:261).

A different characterization of the nature of the superfluid state was proposed by Penrose (1951): the off-diagonal part of the one-particle density matrix in the position-space representation remains finite at large spatial separations.[22] This property, called "off-diagonal long range order" by Yang (1962), has no classical analog but may be viewed as a consequence of the holistic, nonlocal aspect of the quantum description of matter (§3.5). The possibility that a particle may be found to exist in one place interferes with (has a statistical correlation with) the possibility that it exists in another place, no matter how far away those places are. In a Bose-Einstein system below the λ-point, this effect is enhanced to the extent that it affects the macroscopic properties of the system. (Note that the "one-particle density matrix" is derived from the N-particle density matrix, so it reflects indirectly the presence of all the other particles.) Penrose did not give this interpretation, but he did recognize that Bogoliubov's theory leads to a density matrix with off-diagonal order and that this property makes the classical equations of heat transfer and hydrodynamics inapplicable.

In 1957 Keith A. Brueckner and K. Sawada at the University of Pennsylvania rediscovered Bogoliubov's method and his result that weak interactions produce phonon-type interactions in a boson gas. Their paper, and a series of papers by T. D. Lee and C. N. Yang, stimulated a large number of theoretical investigations of the hard-sphere boson gas at temperatures near 0°K. These calculations created considerable confusion about what properties the model really has, and little confidence that it could account for superfluid phenomena other than the simplest ones already known. One weakness was that if the ideal Bose-Einstein gas with 100% occupation of the zero-momentum state is taken as the starting point, it turns out that interactions, treated in the lowest order of approximation, produce a depletion of anywhere from 50% to 270% (Parry & ter Haar 1962; Ostgaard 1971:251–52);

---

[22] This property may be expressed as

$$\langle x'|\rho_1|x\rangle \neq 0 \text{ as } |x - x'| \to \infty.$$

See Yang (1962) and Ziff et al. (1977) for detailed discussion.

there seems little reason to hope that higher approximations will converge to the correct answer. Some theorists began to suspect that the zero-momentum state is not macroscopically occupied at all, and that the condensation is either smeared over a range of low-momentum states (Kraichnan 1959, Girardeau 1962, Goble 1971) or must be described in terms of states of two or more particles (Coniglio & Marinaro 1967; Girardeau 1969; Wong & Huang 1969; Wong 1970, 1971; Nakajima et al. 1971).

In 1966 Pierre C. Hohenberg and P. M. Platzman at Bell Telephone Laboratories proposed that high-energy neutron-scattering experiments could give information about the single-particle momentum distribution of helium atoms, and thereby test the assumption of macroscopic occupation of the zero-momentum state. Such experiments were then done by several groups, with results whose interpretation is still in dispute. H. W. Jackson, at the Ford Motor Company in Dearborn, Michigan, reviewed the evidence in 1974 and concluded that the fraction of particles in the zero-momentum state is much smaller than that predicted by most theories, and may even be zero. G. V. Chester, at Cornell, preferred to believe that the theories predicting 10% occupation of the zero-momentum state at 0°K were correct and that the interpretation of the experiments giving a much smaller value was wrong (1975:21–22).

W.A.B. Evans at the University of Kent, Canterbury, argued that Bose-Einstein condensation cannot be derived from a plausible approximation scheme for an interacting Bose system, so one should accept the conclusion that there is no macroscopic occupation of the zero-momentum state for single particles and turn instead to pairing theories (Evans 1975, 1978; Imry 1970; Evans & Harris 1978, 1979). In pairing theories the superfluid phase could be described by an equation similar to that which arises in the Bardeen-Cooper-Schrieffer theory of superconductivity; "it would also be aesthetically satisfying to have a single theory of *all* superfluids be they of Bose or Fermi statistics" (Evans 1978:41). N. H. March, of Imperial College, London, and Z. M. Galasiewicz, of the University of Wroclaw, Poland, agreed that "superfluidity without a condensate . . . appears possible in principle" but argued that the ground-state wave function in that case "cannot be built from a product of pairs, but must include fundamentally three atom correlations" (March & Galasiewicz 1976:103). But P. Dörre, H. Haug, and D. B. Tran Thoai at Frankfurt University (1979) derived a set of equations which could have as solutions either a condensate theory or a pairing theory, and found that the former has lower free energy, hence would be the thermodynamically stable phase.

The condensation problem seems still unresolved; most theorists still accept the proposition that helium II has a Bose-Einstein condensate even though the occupation of the zero-momentum state may be only a few percent even at 0°K. This condensate somehow makes the entire system superfluid, perhaps by acting as a "pilot wave" dragging along the rest of the fluid, an idea attributed to H. Fröhlich (Cummings 1971; Hyland & Rowlands 1972). It remains to be seen whether a satisfactory theory of condensation of the nonideal Bose gas can be developed.[23]

The most promising theory of the λ-transition now seems to be the lattice model of Matsubara and Matsuda (1957), primarily because its properties can be deduced from results in the general theory of phase transitions. (Recall the situation in the 19th century when Lord Kelvin and James Clerk Maxwell were able to develop electromagnetic field theory by exploiting mathematical analogies with theories of fluid flow and heat conduction.) Vaks and Larkin (1966) pointed out that the phase transition of a Bose-Einstein gas should be similar to that of a lattice of plane dipoles. According to the principles of scaling and universality (§6.7), the critical exponents at the λ-transition should be related to the exponents of a "spin $\frac{1}{2}XY$ model" (Fisher 1967). Thus if the specific heat above T_λ is represented by the equation

$$C_P = \frac{A}{\alpha}[|t|^{-\alpha} - 1] + B \qquad \text{where} \quad t = 1 - T/T_\lambda,$$

and the specific heat below T_λ is represented by a similar formula with A', α', and B', scaling theory predicts $\alpha = \alpha'$, with the same value independent of pressure; and one should have (Eq. 6.7-3):

$$\alpha = \alpha' = 2 - 2\beta - \gamma,$$

where β and γ are the exponents involved in the singularities of spontaneous magnetization and susceptibility for the XY Ising model (Eqs. 6.7-1, 2).

According to Fairbank, Buckingham, and Kellers (1957), c_p has a logarithmic singularity, corresponding to $\alpha = 0$. Their result was not replicated by anyone else for a decade, though Louis Goldstein at Los Alamos (1964) criticized this interpretation of their data. Late in the 1960s,

---

[23] According to a news report in *Physics Today* (June 1980, p. 18), "in contrast to the case of superconductivity, we have at present no satisfactory microscopic theory of superfluid He^4."

Guenter Ahlers at Bell Telephone Laboratories reopened the question, motivated by progress in phase-transition theory[24] which focused attention on the precise values of exponents such as α. In particular, the renormalization-group theory of Kenneth Wilson opened up the possibility of calculating these exponents by starting with a four-dimensional system (which appears to be easier to handle than those of two or three dimensions) and expanding the values for $4 - \epsilon$ dimensions in powers of ϵ (Wilson & Fisher 1972). When Ahlers' data began to suggest very small finite values of α, theorists still under the sway of earlier ideas which dictated ratios of small integers were unhappy—one said to him, "How can you expect me to make a theory which predicts one fiftieth, or even one one-hundredth?" (Ahlers 1978; cf. Fisher 1967:692, where values like 1/8 and 1/16 are suggested). But the rise of renormalization-group theory after 1970 legitimized such values.

In 1973 Ahlers announced that α is probably about -0.02. He then spent a year working with Frank Pobell and Karl Mueller at the Kernforschungsanlage in Jülich, West Germany; their measurements of the thermal expansion coefficient suggested $\alpha = -0.026 \pm 0.04$ (Mueller et al. 1975). Similar results were obtained about the same time by F. M. Gasparini, at SUNY-Buffalo, and M. R. Moldover, at the University of Minnesota (1975); a further refinement of that experiment led to the conclusion that α lies between -0.02 and -0.025 (Gasparini & Gaeta 1978). These are not purely experimental results, since the universality postulate was invoked to require $\alpha = \alpha'$ in fitting the data points; without this constraint one of the exponents might be zero.

Theoretical calculations of α for the λ-transition, based on renormalization-group theory applied to the Matsubara-Matsuda model, suggest that α may indeed have a value of this order of magnitude. Wilson's ϵ-expansion for a two-component order parameter (see §6.7) gives $\alpha \cong -0.02$ (Fisher 1974:607). While there was some skepticism about the reliability of exponents derived from this expansion (Gasparini & Gaeta 1978), the prediction that α should be near zero was a striking result of the new theory.[25] Ahlers quotes a theoretical estimate $\alpha = 0.008$ based on more recent calculations of β and γ by J. C. Le Guillou and J. Zinn-Justin in France (1977). Baker, Nickel, and Meiron (1978) found $\alpha = -0.007 \pm 0.009$, "which is not, in my opinion, significantly different from Ahlers' experimental value" (letter from G. A. Baker to S. G. Brush, 23 August

[24] As Hohenberg (1978:336) noted, Ahlers is an exception to the generalization that "the quantum fluids community has remained somewhat orthogonal of this development."
[25] I am indebted to M. E. Fisher for a letter giving detailed discussion of this and related points.

1979). A recent review of this model, including estimates of critical exponents, is given by Betts and Cuthiell (1980).

If α and α' are really negative numbers (but > -1) and if, as seems likely at this point, $B = B'$, then the specific heat will be continuous at T_λ, though its derivative will be infinite. This would mean that the λ-transition of liquid helium is somewhat more like the transition of the ideal Bose-Einstein gas (c continuous with discontinuous but finite derivative) than had been thought when c was believed to have a logarithmic infinity. If the experimental value of α turns out to be fairly well approximated by a theoretical value based on the Matsubara-Matsuda model, it would seem that the London program based on the Bose-Einstein condensation has at least one solid achievement to its credit. Nevertheless that achievement would depend on having adopted Landau's viewpoint: treating the fluid as a solid with the atoms localized on a lattice, to first approximation. This would be yet another example of the unification of the London and Landau programs, and would only reinforce the general view of physicists that Landau's theory provides the best overall description of superfluidity.

And what about Lev Davidovich Landau himself, the hero of this story, though none of his papers after 1948 has been mentioned? Landau suffered a fate even more bizarre than that of helium cooled down to absolute zero: he died twice. On January 7, 1962, he was killed in an automobile accident, but was resurrected from a state of clinical death by the heroic afforts of an international group of physicians and physicists (Dorozynski 1965). He never recovered his full mental capacities during the remaining six years of his life (Lifshitz 1969) but managed to attend the ceremony in Moscow at which he was awarded the 1962 Nobel Prize in physics "for his pioneering theories for condensed matter, especially liquid helium."

V. INTERATOMIC FORCES AND THE CHEMICAL BOND

5.1 EARLY IDEAS ABOUT INTERATOMIC FORCES

A recurrent theme in the physical science of the past three centuries has been provided by a research program first explicitly formulated by Isaac Newton in his *Principia* (1687): from the phenomena of nature to find the forces between particles of matter, and from these forces to explain and predict other phenomena. The success of this program in the case of long-range gravitational forces encouraged Newton and his successors to apply it to the short-range forces that presumably determine the properties of solids, liquids, and gases.

Newton himself showed that Boyle's law, PV = constant, could be derived by assuming a gas to be composed of particles that *repel each other with a force inversely as the distance* between them, provided that this force acts only between neighboring particles. (He was aware of the mathematical fact that summing the $1/r$ force between *all* pairs of particles would give a result that depends on the shape of the substance, and diverges as the volume goes to infinity; his successors often ignored this restriction.) Newton's gas model was essentially static—the pressure or elasticity of the gas was attributed to the constant repulsive force between the particles in a given configuration, rather than to collisions of moving particles with the sides of the container as in the later kinetic theory. Newton himself did not claim to have proved that gases are actually composed of such particles—he called that a "physical question" to be settled by others—but only that *if* gases have this structure, *then* the force law must have this form. Nevertheless many scientists in the 18th and early 19th centuries, including John Dalton and Thomas Young, thought he had proved that gases do consist of mutually repelling particles.

Newton suggested a rather different atomic theory in his *Opticks*: atoms are hard and impenetrable, but have short-range *attractive* forces diminishing with distance.

In 1758 the Yugoslavian scientist Roger Boscovich [Rudjer Boskovic] proposed that atoms are nothing but point-centers of force. He postulated a rather complicated force law, oscillating several times between attraction and repulsion before merging with the inverse square law of gravita-

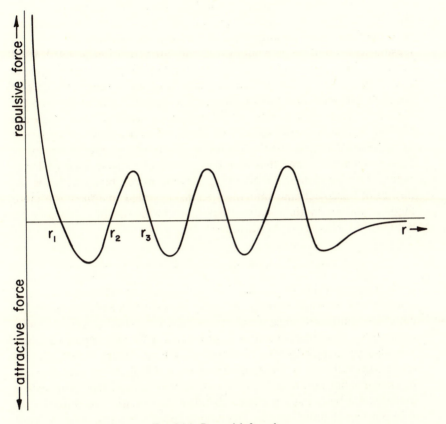

FIG. 5.1.1. Boscovich force law.

tional attraction at very large distances. Impenetrability was preserved by having the repulsion become infinite as the interatomic distance r goes to zero. The existence of definite equilibrium distances corresponding to solids and liquids was explained by the multiple crossings of the axis by the force curve; the point r_1 corresponds to stable equilibrium because any displacement pushing the atoms closer together will be counteracted by a repulsive force, while any displacement moving them apart will be counteracted by an attractive force. This point could thus be identified with the solid state. Similarly r_3 would correspond to the liquid, but r_2 is unstable.

During the 19th century it was common to postulate short-range attractive forces in order to explain phenomena such as surface tension and capillarity; such theories were developed quantitatively by Laplace and later by J. D. van der Waals (1873) in his theory of the liquid-gas transition. At the same time it was tacitly or explicitly assumed that there must be some kind of short-range repulsive force to prevent matter from collapsing to a point, as it would do if only attractive forces acted at all distances. Whether this repulsive force was described by a curve such as that suggested by Boscovich, or by postulating an impenetrable core of the atom, depended on one's philosophical commitments, at least on the Continent. (British scientists seemed able to go back and forth between the two models using criteria of convenience rather than metaphysical principles.) The mechanistic Cartesian tradition suggested that atoms have a definite size, and that no forces can act at a distance beyond that radius. The Newtonian or Leibnizian tradition, combined with the influence of Boscovich and Immanuel Kant, suggested that *force* is a more fundamental entity than *matter*; that is, all properties of matter can be reduced to forces acting between particles.

5.2 Force Laws in 19th-century Kinetic Theory

With the revival of the kinetic theory of gases in the 1850s, the assumption of a repulsive force varying inversely as the distance between particles could be eliminated, since pressure was now attributed to the collisions of particles. (The idea that *heat* is associated with such repulsive forces was also given up.) In his first paper on kinetic theory (1860a), James Clerk Maxwell used a simple model of billiard-ball atoms with impenetrable cores but no attractive forces at all, and showed that from such a model could be derived not only the ideal-gas equation of state ($PV = RT$) but also definite testable predictions about the transport properties of gases such as viscosity, heat conduction, and diffusion. In particular, he inferred from the model that the viscosity coefficient should be (1) independent of pressure or density, and (2) proportional to the square root of absolute temperature. Since both predictions contradicted intuitive ideas about the behavior of fluids (based on experience with liquids) and (1) seemed to be contradicted by the only known experiment on the viscosity of air, Maxwell apparently thought the kinetic theory could be refuted and did not at that time try to develop a more elaborate or "realistic" model of atoms.

A few years later, having found by his own experiments that the viscosity of air *is* independent of pressure over a wide range, and that it increases with temperature, Maxwell decided that the kinetic theory deserved further elaboration. He worked out a general theory of transport processes in gases on the assumption that the particles are *point-centers of repulsive force*, with a potential function of the form r^{-n} (thus the force would be proportional to r^{-n-1}). He discovered that an important simplification in the mathematical theory is achieved if n is given the special value of 4; in this case it is not necessary to know the velocity-distribution for the gas (which would in general be different from the equilibrium Maxwellian distribution). Moreover it turned out that for this special case the viscosity coefficient is directly proportional to the absolute temperature, in agreement with Maxwell's own experimental results for air. Thus in 1866 Maxwell arrived at the conclusion that the interatomic force law can conveniently be represented by a *pure repulsion varying inversely as the fifth power of the distance.* He did not of course believe that this law was the ultimate solution of the problem, since other properties of matter would obviously require adding some kind of attractive force.

J. D. van der Waals (1873) derived his famous equation of state,

$$\left(p + \frac{a}{V^2}\right)(V - b) = RT,$$

by assuming both attractive and repulsive forces. The repulsive forces are approximated by an impenetrable core of diameter d, so that $b = (2\pi d^3/3)N$. (The excluded volume is actually four times the volume occupied by each of the N particles.) There was some confusion about what kind of attractive force between atoms is associated with the term a/V^2. Van der Waals himself simply asserted that the results of the Joule-Thomson experiment[1] showed that there are attractive forces which must vanish beyond some distance, whereas a repulsive force must dominate at much shorter distances. He did not propose any definite law for the attractive force, so his result a/V^2 must be regarded as a lucky guess based on physical intuition rather than the consequence of a mathematical derivation. Boltzmann, in 1898, objected to van der Waals' assumption that the attractive forces vanish at large distances, and claimed instead that the van der Waals equation could be derived only on the assumption that the force has a small but *constant* value at large distances. (Van der Waals refused to accept this claim, but it is essentially correct.)

Other force laws, chosen primarily for ease of calculation, were used in the 19th century. For example William Sutherland in 1893 showed that one could take account of short-range attractive forces in calculating the viscosity coefficient by specifying the value of the attractive potential at the point when the two molecules, regarded as elastic spheres, are in

---

[1] J. P. Joule and William Thomson (1854, 1862) showed that when a gas is allowed to expand into a vacuum there is in general a small cooling effect.

contact; the result would be a temperature-dependent "effective diameter." In other papers Sutherland argued that the attractive force law itself could best be represented by an *inverse fourth power* (potential r^{-3}).

5.3 ELECTROSTATIC FORCES AND ANOTHER TRY WITH KINETIC THEORY

After 1900, when it was generally believed that the atom was composed of electrically charged particles, several attempts were made to derive force laws from assumed charge distribution. If the atom as a whole is electrically neutral (the case of charged ions raises special difficulties that will be discussed later), the simplest assumption would be an *electric dipole*. The potential energy of interaction of two dipoles is inversely proportional to the *third power* of their distance, multiplied by an angle-dependent term.[2] If all relative orientations of the two atoms were equally probable, the average force would be zero, since the force between atoms is attractive for some orientations and repulsive for others. However, according to statistical mechanical principles, the attractive orientations will be more likely; their probability is given by the Boltzmann factor, depending on E/kT where E is the attractive potential energy. This effect was studied by J. H. van der Waals, Jr., and W. H. Keesom; the result, according to Keesom (1912), is an average attractive potential

$$V = -\frac{2}{3}\frac{\mu^4}{kTr^6} \quad \text{(dipole-dipole)}$$

(μ = dipole moment). This was subsequently known as Keesom's "direction effect."

Peter Debye pointed out in 1920 that these dipole-orientation forces cannot be sufficient to explain the physical properties of gases, since they must vanish at very high temperatures; he argued that the experimental data on gas properties, especially the equation of state, require a temperature-independent attractive force. This force must originate in the fact that the electric charge distribution in an atom is not fixed but mobile, as shown by the fact that every gas has a refractive index different from unity and hence has a finite electric polarizability. He suggested that a

---

[2] If a dipole consists of a charge $+e$ at the origin and a charge $-e$ at $x = d$, then the electric potential at a distance R along the x axis is $(e/R - e/[R - d]) = -ed/R(R - d)^2$. In the limit $d \to 0$, $e \to \infty$, such that $ed = \mu$ is fixed, one obtains the dipole potential μ/R^2. This would give the force on a point charge at R; the force on another dipole can be found by the same procedure and gives a potential proportional to R^{-3} and a force proportional to R^{-4}.

dipole moment of one atom can *induce* a dipole moment in the other. The result would be an interaction energy varying *inversely as the sixth power of the distance.* When this "induction" force is insufficient, for example when none of the atoms or molecules has a dipole moment to begin with, then he postulated a quadrupole moment which could thus induce a dipole moment in other atoms. This would give an *inverse eighth-power* attractive energy.

After quantum mechanics was developed in 1925–26, it was found that Debye's theory was still insufficient to account for the existence of attractive forces, since most atoms and molecules have dipole and quadrupole moments that are either zero or too small to give the required attraction.

In the meantime some scientists tried to determine the interatomic force law indirectly, by working backward from measured properties such as gas viscosity. Lord Rayleigh had shown by dimensional analysis in 1900 that if the viscosity coefficient is independent of density, then it must be proportional to $T^{(n+3)/(2n-2)}$, where the repulsive force law is r^{-n}. (For the two special cases investigated by Maxwell, hard spheres [$n = \infty$] and $n = 5$, this gives $T^{1/2}$ and T respectively.) At that time experimental data indicated that the viscosity of air varies as $T^{0.77}$, whereas the choice $n = 8$ corresponds to $T^{0.79}$. This would suggest something like an *inverse seventh-power repulsive potential.*

In 1916 Sydney Chapman in England and David Enskog in Sweden developed methods for solving the Maxwell-Boltzmann transport equations for more general force laws such as combinations of attractive and repulsive potentials. Thus there was some hope that with sufficiently accurate measurements of the temperature-dependence of the viscosity coefficient it might be possible to determine the complete force law. In addition, properties such as the second virial coefficient (deviation from ideal gas equation of state) and the compressibility of crystals depended, theoretically, on the interatomic force law.

Around 1920, Max Born and other physicists began to use force laws represented by potential functions of the form

$$V(r) = -\frac{a}{r^n} + \frac{b}{r^m},$$

where the first term represents attraction and the second repulsion. In 1924 J. E. Jones, formerly a student of Chapman at Manchester, began an intensive program to determine the parameters of the force law in this form. (After marrying Kathleen Lennard in 1925 he changed his name to Lennard-Jones.) His early results indicated that n is probably 4 (inverse fourth-power attractive potential), while m may vary from 8 to 20. Sim-ilar results were reported by H. R. Hassé and W. R. Cook (1929). The value $n = 4$ seemed to be in agreement with early quantum-theory results found by Born and Heisenberg in 1924.

5.4 The Quantum Theory of Dispersion Forces

Shortly after Schrödinger published his papers on wave mechanics, Debye was visiting at Columbia University, where he met a Chinese graduate student, Shou Chin Wang. At Debye's suggestion, Wang cal-culated the interaction energy between two hydrogen atoms using the new wave mechanics (1927). He worked with an approximate solution of the Schrödinger equation, using as the potential energy the instantaneous dipoles formed by the electrons with their nuclei, and assuming that both dipoles lie in a plane. He found that the energy levels for the two-atom system vary as the *inverse sixth power* of the distance between the nuclei.

Wang suggested that the attractive force he had calculated was related in some way to the Debye induction effect; his formula for the potential energy, $-(243/28)(e^2 a_0^5/R^6)$, could be identified with Debye's formula $-2\alpha\mu^2 R^6$, if one substitutes for the polarizability α its value determined from the theory of the quadratic Stark effect, $\frac{9}{2}a_0^3$, and gives the dipole moment the value $\mu = 0.982ea$. Slater (1928) described Wang's effect as "in a certain way, a Debye attraction of the variable electric moment of one atom (the variable terms are not zero, even in the ground state, al-though the diagonal term is) for the dipole moment it induces in the second atom" (p. 354).

A more accurate calculation of Wang's force was published by R. Eisenschitz and F. London in 1930; they found the same general form but a numerical coefficient about 26% smaller.[3] The attractive potential is

$$V = -6.47 \frac{e^2 a_0^5}{R^6} \quad (a_0 = \text{Bohr radius}).$$

London called this the *dispersion force*, since it is due primarily to the outer electrons in the atom that are responsible for the dispersion of light. This is not a very illuminating name. The standard explanation of disper-sion forces, following Slater and London, is that the instantaneous dipole moment in one atom induces a dipole moment in the other atom, and these attract each other; the attraction itself will produce correlations in

---

[3] See also the Debye-London correspondence on this subject in the London papers at Duke University.

the motions of the electrons, which further enhance the effect. The connection with optical dispersion does allow one to express the force constant in terms of "oscillator strengths" for spectral line intensities, which can be observed experimentally.

At short distances there will be additional kinds of forces between neutral atoms. One of these is the *exchange force* studied by Heitler and London in 1927 (see §5.8), which may lead to chemical bonding or to repulsion. Since the formation of molecules from atoms introduces further complications into statistical mechanical calculations, it has become customary to develop and test the theory primarily for rare-gas atoms. The simplest case is therefore helium, and this was first studied using quantum mechanics by J. C. Slater in 1928.

Slater recalled in his autobiography that he shifted his interest to the helium atom in 1927:

> This in a sense was going back to the question that had led me to wave mechanics in the first place: the compressibility of the alkali halides. . . . when I was working on that problem for my Ph.D. thesis, I concluded that nothing in the quantum mechanics as known in that period [1922–23] would explain the repulsive forces felt at short distances between inert gas atoms, helium, neon, argon, and so on. The simplest case was helium. Why were they relatively hard and impenetrable, as kinetic theory showed them to be? I felt that if one could first investigate the wave functions of a helium atom, and then their interaction, one should be able to throw light on this problem. It was with this in mind that I started work on the helium atom. (1975: 151)

Slater calculated the energy of two helium atoms at short distances, using a wave function which explicitly took account of the interaction between the two electrons. He found a repulsive force, described by a potential that is an exponential function of the distance between the two nuclei. Assuming that the long-range attraction has the same form as Wang had found for hydrogen, but with a different magnitude based on the different size and polarizability of the helium atom, Slater proposed in 1928 the following potential function:

$$\text{Energy (ergs)} = 7.7 \times 10^{-10} e^{-2.43(R/a_0)}$$
$$- .67 \times 10^{-10}/(R/a_0)^6 \quad \text{(helium)}.$$

The energy has a minimum at $R/a_0 = 5.6$, that is, at $R = 3.0 \times 10^{-8}$ cm. This is fairly close to the results found by later calculations and experiments (Margenau & Kestner 1971, §4.2).

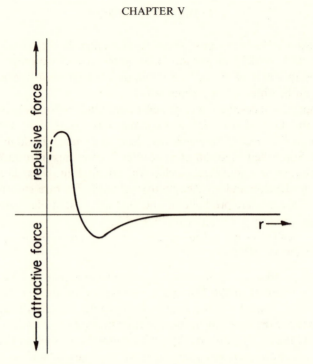

Fig. 5.4.1. Slater's Model (1928), usually called "exp-6" potential. $F(r) = Ae^{-kr} - B/r^7$ for all r, except that various modifications are used to eliminate the negative infinity at $r = 0$.

5.5 What is the True Interatomic Force Law?

Lennard-Jones, who had participated in some of the quantum-mechanical calculations and worked out a more efficient way of getting the Eisenschitz-London result for the Wang force, presented a general survey of the subject of interatomic forces in a lecture to the Physical Society of London in 1931. He abandoned the r^{-4} attractive potential, which he had previously inferred from gas properties, in favor of the quantum-theoretical r^{-6} potential. But, rather than adopting the exponential form for the repulsive potential as suggested by Slater and others, he retained the r^{-m} form. One justification for this decision was that with a suitable choice of constants both functions give about the same force "over the range which is most effective in atomic collisions." The value of m could be estimated from data on solids—the interatomic distance and heat of sublimation—but values between 9 and 12 seemed to give about equally good agreement for neon, argon, and nitrogen. He seemed to have a slight preference for $m = 12$.

In 1936 Lennard-Jones advocated an eighth-power repulsive potential, but in a series of papers with A. F. Devonshire on the equation of state and critical phenomena in 1937–39, he adopted what has subsequently become known as the "Lennard-Jones 6–12 potential":

$$V = -ar^{-6} + br^{-12}.$$

The main competitor was the "exp-6" potential proposed by Slater for helium,

$$V = -ar^{-6} + ce^{-r/d}.$$

Some theoretical calculations, such as second virial coefficients, could be done more easily with the 6–12 potential, because the integrals involved are simplified when the repulsive index is twice the attractive index. Transport coefficients were calculated for the 6–12 potential by four

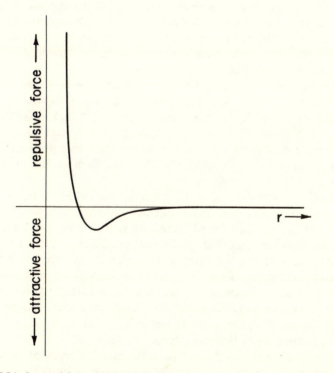

FIG. 5.5.1. Lennard-Jones (1931) "6, 12" potential. $F(r) = -(a/r^7) + (b/r^{13})$ for all r; a positive value of $F(r)$ means that the force is repulsive, negative means attractive.

separate groups during the 1940s: Kihara and Kotani in Japan; Hirschfelder, Bird, and Spotz in the United States; de Boer and van Kranendonk in Holland; and Rowlinson in England. The fact that all these calculations gave essentially the same results produced confidence in them; but perhaps more significantly, the fact that all four groups chose to use the same potential created the impression that the 6–12 function was now the preferred one. At least it was the only "realistic" potential for which the parameters were established for a large number of gases. At the time when Hirschfelder, Curtiss, and Bird published their massive treatise, *Molecular Theory of Gases and Liquids* (1954), calculations with the exp-6 potential were only just beginning, so that the postwar generation of scientists who used *Molecular Theory* as their chief source of information may well have gotten the impression that the 6–12 potential had been generally accepted by the experts.

A strict comparison of results based on the exp-6 and 6–12 models was not possible until virial coefficients had been computed for both of them. The calculations are somewhat simpler for the 6–12 potential, so it had a head start despite the theoretical prestige of the exponential repulsion. Moreover, it was necessary to adopt some kind of short-range modification of the latter, since otherwise the inverse sixth-power attraction would overcome the exponential repulsion and produce an "unphysical" negatively infinite potential at $r = 0$.

By the end of 1954, just after the publication of *Molecular Theory*, E. A. Mason and W. E. Rice had carried out direct comparisons of the 6–12 and exp-6 models for several gases. They found that while the exp-6 was definitely better when applied to hydrogen and helium, on the whole neither model was markedly superior to the other in reproducing virial coefficients and viscosity coefficients. Thus gas theory could be used to select the correct form of the force law.

While the Lennard-Jones 6–12 potential continued to be the most popular for statistical mechanical calculations when a "realistic" model was needed, it has been found increasingly inadequate. In 1965, A. E. Kingston reported that the coefficient of the r^{-6} attractive term, calculated theoretically from quantum mechanics, is only about half of the value determined from second virial coefficients and viscosity. Although the discrepancy, once it was known to exist, could be removed by reinterpretation and extension of the experimental data, Kingston had to conclude that "in general the Lennard-Jones (12–6) potential is not a good representation of the true interatomic potential."

At very large distances the r^{-6} potential must be replaced by an r^{-7}, according to the results of a theoretical calculation by H.B.G. Casimir and D. Polder in 1946.[4] The physical reason, as explained by Margenau and Kestner, is:

> Dispersion forces arise from the interaction of one instantaneous dipole inducing another dipole in a neighboring molecule. This new induced dipole then interacts with the original dipole. Because of the finite velocity of electromagnetic radiation the total time elapsed is $2R/c$ (c being the velocity of light). During this time the original dipole has changed its orientation. The returning field is then "retarded" with respect to the initial field, hence the name "retarded dispersion forces." Since the maximum interaction occurs for no retardation or no change in orientation, the corrected interaction must be weaker. It is weaker by a factor depending on R at large separations. (1971:217)

For a recent review of this topic see Sucher (1977).

At very short distances, the dominant force between two atoms is the Coulomb repulsion between the nuclei rather than either an r^{-6} or $e^{-r/d}$ potential. This rather obvious fact seems to have been first incorporated into interatomic potential functions used to interpret experiments by A. A. Frost and J. H. Woodson, and independently by R. A. Buckingham, in 1958. If the atoms are so close together that this effect is important, as for example in matter compressed to a small fraction of its normal volume, then the atom itself can no longer be considered an independent unit; the electrons will be squeezed out ("pressure ionization") and form a uniform Fermi-Dirac gas, in which the nuclei move as point-charges.

Even without going to extremely dense matter, one can find evidence that the original goal of Newton's research program, the "true interatomic force law," is not a very useful concept. It is of course possible to determine the force between two atoms by scattering experiments, at least under circumstances where it can be assumed that the potential is a monotonic

---

[4] According to E. J. Verwey (1967:171), the Casimir-Polder theory originated in problems encountered in an industrial laboratory. "Before the war Hamaker and I worked on suspensions in various media, and we had to pay special attention to the problem of stability of these suspensions in connection with their practical applications (mainly electrophoretic deposition of various materials upon an electrically conducting substrate). Accordingly Hamaker started to calculate the van der Waals–London attraction between the particles. . . . However, the complete theory . . . led to a serious deviation from practical experience for the case of larger particles (as present in suspension), suggesting that the calculated van der Waals–London attraction for larger distances was too large. Then [J.T.G.] Overbeek had the idea that it might be the retardation which had to be taken into account, and Casimir and Polder helped us out of the difficulties by developing their theory, which gave better equations for the interaction of larger particles. . . ."

function of distance. A considerable amount of information of this type has been collected over the past few decades and can be represented by assuming that the short-range repulsive potential has the form r^{-m}, where m ranges from 3 to 20 for various atoms and molecules. But if an interatomic potential is to be of any use in explaining and predicting properties of bulk matter, it must be applicable to systems of more than two atoms, and it must give an adequate description of attractive as well as repulsive forces. There is now considerable doubt whether these criteria are satisfied by *any* of the currently used potentials.

Since experimental data on crystals played a significant role in the original determination of the exponents 6 and 12 in the Lennard-Jones potential (especially the 12), one might at least expect that this potential would be useful in the theory of solids. For metallic solids or ionic crystals such a potential would of course be irrelevant; no one nowadays would attribute a major share of the binding energy of such solids to the forces known to act between neutral atoms or molecules. However, there is a small but interesting class of substances of low melting point, such as the solids of the rare gases, which are believed to be held together by essentially the same forces which produce the deviations from the ideal gas law at higher temperatures, the so-called van der Waals lattices. Not surprisingly, in view of the results mentioned above, the lattice energy of these solids can be fitted fairly well by either the 6–12 or the exp-6 potential. But one also expects to be able to explain theoretically the actual crystal structure of such a solid, and here one encounters another failure of the interatomic-potential concept. It turns out that any physically plausible potential of the 6–12 or exp-6 type leads to the prediction that a face-centered cubic lattice has a *higher* energy (is less stable) than a hexagonal close-packed lattice, whereas in reality all the rare gases crystallize in the face-centered-cubic form except for helium, where the quantum zero-point energy is dominant (see L. Meyer 1969).

It now appears, as a result of research in the 1960s, that the interatomic-force concept is inadequate because the interaction energy of a system of three or more atoms or molecules cannot be written accurately as a sum of two-particle energies. For example, an elaborate study of the properties of rare-gas solids, published by M. L. Klein and R. J. Munn in 1967, using a complicated ten-parameter potential said to be the best available function with constants determined from virial-coefficient and transport data, concluded that "if pairwise additivity is assumed for the interatomic forces in the solid, the agreement with experiment is poor." (In this case "poor" means about 20% too high.) But the agreement could be improved significantly by adding a three-body potential to the energy of the system.

Similar conclusions emerged from experiments on the shock compression of argon done by M. Ross and B. J. Alder in 1967. They used both the Lennard-Jones and exp-6 potentials to calculate the shock ("Hugoniot") equation of state, and found that these theoretical pressures (based on the assumption of pairwise additivity) were different from the experimental pressures by up to 30% for high-density states. Other studies by T. Halicioglu and O. Sinanoglu, and by D. A. Copeland and N. R. Kestner, both in 1968, showed that even in liquids and gases some kind of density-dependent three-body interaction must be included in order to interpret experimental results.[5] The quantum-mechanical basis for such a three-body force is well understood and was investigated in detail by A. Shostak in 1955 for some simple systems. (See Margenau & Kestner 1971, chap. 5, for a review of this topic.)

It was probably inevitable, once it became clear that the "atom" of 19th-century chemistry and physics is composed of smaller particles, that the law of force between atoms would lose its fundamental significance in science.[6] For the 20th century the proper question is: What is the law of force between elementary particles? For the 21st century, it will probably be: What is the law of force between the constituents of those "elementary" particles (quarks or whatever)? And so forth.

5.6 The Exclusion Principle and the Electronic Structure of Atoms

Early in the 1920s it had already become clear that the Bohr model of the hydrogen atom might be extended to a more general theory of atomic structure. An atom of atomic number Z (nuclear charge $Q = Ze$) would contain Z electrons in the neutral state; these electrons would occupy the various energy levels available to the single electron in the hydrogen atom. However, it was necessary to adopt some apparently arbitrary assumptions about how many electrons could occupy each energy level. The lowest level, whose quantum number is $n = 1$, could hold 2 electrons; the next one, $n = 2$, could hold 8; the next, $n = 3$, could hold 18, and so forth (except that there was no general formula by which one could predict these numbers for higher n). It was suggested that the different electrons occupying the same energy level might be distinguished by assigning them other quantum numbers. For example, a quantum number might

---

[5] I am indebted to Hugh DeWitt of Lawrence Livermore Laboratory for correspondence about recent research showing the need to include three-body interactions.

[6] Schlier (1969:195) said that the very existence of the periodic table should demonstrate that no "universal reduced potential" for all atoms should exist.

determine the eccentricity of an elliptical orbit. Ordinarily the elliptical orbit might have the same energy as a circular orbit, but it might be affected differently by a magnetic field; this idea was suggested by the splitting of spectral lines in magnetic fields (Zeeman effect).

In 1925, Wolfgang Pauli suggested that all possible electron states could be accounted for by assigning each electron a set of four quantum numbers:

$$n = 1, 2, 3, 4, 5, \ldots$$
$$l = 0, 1, \ldots, n - 1$$
$$m = -l, -l + 1, \ldots, l - 1, l$$
$$m_s = -\tfrac{1}{2}, +\tfrac{1}{2}.$$

Note that I am using the modern definitions of these quantum numbers, not the definitions originally proposed by Pauli; but the basic idea is the same. The first three quantum numbers could be related to some conceivable physical property of the electron orbit—energy, angular momentum, etc. Moreover, it was found that they could be *derived* from the Schrödinger equation, so they did not have to be postulated ad hoc (see §3.4). However, the fourth quantum number, m_s, could only be attributed to some mysterious "two-valuedness" of the electron.[7]

Using this set of quantum numbers, Pauli could then postulate his "exclusion principle"—no two electrons in an atom may occupy the same state defined by values of these quantum numbers. As we have seen in §4.5, Fermi and Dirac successfully applied this principle to the free electron gas, and their theory was subsequently developed into the modern electron theory of metals.

The fact that only two electrons can occupy the lowest level, $n = 1$, now follows immediately, since l and m must both be zero, and m_s can have the two values $m_s = +\tfrac{1}{2}$ or $-\tfrac{1}{2}$. For $n = 2$, l may be either 0 or 1, and for $l = 0$ you can have only $m = 0$, whereas for $l = 1$ you can have three values of m: -1, 0, or $+1$. This gives four possible combinations of (l, m) values, and for each of these there are two possible values of m_s; so the total number of states is 8.

From this set of rules physicists could immediately derive the "electron-shell" theory of the atom. A shell consists of all the electrons that have a

---

[7] Although Pauli would not say what the fourth quantum number is, others were not so cautious. Two young Dutch physicists, G. Uhlenbeck and S. Goudsmit, proposed in 1926 that it corresponds to the "spin" of the electron. A few years later P.A.M. Dirac in England found that "spin" comes naturally out of a relativistic form of quantum mechanics.

given value of n. Thus the first shell is filled by 2 electrons, the second by 8, the third by 18, and so forth. A closed shell corresponds to a "rare-gas" atom—thus helium has 2 electrons, neon has $2 + 8 = 10$. Things get a little more complicated after that, because the energies of the electron states of a many-electron atom are not quite the same as those for hydrogen. In fact it is surprising that they are similar at all, since the potential energy of interaction between electrons is of the same order of magnitude as that between an electron and the nucleus. However, it was possible to work out a fairly good approximate theory of the electronic structure of all the elements, and to predict correctly the order in which the various states would be filled as more electrons were added.

The chemical similarity between elements in the same column of the periodic table now had an obvious explanation. Lithium and sodium, for example, both contain one electron outside a "core" of closed shells; only this outer electron is involved in chemical reactions. Since the outside electron has an orbit almost entirely outside the core, it is not acted on simply by the entire attractive force of the positive nucleus (charge $Q = Ze$); instead, most of this force is canceled by the core of $Z - 1$ electrons which repel it. So the outer electron is fairly likely to be removed from the atom under favorable circumstances. Conversely, atoms like fluorine ($Z = 9$) are one electron short of a closed shell, and tend to want to pick up that electron.

After the establishment of quantum mechanics had provided a firm foundation for the above semiquantitative account of atomic structure, physicists and a few theoretical chemists turned to the problem of molecular structure and chemical bonds. Four outstanding problems may be mentioned:

(1) How can one explain "saturation"? In Avogadro's theory of diatomic molecules, for example, it wasn't clear (as Dalton and others pointed out) why—if there *are* attractive forces that can hold atoms together—only *two* atoms of the same kind actually cohere, rather than three or four.

(2) Stereochemistry (the three-dimensional structure of the molecules, in particular the spatial directionality of bonds as in the "tetrahedral" carbon atom) needed to be treated.

(3) One should be able to calculate the "bond length"—there seems to be a well-defined equilibrium distance between two atoms, which can be accurately determined by experiment.

(4) Why do some atoms (for example helium) form no bonds, while others can form one or more? (These are the empirical rules of "valence.")

5.7 THE LEWIS-LANGMUIR THEORY

Soon after J. J. Thomson's discovery of the electron (1897) there were several attempts to develop theories of chemical bonds based on electrons. The most successful was the "Lewis-Langmuir" theory. It was proposed originally by the American chemist Gilbert Newton Lewis in 1916 and popularized by another American chemist/physicist, Irving Langmuir, starting in 1919. Lewis, starting in 1902, had developed the idea that an atom consists of a series of concentric cubes, with possible positions for electrons at the eight corners of each cube. In contrast to other theories, such as that of J. J. Thomson (1907), which regarded all chemical bonds as essentially polar—involving the transfer of one or more electrons from one atom to another—Lewis proposed the concept of a *shared-electron bond*. In his later writings, Lewis described polar bonds as a special case of the shared-electron bond and argued that in reality there is a continuum of possibilities with more or less equal sharing of electrons between atoms.

The Lewis-Langmuir theory successfully accounted for most of the known facts about valence. The emphasis on shared *pairs* of electrons was justified by the fact that almost all molecules have an even number of electrons. The *static* character of the cubic model made it possible to account for the definite spatial arrangements of atoms in molecules (stereochemistry), although some modifications were needed to deal with

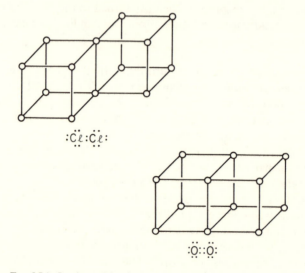

FIG. 5.7.1. Lewis models of molecules (single and double bonds).

FIG. 5.7.2. Lewis model of tetrahedral atom.

carbon compounds. To explain the tetrahedral carbon atom Lewis moved his electrons in from the corners of the cube to the edges.

In representing molecules by combinations of ordinary symbols on a two-dimensional piece of paper rather than by combinations of cubes in three-dimensional space, Lewis gradually liberated his theory from the constraints of the original "cubic" atom. Chemists could adopt his system of placing dots around the atomic kernels to represent the valence electrons, without committing themselves to a particular three-dimensional structure.

Lewis's theory was clearly in conflict with the Bohr atomic model of 1913. From the standpoint of classical electrostatics, no static model could be stable if the forces are purely those given by Coulomb's law (Earnshaw theorem). Lewis avoided this criticism by suggesting that the forces between electrons deviate from Coulomb's law at short distances. In Bohr's model Coulomb's law is assumed in order to compute the relation between the energy, speed, and position of an electron in a particular orbit, but, as Lewis pointed out, the laws of electromagnetic theory requiring radiation from moving charged particles are ignored. As Lewis put it,

> Planck in his elementary oscillator which maintains its motion at the absolute zero, and Bohr in his electron moving in a fixed orbit, have invented systems containing electrons of which the motion produces no effect upon external charges. Now this is not only inconsistent with the accepted laws of electromagnetics but, I may add, is logically objectionable, for that state of motion which produces no physical effect whatsoever may better be called a state of rest. (Lewis 1916, as quoted by Lagowski 1966:103)

By 1923 Lewis decided that his conception of the chemical atom could probably be reconciled with Bohr's latest views on the physical atom, especially "if we regard as the important thing the orbit as a whole, and not the position of the electron within the orbit" (Lewis 1923:56). Although he was not yet ready to propose a fundamental explanation for the tendency of electrons to form shared pairs, he noted that atoms and molecules with an odd number of electrons possess magnetic moments, while most of those with an even number generally have no moment. He had already noted with some approval the suggestion of A. L. Parsons that electrons may be small magnets (Lewis 1916:773). He now suggested that the coupling of two electron orbits, with a resulting neutralization of their magnetic fields, may be the basis of chemical phenomena.

It was well known that a circulating electric current produces a magnetic dipole. By invoking this effect it might appear that Lewis was going over to Bohr's dynamical theory, in which the orbital motion of the electron around the nucleus was associated with a magnetic moment. Yet Lewis still insisted that his electrons were localized at the corners of a cube, though he allowed some "distortion" to produce his tetrahedral atoms. He seemed to have entertained the possibility of a magnetic moment arising from some kind of motion of the electron around its equilibrium position. But he did not (at least in public) pursue this idea far enough to arrive at the concept of electron *spin*, suggested in 1925 by Goudsmit and Uhlenbeck. Nevertheless electron spin turned out to be the reason why much of the Lewis-Langmuir theory is valid.

(Additional references for this section: Bykov 1965; Kohler 1974, 1975a, 1975b; Palmer 1965; Russell 1971. For Bohr's early ideas on the chemical bond see Kragh 1977.)

5.8 RESONANCE AND THE HEITLER-LONDON THEORY

The key to the solution of many-body problems in general, and the nature of the chemical bond in particular, is the phenomenon of quantum-mechanical *resonance*. It was first described by Heisenberg (1926a, 1927) using both his own matrix mechanics and Schrödinger's wave mechanics. Resonance is related to the requirement that the wave function must be symmetric or antisymmetric in the coordinates of identical particles, as pointed out independently by Dirac a few months later[8] in his discussion of quantum statistics (see above, §4.5).

In classical mechanics resonance occurs in a mechanical system consisting of two harmonic oscillators which, if they did not interact, would each have the same frequency v_0. If the interaction is characterized by a small parameter λ, the potential energy for the system can be written in the form (using the notation of Pauling & Wilson 1935, which is a

---

[8] In a footnote to this 1926 paper (p. 670) Dirac said that Born "has informed me that Heisenberg has independently obtained results equivalent to these" (on the symmetry of wave functions), and he added in proof the reference to Heisenberg's published paper in the *Zeitschrift für Physik*.

convenient reference for this topic)

$$V(x_1, x_2) = 2\pi^2 m v_0^2 x_1^2 + 2\pi^2 m v_0^2 x_2^2 + 4\pi^2 m \lambda x_1 x_2,$$

where x_1 and x_2 are the coordinates of the two oscillators, each of mass m. The standard procedure for such problems is to transform to new variables ("normal coordinates") X and Y, defined as

$$X = \frac{1}{\sqrt{2}}(x_1 + x_2), \qquad Y = \frac{1}{\sqrt{2}}(x_1 - x_2),$$

the reason being that the potential and kinetic energies are now simply sums of squares,

$$V(X, Y) = 2\pi^2 m (v_0^2 + \lambda) X^2 + 2\pi^2 m (v_0^2 - \lambda) Y^2$$
$$T = \tfrac{1}{2} m \dot{x}_1^2 + \tfrac{1}{2} m \dot{x}_2^2 = \tfrac{1}{2} m \dot{X}^2 + \tfrac{1}{2} m \dot{Y}^2$$

The system is thus equivalent to two independent harmonic oscillators with frequencies $\sqrt{v_0^2 + \lambda}$ and $\sqrt{v_0^2 - \lambda}$. If λ/v_0^2 is small, one finds that the actual interacting oscillators (coordinates x_1 and x_2) carry out approximately harmonic oscillations with frequency v_0 but with amplitudes that change with time, in such a way that the total energy of the system gradually shifts back and forth between them. The period of the resonance (time between successive maxima of x_1) is approximately v_0/λ. (For details see Pauling & Wilson 1935:316–17.)

Whereas in classical mechanics resonance is relatively uncommon— real oscillators generally do not have exactly the same frequency, and their interaction is not usually given by the linear form above—in quantum mechanics it turns out to be very important. This is because atoms and molecules consist of particles most of which are identical (at least the electrons) and the basic equations of the theory are linear. In the simplest case of two-electron atoms, first discussed by Heisenberg from this viewpoint, the wave function of the system corresponding to one electron in state n and the other in state m can be written as

$$\psi = \frac{1}{\sqrt{2}} [\psi_n(1)\psi_m(2) \pm \psi_n(2)\psi_m(1)].$$

Instead of saying that the first electron is in state n while the second electron is in state m, we have to say that this configuration is mixed with another in which the first is in m while the second is in n. The two possible mixtures corresponding to the $+$ and $-$ signs are called symmetric and antisymmetric; these two states have slightly different energies. Heisenberg attributed the splitting of spectral lines in helium to this effect, which he described as a continual interchange of the electrons.

Heisenberg's resonance theory was applied to the simplest[9] bond, between two hydrogen atoms, by Walter Heitler and Fritz London in 1927 (see the reminiscences by Heitler 1967). As a first approximation, one can construct a wave function for the system from the ground-state wave functions of the hydrogen atom; thus for the first electron, in atom a, one has

$$\psi_a(1) = \frac{1}{\sqrt{\pi a_0{}^3}} \exp\left\{-r_{a_1}/a_0\right\}$$

$$\left[a_0 = \frac{h^2}{4\pi^2 me^2} = 0.529\text{Å, "Bohr radius"}\right],$$

where r_{a1} is the distance between the nucleus of a and the first electron. Since the system resonates between a state with the first electron in atom a and the second in b, and a state with the first in b and the second in a, the wave function for the system is written

$$\psi(1, 2) = c_1\psi_a(1)\psi_b(2) + c_2\psi_b(1)\psi_a(2). \tag{1}$$

The nuclei are assumed to be fixed at a distance R; the energy of the system is calculated for various values of R, and it is assumed that the equilibrium distance between nuclei in the hydrogen molecule is then one for which the energy is a minimum. (The procedure of holding the nuclei fixed in calculating the electronic wave function was justified by Max Born and the American physicist J. Robert Oppenheimer, in a paper published the same year, 1927, though Heitler and London do not cite it.)

The calculation shows that the symmetric wave function ($c_2 = +c_1$) gives a minimum energy at $R = 0.80\text{A}$, compared to the experimental value 0.740Å. Without resonance, there would be a much weaker minimum; and the antisymmetric function ($c_2 = -c_1$) has no minimum at all. The error in the total electronic energy is about 5%.

Whether the two atoms will be in the symmetric or antisymmetric state with respect to the electronic wave function depends on the *spins* of the

[9] Strictly speaking the hydrogen molecule-ion, formed from two protons and *one* electron, is even simpler; the first quantum-mechanical treatment of this system was published by a Danish physicist, Ø. Burrau, in 1927. See Pauling (1928) for a discussion of his result.

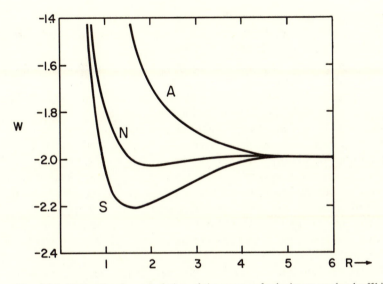

FIG. 5.8.1. Heitler-London calculation of the energy of a hydrogen molecule; W in units of $e^2/2a_0$, R in units of a_0. Curve A is for an antisymmetrical wave function, S for symmetrical, N for a function without resonance.

two electrons. According to the Pauli principle as interpreted by Heisenberg and Dirac, the complete wave function (including spin functions) must be antisymmetric for the interchange of two electrons. Since the total wave function is a product of the orbital and spin functions, we have the general rule that one factor must be symmetric and the other antisymmetric. An antisymmetric spin function means that the two electrons have opposite spins; it can be written $\alpha(1)\beta(2) - \alpha(2)\beta(1)$, where α corresponds to spin $+\frac{1}{2}$ and β corresponds to spin $-\frac{1}{2}$. Thus the Heitler-London calculation shows that if two hydrogen atoms approach each other, they can form a molecule if the electron spins are opposite (curve S in Fig. 5.8-1), but they will repel at all distances if the electron spins are the same (curve A). This is consistent with the fact that if we united the two hydrogen atoms to form a helium atom, the two electrons could not both be in the ground $(1s)$ state unless they had opposite spins.

Heitler and London applied a similar procedure to the interaction of two helium atoms. In this case application of the antisymmetry requirement excludes the only combination of orbital wave functions that could lead to binding; the only allowed wave function composed of ground-state electronic functions leads to repulsion at all distances. Thus quantum mechanics predicts that while a stable H_2 molecule can be formed, a stable He_2 molecule probably cannot (one would have to bring in excited states of one or more electrons).

A general conclusion from the early work of Heitler and London was that the formation of covalent bonds depends on "resonance energy," the contribution to the electronic energy from integrals like $\int \psi_a(1)\psi_b(2)\psi_b(1)\psi_2(2)/r_{12}$, in which the electrons change places. The classical analogue of this energy might be considered to be the reduction of the frequency of an oscillator from v_0 to $\sqrt{v_0{}^2 - \lambda}$ when it interacts with another oscillator. However, the identification of some portion of the total energy of a system as "resonance energy" is artificial, since it depends on our choice of the one-electron atomic wave functions as the first approximation. One could start from some other first approximation and get a similar final result, without making use of the particular integral that was associated with the resonance energy (cf. Pauling & Wilson 1935:321–25).

The Heitler-London approach to chemical bonds was rapidly developed by two Americans, John C. Slater and Linus Pauling. Slater in 1929 proposed a simple general method for constructing many-electron wave functions that would automatically satisfy the Pauli exclusion principle by being antisymmetric on the interchange of any two electrons. He did this by incorporating the spin into the electron wave function, and then wrote the entire wave function as a determinant whose elements are the wave functions for the individual electrons, with all possible permutations of the spatial and spin coordinates. Prior to Slater's work, it was necessary to use the mathematical theory of groups to accomplish the same thing; since most physicists and chemists were at that time unfamiliar with group theory, it made the subject inaccessible to all but a few theorists. As Slater recalls in his autobiography,

> It was at this point that Wigner, Hund, Heitler, and Weyl entered the picture, with their "Gruppenpest": the pest of the group theory, as certain disgruntled individuals who had never studied group theory in school described it. The mathematicians have made a great study of what is called the symmetric group, or the permutation group. This is a study of the effect of the $N!$ possible permutations of a set of N objects. The authors of the "Gruppenpest" wrote papers which were incomprehensible to those like me who had not studied group theory, in which they applied these theoretical results to the study of the many-electron problem. The practical consequences appeared to be negligible, but everyone felt that to be in the mainstream of quantum mechanics, one had to learn about it. Yet there

were no good texts from which one could learn group theory. It was a frustrating experience, worthy of the name of a pest.

I had what I can only describe as a feeling of outrage at the turn which the subject had taken . . . . (1975:60–61)

After publishing his 1929 paper on the determinantal method, Slater recalled, "it was obvious that a great many other physicists were as disgusted as I had been with the group-theoretical approach to the problem. As I heard later, there were remarks made such as 'Slater had slain the "Gruppenpest." ' I believe that no other piece of work I have done was so universally popular" (1975:62).

Slater's victory over group theory proved to be short-lived. Thirty years later, it was generally admitted that theoretical chemists as well as theoretical physicists needed to have a sound working knowledge of group theory in order to understand and apply existing theories of the atom as well as to develop new ones.

5.9 VALENCE BOND VERSUS MOLECULAR ORBITAL

Pauling was one of the first theorists to apply the Heitler-London method to chemical bonds; this method became known as the "valence bond" method because it involved picking out one electron in each of the two combining atoms and constructing a wave function representing a paired-electron bond between them. As early as 1928 Pauling reported that "it has further been found that as a result of the resonance phenomenon a tetrahedral arrangement of the four bonds of the quadrivalent atom is the stable one" (1928b:361). He did not publish the details until 1931, at which time he acknowledged that Slater had independently discovered "the existence, but not the stability, of tetrahedral eigenfunctions" in 1930 (1931:1367). Pauling proposed that the carbon atom, which in its ground state has only two unpaired electrons in the valence shell (this shell is usually described as $2s^2 2p_x 2p_y$), can be quadrivalent if one of the two $2s$ electrons is excited to the $2p_z$ orbital. Although it takes some energy to do this, the expenditure of energy is more than compensated by the energy gained in forming four bonds. Moreover, Pauling argued that the four bonds will all be equivalent rather than one of them being different, as might be expected from the fact that the carbon atom has one $2s$ and three $2p$ electrons available for bonding. The bond eigenfunctions will actually be mixtures of s and p orbitals, now known as *hybrid* bonds; and the bonds will be directed toward the corners of a regular tetrahedron.

At about the same time as Pauling was developing the valence-bond method, another American, Robert S. Mulliken, was developing an alternative technique based on what he called "molecular orbitals." This term was first used by Mulliken in 1932 (1975:458), but the concept had been introduced a few years earlier. Friedrich Hund, beginning in 1926, and Mulliken, in 1928, proposed schemes for assigning quantum numbers to electrons in molecules, based on atomic quantum numbers and the interpretation of molecular spectra. The first person to write an explicit wave function for a molecular orbital was J. E. Lennard-Jones, in 1929. He suggested a function of the form

$$\Phi(1) = \psi_A(1) \pm \psi_B(1)$$

for an electron in a diatomic molecule; $\psi_A(1)$ refers to an atomic wave function centered on nucleus A, and $\psi_B(1)$ to one on nucleus B. Lennard-Jones described his wave function as representing a "resonance" of the electron between two possible states. Note that this is not the same kind of resonance as that discussed by Heisenberg and by Heitler and London, which involves the interchange of *two* electrons between two possible states. (In the later literature of quantum chemistry the Heisenberg form of resonance is usually called "exchange.")

In 1931, Erich Hückel applied both the valence-bond and the molecular-orbital methods to benzene and related compounds. Mulliken (1932) argued that the molecular-orbital method is more useful than the valence-bond method, since it is not restricted to electron-pair bonds. It can be used to describe one-electron bonds (though these are admittedly rather rare). More important, it can represent an electron as distributed over more than two nuclei. G. W. Wheland, who had worked with Pauling at Caltech, compared the two methods as applied to aromatic molecules in 1934; he concluded that while the valence-bond method gives better agreement with experiment, the molecular-orbital method can be applied to a wider variety of problems.

In treating benzene by the valence-bond method in 1933, Pauling and Wheland constructed a wave function which was a linear combination of

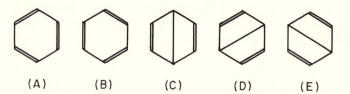

(A)   (B)   (C)   (D)   (E)

FIG. 5.9.1. Benzene structures: Kekulé (A, B) and Dewar (C, D, E).

five possible structures, that is, five possible arrangements of double and single bonds. Two of them are the structures proposed by Kekulé in 1865 (A and B); the other three (now called "Dewar structures") were suggested by Wichelhaus in 1869 (C, D, E). Although the Dewar structures have a lower statistical weight in the total wave function, their inclusion is justified by the result that they contribute about 20% to the "resonance energy." Other possible structures based on a plane hexagonal arrangement (suggested by Claus and others) can be considered as included in the wave function even though they are not listed explicitly, since they can be expressed as linear combinations of the Kekulé and Dewar structures.

The valence-bond method, with its emphasis on resonance between different structures as a means of analyzing aromatic molecules, dominated quantum chemistry in the 1930s. The method was comprehensively presented and applied in Linus Pauling's classic treatise, *The Nature of the Chemical Bond* (1939), and in G. W. Wheland's book, *The Theory of Resonance and its applications to organic chemistry* (1944). One reason for its popularity was that ideas similar to resonance had been developed by organic chemists—especially F. G. Arndt in Germany and C. K. Ingold in England—independently of quantum theory in the late 1920s. Ingold preferred the term *mesomerism* to resonance, to convey the idea that the actual structure of the molecule is somewhere in between the various classical structures; some chemists complained that "resonance" implied that the molecule actually switched back and forth between the different structures, leading to confusion with tautomerism (in which the structures can actually have separate existence).

After World War II, there was a strong movement away from the valence-bond method toward the molecular-orbital method, led by Mulliken in the U.S. and by Charles Coulson, J. E. Lennard-Jones, H. C. Longuet-Higgins, and M.J.S. Dewar in England. The MO advocates argued that their method was simpler and easier to apply to complicated molecules, since it allowed one to visualize a definite charge distribution for each electron. The fact that this picture left out exchange effects—the indistinguishability of electrons which required using a wave function antisymmetric for the interchange of any two electrons—was not a fatal objection, since equivalent effects could be added in higher approximations, and the level of accuracy of the MO method at any given level of approximation was about as good as that of the VB method.

In 1949, as Western theoretical chemists were beginning to abandon the VB method in favor of the MO method, the resonance concept began to come under strong attack in the Soviet Union for ideological reasons. This was part of the same movement that established Lysenko's theories in genetics and temporarily displaced the Copenhagen interpretation of qunatum mechanics. According to dogmatic Marxist-Leninists, resonance theory was tainted with idealism because it described molecular structure in terms of forms that were physically impossible but had only a theoretical existence. Although there was much rhetoric about the dangers of "Pauling-Ingoldism," it appears that (unlike the Lysenko controversy) this particular attempt to impose the party line on science resulted in not much more than verbal modifications. Russian theoretical chemists could do their calculations however they thought best as long as they did not use the word "resonance."

The apparent absurdity of this ideological critique has obscured a significant difference between the VB and MO methods. The valence bond could be interpreted in the spirit of Copenhagen: the wave function is a mixture of two states, $\psi_A(1)\psi_B(2)$ and $\psi_B(1)\psi_A(2)$, and it is not possible to say that the first electron is *really* in the state ψ_A (that is, in an orbital centered on nucleus A) or ψ_B. Only by a rather violent "observation"—which would involve pulling the molecule apart into two atoms—could one force the electron to decide which atom it belongs to. This is something like a Schrödinger cat paradox. In the case of the resonance of the benzene molecule between five different valence-bond structures the situation is even worse: there is no observation that could determine which structure really is present. Thus one has to abandon the idea that the electron is *in* a particular state. The MO approach permits a more "realist" interpretation in this respect, since it assigns each electron to a particular state, though that state may be spread out over the entire molecule. As Dewar expressed it (1949:1), "ψ^2 may be taken as the measure of density of an 'electron gas' representing the average electron distribution over a comparatively long period," and since this distribution represents all we can know about the electron, "the failure to localize the electrons more exactly is of little practical importance."

[Additional references for this section: Graham 1964, Nachtrieb 1975, Slater 1967, Paradowski 1972, Pauling 1970, Rich & Davidson 1968, van Vleck 1970.]

5.10 THE REDUCTION OF CHEMISTRY

The development of quantum chemistry has led to a general impression that all of chemistry has now been reduced to physics. Dirac wrote in 1929 that

The general theory of quantum mechanics is now almost complete.
... The underlying physical laws necessary for the mathematical

theory of a large part of physics and the whole of chemistry are thus completely known . . . . (1929:714)

All problems of chemistry could be solved "in principle" if one could determine the wave function. This would of course involve solving the corresponding Schrödinger equation in $3N$ variables for a system of N particles; for typical molecules of interest to chemists, N ranges from 20 to 200, so this is a very difficult task even with modern computers. Nevertheless, according to Heitler (1967:35),

> within 8 years after the advent of quantum mechanics all the fundamental facts of chemistry had been understood in the sense that they could be reduced to the laws of atomic physics.

Insofar as chemists believe that their subject is reducible to physics, the result has been rather damaging. It seems that chemistry is no longer an independent science but "only" a subfield of applied physics. If this were true it would no longer be possible for anyone to propose a precise chemical theory with any hope of obtaining unambiguous testable predictions. As long as such a theory is regarded as only an approximation to the Schrödinger equation, one can always—if some of its predictions don't agree with experiment—go back and fiddle with some of the parameters. Chemical theories would thus lose the property of falsifiability, which (according to Karl Popper) is essential to scientific theories and hypotheses.

A further consequence of this belief in the reducibility of chemistry is the tendency to suppress history in teaching chemistry. Chemistry is defined as primarily an *experimental* science, having no theory of its own other than what it borrows from atomic physics. Chemistry is to be studied for its practical value, as a collection of experimental facts tied together by a few correlation rules rather than as an intellectually stimulating discipline. The history of science, on the other hand, has generally concentrated on concepts and theories, and scientists who are proud of the theoretical achievements of their own discipline are often interested in finding out where they came from.

Recently some philosophers of science (for example David Bantz at the University of Illinois, Urbana, and Paul Bogaard at Mt. Allison University in Canada) have challenged the idea that chemistry has been or can be reduced to physics. Thus Bogaard points out that the explanation of chemical phenomena does not follow directly from the Schrödinger equation unless one adds the Pauli exclusion principle[10] as

---

[10] Thus Lennard-Jones pointed out that the Pauli principle does not follow from wave mechanics but must be added because of our "experience" that "all the electrons of an atom do not assume the same pattern" (1931:462).

well as a practical approximation method. Yet the Pauli principle itself was originally deduced at least in part from the periodic system of elements. And the use of approximation methods in quantum chemistry has always been guided by chemical knowledge; it is doubtful that explanations of such phenomena as covalent bonds and tetrahedral carbon atoms would have emerged as pure *predictions*. Even the concept of resonance, which is sometimes presented as a purely quantum effect, was to some extent anticipated in the chemical theories of Arndt and Ingold.

VI. PHASE TRANSITIONS
AND THE CRITICAL POINT

6.1 THE VAN DER WAALS THEORY OF THE
GAS-LIQUID CRITICAL POINT

In §1.12 we discussed the earlier work of J. D. van der Waals on the continuity of gaseous and liquid states. We must now consider what his theory has to say about the precise nature of the singularity at the critical point, since it is here that the essential features of phase transitions in general are revealed. As the first qualitatively successful theory of phase transitions, the van der Waals theory provided a model for theories of other kinds of transitions; as one might expect, it eventually had to be replaced by more sophisticated approaches.

In 1893 van der Waals became interested in the behavior of fluids very near the critical point in connection with experiments on surface tension by de Vries in Leiden and by W. Ramsay and J. Shields in England. According to his equation of state, the difference between the densities of coexisting liquid and gas phases (ρ_L and ρ_G) should go to zero at the critical point; in particular,

$$\rho_L - \rho_G \propto (\Delta T)^\beta,$$

where $\Delta T = T_c - T$ and β is a positive number which he found to be equal to $\frac{1}{2}$.

At the same time van der Waals showed that the surface tension should vanish at the critical point according to the equation

$$\sigma \propto (\Delta T)^\mu,$$

where $\mu = \frac{3}{2}$ if the van der Waals equation is valid. Older theories of Laplace and Gauss predicted $\mu = 1$. Experimental data indicated $\mu \sim$ 1.23 to 1.27; van der Waals did not seem too concerned about the discrepancy in view of the difficulty of determining accurate values very close to T_c. (See J.M.H. Levelt Sengers 1976:324–26.)

Shortly after this the Belgian physicist J. E. Verschaffelt, working in Kamerlingh Onnes' laboratory at Leiden, began to obtain more accurate experimental values for critical point exponents. By 1900 he had found strong evidence that β is not equal to the van der Waals value $\frac{1}{2}$, but is probably about 0.34. Although he continued to publish extensive data supporting this conclusion for the next few decades, he did not succeed in shaking the faith of physical chemists in the validity of van der Waals' equation; in the 1960s his work had been largely forgotten, and many of his conclusions were rediscovered by others (see J.M.H. Levelt Sengers 1976 for extensive discussion of this situation).

6.2 THE SOLID-LIQUID TRANSITION

Turning now to research on the liquid-solid transition, we find that in the 19th century comparatively little progress was made toward a molecular theory, although thermodynamics was applied rather successfully to the macroscopic properties of the transition. In fact it was James Thomson's theoretical prediction in 1849 that the freezing point of water would be lowered by increased pressure that helped persuade his brother William Thomson of the validity of the "second law of thermodynamics" (Carnot's principle).[1] William Thomson had originally thought that it would be possible to violate Carnot's principle with an engine that used the expansion of water on freezing to perform mechanical work; it appeared that the cycle of transformation of heat to work could be indefinitely repeated by converting back and forth from solid to liquid at the same temperature without ever having heat pass from a hotter to a colder body. James Thomson reasoned that the violation could be avoided if the melting point of ice is lowered by pressure; then as soon as the expanding substance starts to perform mechanical work it will be at a higher temperature, hence will stop freezing and will not be able to do work. William Thomson quickly confirmed by experiment that the melting point of ice *is* lowered by pressure, thus confirming in an unexpected way the validity of Carnot's principle.

Rudolf Clausius, soon after the publication of William Thomson's paper in 1850, showed that the equation relating the effect of pressure on the boiling point to the latent heat of vaporization and the volume change, derived by him from thermodynamics on the basis of earlier work by Clapeyron, also applied to the solid-liquid transition. The equation, now known as the "Clausius-Clapeyron equation," is

$$\frac{dp}{dT} = \frac{L}{T\,\Delta V}$$

---

[1] James at that time accepted "the almost universally adopted supposition of the perfect conservation of heat," as did his brother until a few months later.

(L = latent heat, $\Delta V = V_L - V_S$, T = absolute temperature, p = melting pressure). For water, ΔV is known to be negative; hence one expects dp/dT to be negative since L is positive; and the numerical value for dT/dp calculated from this equation using known values of L and ΔV is within about 2% of the experimental value found by William Thomson.

In order to arrive at some rudimentary understanding of melting and freezing on the molecular level, one should know whether the behavior of water is "normal" or exceptional. Do most substances expand or contract when they melt? It seems to have been generally believed for some time that most substances expand, but the question was by no means definitely settled in the 19th century. For example, in the debate about the internal structure and evolution of the earth, one side (J. H. Pratt, William Thomson, and others) argued, on the basis of experiments in the 1840s by Gustav Bischoff, that the liquid material forming the primitive earth would contract on solidification. Thus if a solid crust formed on the surface because of the lower temperature there, it could not be stable but would break up into pieces which must sink toward the center because they are denser than the liquid interior. Moreover, according to the Clausius-Clapeyron equation, the presumed higher temperature of the interior of the earth might not be sufficient to keep material liquid there, since the higher pressure in the interior would raise the melting point. Hence Pratt and Thomson concluded that the earth would have solidified from the center outward, and the presence of a solid crust indicates that it is entirely solid inside. The other side (including most geologists) argued that rocks and iron expand on freezing, like water; they pointed to observations of solid lava floating on top of liquid, and solid iron on top of liquid iron. From these observations they concluded that a stable solid crust could form around a hot liquid earth, and that the earth is even now partly or mostly liquid inside. The high pressure inside the earth would then reinforce the tendency toward liquidity.

Since the effect of pressure on the melting point is generally much smaller than its effect on the boiling point, 19th-century experiments could do little more than attempt to establish the initial slope of the melting curve, and in some cases, such as iron and rock, even the sign of this slope was in doubt. In the 1890s a Russian-German physical chemist, Gustav Tammann, argued on the basis of his experiments that the melting curve does not end in a critical point but, even if it initially has positive slope, must eventually bend back toward the zero-temperature axis at high pressures. Thus it would never be possible to go continuously from the solid to the liquid state, but all substances must melt at sufficiently high pressures. This theory was strongly criticized by P. W. Bridgman (see for example 1931:204–6) and seems to have been generally rejected by

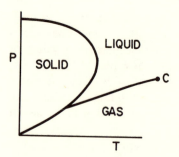

FIG. 6.2.1. Phase diagram showing a melting curve that bends back.

experimentalists. According to 20th-century theories of atomic structure, all substances must change from solid to fluid at high enough pressures, no matter how low the temperature, though the change may be either continuous or discontinuous; and phase-transition theorists still believe that there cannot be a solid-liquid critical point (see for example Fisher 1972). So Tammann's theory cannot yet be rejected.

Tammann's theory of the melting curve was used by the Swedish physical chemist Svante Arrhenius in 1900 to support the theory that the center of the earth is gaseous. It had been suggested in 1878 by the German astrophysicist August Ritter that the temperature inside the earth (estimated by extrapolating its increase with depth near the surface) is above the critical temperature for all known substances. A number of German geophysicists in the the 1880s and 1890s accepted the idea that the central part of the earth is a supercritical fluid or gas. Tammann's theory was taken to imply that a substance could never be solidified by pressure at a temperature above its gas-liquid critical point. However, in 1903 Tammann found that phosphonium chloride (PH_4Cl) could be solidified by pressure at a temperature above its critical point, and later Bridgman found that the same is true for carbon dioxide. (For further details on these geophysical debates see Brush 1979.)

6.3 MAGNETISM AND THE CURIE TEMPERATURE

A rather different kind of phenomenon, the loss of permanent magnetism at high temperature, was not recognized as a "phase transition" similar to those between gases, liquids, and solids until near the end of the 19th century. The phenomenon was known in China by the 11th century A.D. (Needham 1962:252–53) and was mentioned in Chapter IV of William Gilbert's *De Magnete* as a fact well known in the West at that time (1600).

In 1889 John Hopkinson, an English engineer, suggested that the temperature at which magnetism disappears should be called the "critical temperature" but did not explicitly state that he intended this term to suggest an analogy with the gas-liquid critical point.

The most important 19th-century publication on the magnetic properties of matter was the doctoral dissertation of Pierre Curie, the French physicist who is better known as the husband of Marie Curie. In his 1895 dissertation Pierre Curie presented an extensive quantitative investigation of the dependence of magnetization on both temperature and external magnetic field. He defined and described the three basic kinds of magnetism now recognized:

(1) ferromagnetism, for example of iron, which may remain in the absence of the external field up to a certain temperature characteristic of the substance (Hopkinson's "critical temperature" now known as the "Curie temperature");[2]

(2) paramagnetism, a much weaker magnetization which is produced by an external field but vanishes with it (the paramagnetic susceptibility decreases with temperature);

(3) diamagnetism, a very weak magnetization in the direction opposite to that of the applied field, and independent of temperature.

He suggested that diamagnetism is a property of individual atoms, whereas ferro- and paramagnetism depend on interactions between atoms.

Curie explicitly proposed that there is an analogy between magnets and gas-liquid phase transitions; the way in which the intensity of magnetization varies with temperature and external field is similar to the way in which the density of a fluid varies with temperature and pressure. To demonstrate this analogy he presented on facing pages the following diagrams (Figs. 6.3.1 and 6.3.2), one showing the density of CO_2 as a function of temperature at various pressures (as measured by the French physicist Émile Amagat) and the other showing the magnetization of iron as a function of temperature at different magnetic field strengths.

Curie did not attempt to develop a detailed molecular theory of magnetism in this paper, and shortly afterward he was diverted into the study of radioactivity with his wife Marie. He left only a brief remark at the end of his 1895 paper:

From the point of view of molecular theories, one can say, by analogy with the hypotheses which one makes about fluids, that the rapid increase of the intensity of magnetization is produced when the

[2] As pointed out by Domb (1971), Curie did not identify a single temperature at which ferromagnetism disappears; the term "Curie temperature" was first used by Weiss and Kamerlingh Onnes in 1910.

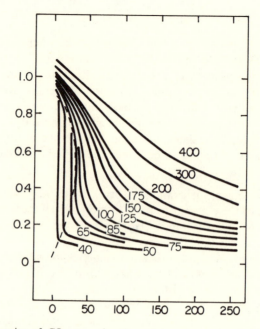

Fig. 6.3.1. Density of CO_2 as a function of temperature at various pressures (Amagat).

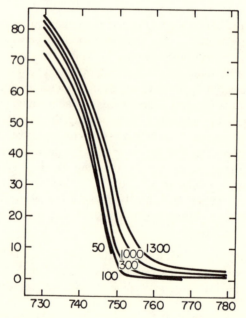

Fig. 6.3.2. Magnetization of iron as a function of temperature at various fields (Curie).

intensity of magnetization of the magnetic particles is so strong that they can react on each other. (1895:404)

The molecular theory of magnetism was developed by two other French physicists, Paul Langevin[3] and Pierre Weiss. In 1905 Langevin applied statistical mechanics to a gas of molecules with permanent dipole moments and showed that at low fields and high temperatures the system would have a paramagnetic susceptibility inversely proportional to the absolute temperature, in agreement with Curie's experimental results. Weiss in 1907 extended Langevin's theory by introducing an internal molecular field, assumed to be proportional to the intensity of magnetization of the substance. The force tending to orient each molecular magnet would be the sum of the external field and the molecular field. The system would then have a spontaneous magnetization even in zero external field below the Curie temperature. The direction of this magnetization would depend on the previous magnetic history of the sample, and might also be different for different parts of the system (later known as "Weiss domains").

As mentioned earlier (§3.3), Niels Bohr's early work on electrons was stimulated in part by the Langevin-Weiss theory; Bohr proved that this theory could not be completely consistent with classical statistical mechanics, but that it was necessary to introduce some nonclassical rigidity in the states of electrons within atoms in order to justify the assumption of a fixed magnetic moment.

6.4 THE LENZ-ISING MODEL

It is now believed that many physical systems can be fairly well described by means of the following idealized model: each node of a regular space-lattice is characterized by a two-valued parameter, called for convenience the "spin" of node i: $\sigma_i = \pm 1$, $i = 1$ to N. Each of the 2^N configurations of the lattice, specified by a particular set of values of the N spins, is assigned an energy which may be written in the form

$$E = \sum_{n.n} U\sigma_i\sigma_j + \mu H \sum_i \sigma_i, \tag{1}$$

where U is a positive or negative constant characterizing the interaction between neighboring spins, and the product μH may be assigned a fixed value (frequently zero) or may be allowed to vary. The first sum is meant

[3] It was Langevin who apparently became Marie Curie's lover after Pierre's death; see Reid (1974, chaps. 15–17) for details.

to be taken over all pairs of nodes (i, j) that are "nearest neighbors" on the lattice.

Less abstractly, one may identify the parameter σ with the direction ("up" or "down") of the actual spin of a magnetic atom, μ with its magnetic moment, and H with the external magnetic field; one then has a very crude representation of a magnetic system, corresponding to a theory of ferromagnetism that is not generally accepted at present. A second possibility is to let σ indicate that either of two kinds of atoms is situated at the lattice node, so that the model describes a two-component mixture or alloy with possible ordering effects. In this case μH will be assigned a value appropriate to ensure that the statistical average of $\sum \sigma_i$ will correspond to the specified composition of the system. A third physical system that can be represented by this model is the "lattice gas," a mixture of atoms and "holes" (vacant lattice nodes). The properties of the model with positive or negative values of U, using μH as a parameter chosen to determine the density, show a rough similarity to gas-liquid or liquid-solid equilibrium.

This model is commonly referred to as the "Ising model," although one has only to read Ising's original paper (1925) to learn that it was originally proposed by Ising's research director, Wilhelm Lenz, in 1920. Lenz suggested that dipolar atoms in a crystal may be free to rotate around a fixed rest position in a lattice. To understand the physical basis of his assumptions, it would be necessary to examine the development of the "old quantum theory" in somewhat more detail than we have done here; instead of doing this, we shall simply quote the remarks of Ising and Lenz.

If α is the angle by which an atomic dipole is rotated from a rest position, then, says Lenz,

> In a quantum treatment certain angles α will be distinguished, among them in any case $\alpha = 0$ and $\alpha = \pi$. If the potential energy W has large values in the intermediate positions, as one must assume taking account of the crystal structure, then these positions will be very seldom occupied, *Umklapp*-processes will therefore occur very rarely, and the magnet will find itself almost exclusively in the two distinguished positions, and indeed on the average will be in each one equally long. In the presence of an external magnetic field, which we assume to be in the direction of the null position for the sake of simplicity, this equivalence of the two positions will disappear, and one has, according to the Boltzmann principle, a resulting magnetic moment of the bar-magnet at temperature T:

$$\bar{\mu} = \mu \frac{e^a - e^{-a}}{e^a + e^{-a}}; \qquad a = \frac{\mu H}{kT}.$$

For sufficiently small values of a, this reduces to

$$\bar{\mu} = \frac{\mu^2 H}{kT},$$

i.e. we obtain the Curie law . . . .

For ferromagnetic bodies, in addition to the temperature-dependence of the susceptibility, one has to explain first of all the fact of spontaneous magnetization, as is observed in magnetic and pyrites . . . . If one assumes that in ferromagnetic bodies the potential energy of an atom (elementary magnet) with respect to its neighbors is different in the null position and in the π position, then there arises a natural directedness of the atom corresponding to the crystal state, and hence a spontaneous magnetization. The magnetic properties of ferromagnetics would then be explained in terms of nonmagnetic forces, in agreement with the viewpoint of Weiss, who has by calculation and experiment established that the internal field which he introduced, and which generally gives a good representation of the situation, is of a nonmagnetic nature. It is to be hoped that one can succeed in explaining the properties of ferromagnetics in the manner indicated.

Lenz was at Rostock University in 1920, but the following year he was appointed Ordinary Professor at Hamburg. One of his first students was Ernst Ising, who worked out the theory described above on a quantitative basis for a one-dimensional lattice. The partition function of the one-dimensional model could be obtained exactly, and the analysis showed that there was no transition to a ferromagnetic ordered state at any temperature; Ising then used some approximate arguments to extend this conclusion to the three-dimensional model.

According to a letter from Ising to S. G. Brush,

At the time I wrote my doctor thesis Stern and Gerlach were working in the same institute on their famous experiment on space quantization. The ideas we had at that time were that atoms or molecules or magnets had magnetic dipoles and that these dipoles had a limited number of orientations. We assumed that the field of these dipoles would die down fast enough so that only interactions should be taken into account, at least in the first order. . . . I discussed the result of my paper widely with Prof. Lenz and with Dr. Wolfgang Pauli, who at that time was teaching in Hamburg. There was some disappointment that the linear model did not show the expected ferromagnetic properties.

Heisenberg, when he proposed his own theory of ferromagnetism in 1928, began by saying that "Ising succeeded in showing that also the assumption of directed sufficiently great forces between two neighboring atoms of a chain is not sufficient to explain ferromagnetism." Thus Heisenberg used the supposed failure of the Lenz-Ising model to explain ferromagnetism as one justification for developing his own theory based on a more complicated interaction between spins. In this way the natural order of development of theories of ferromagnetism was inverted; the more sophisticated Heisenberg model was exploited first, and only later did theoreticians return to investigate the properties of the simpler Lenz-Ising model.

Indeed, as a result of Ising's own rejection of the model, we might never have heard any more about it, if it had not been for developments in a different area of physics: order-disorder transformations in alloys. Tammann had proposed in 1919 that the atoms in alloys may be in a definite ordered arrangement, and a number of workers had developed the idea that order-disorder transformations result from the opposing effects of temperature and of the lower potential energy of order. In 1928, Gorsky tried to construct a statistical theory on this basis, but the first satisfactory theory was that of Bragg and Williams (1934, 1935). Like Gorsky, they assumed that the work needed to move an atom from a position of order to one of disorder is proportional to the degree of order already existing; they note that Gorsky's formula is incorrect,

> since in his statistical treatment by the Boltzmann principle, he overlooked the factor for the number of places to which an atom could be moved; but the principle is the same . . . we wish to acknowledge this previous formulation of an essential feature of the order-disorder transformation (Bragg & Williams 1935).[4]

In 1935, Hans Bethe showed how the Bragg-Williams theory could be improved by taking account of the short-range ordering produced by in-

---

[4] According to a letter from C. Domb to S. G. Brush (29 Oct. 1971), "Bragg told me that he had given a colloquium to metallurgists in Manchester describing his experimental work and giving a general theoretical explanation. Williams was in the audience and sat down afterwards and worked the theory out from scratch and presented it to Bragg. Although Bragg was most impressed he felt sure that such a straightforward piece of work must have been done by someone else before. He consulted the metallurgists, but none of them knew of anything so he and Williams went ahead and published their work. Bragg was most embarrassed to hear later of the previous work of Gorsky and Borelius, particularly as he found a reprint on his desk of the work of the latter which he had not been able to read so far. He told me that he always wanted to give due credit to his predecessors and was not too happy that he and Williams were always quoted with no reference to previous work. He felt that the reason for this was probably because his account with Williams was much clearer and more intelligible than the others."

teractions between neighboring atoms. He did this by constructing an approximate combinatorial factor based on configurations of the first shell of lattice sites around a central one. In the same year, E. A. Guggenheim developed a theory of liquid solutions in which nearest-neighbor interactions were taken into account by what is known as the "quasi-chemical" (QC) method. In the QC method one constructs an approximate combinatorial factor by counting configurations of neighboring pairs or larger groups of atoms, assuming that these groups can be treated as independent statistical entities. (This is of course not strictly true, since each atom belongs to several groups, so that the configurations of these groups cannot really be independent.) The QC method was improved and extended by Rushbrooke (1938) by means of Bethe's method. Guggenheim (1938) then showed that the QC method is really equivalent to the Bethe method and more convenient to use in many cases, such as those involving complicated lattices. In 1940, Fowler and Guggenheim published a general formulation of the QC method and applied it to alloys with long-range order.

The concept of the "Ising model" as a mathematical object existing independently of any particular physical approximation seems to have been developed by the Cambridge group led by R. H. Fowler. Fowler discussed rotations of molecules in solids in a paper published in 1935, where he stated that the need for a quantitative theory of such phenomena

> was first brought clearly to my notice at a conference on the solid state held at Leningrad in 1932. As will appear, however, an essential feature of the theory is an application of the ideas of order and disorder in metallic alloys, where the ordered state is typically cooperative, recently put forward by Bragg and Williams. As soon as their ideas are incorporated the theory "goes."

In 1936, R. Peierls published a paper with the title "On Ising's model of Ferromagnetism" in which he recognized the equivalence of the Ising theory of ferromagnetism, the Bethe theory of order-disorder transformations in alloys, and the work of Fowler and Peierls on absorption isotherms. Peierls gave a simple argument purporting to show that (contrary to Ising's statement) the Ising model in two or three dimensions *does* exhibit spontaneous magnetization at sufficiently low temperatures. He pointed out that each possible configuration of $(+)$ and $(-)$ spins on the lattice corresponds to a set of "boundaries" between regions of $(+)$ spins and regions of $(-)$ spins. If one can show that, at sufficiently low temperatures, the average $(+)$ area or volume enclosed by boundaries is only a small fraction of the total area or volume, then it will follow that the majority of spins must be $(-)$, which corresponds to a system with net magnetization. Unfortunately Peierl's proof is not rigorous because of an incorrect step, which was pointed out by M. E. Fisher and S. Sherman; the damage was repaired by R. B. Griffiths in 1964.

The next advance was made by J. G. Kirkwood (1938), who developed a systematic method for expanding the partition function in inverse powers of the temperature. His method was based on the semi-invariant expansion of T. N. Thiele, used in statistics to characterize a distribution function by its moments. Since only a small number of terms in the expansion could actually be computed, the result was "essentially equivalent to Bethe's in its degree of approximation, but somewhat less unwieldy in form" (1938:70).

The first exact quantitative result for the two-dimensional Lenz-Ising model was obtained by H. A. Kramers and G. H. Wannier in 1941. They located the transition temperature by using the symmetry of the lattice to relate the high and low temperature expansions of the partition function. They showed that the partition function can be written as the largest eigenvalue of a certain matrix (some of their ideas were attributed to Elliott Montroll, who subsequently published a similar method). Kramers and Wannier developed a method that yields the largest eigenvalue of a sequence of finite matrices and would therefore converge to the exact solution, but did not succeed in obtaining this exact solution in closed form.

A major breakthrough, called a "revolution" by Domb (1971), was achieved by the Norwegian-American theoretical chemist Lars Onsager. Onsager announced his solution of the two-dimensional Lenz-Ising problem in 1942, and the details were published two years later. His method was similar to that of Kramers and Wannier (and of Montroll) except that he emphasized "the abstract properties of relatively simple operators rather than their explicit representation by unwieldy matrices." We quote Onsager's summary of his method:

> The special properties of the operators involved in this problem allow their expansion as linear combinations of the generating basis elements of an algebra which can be decomposed into direct products of quaternion algebras. The representation of the operators in question can be reduced accordingly to a sum of direct products of two-dimensional representations, and the roots of the secular equation for the problem in hand are obtained as products of the roots of certain quadratic equations. To find all the roots requires complete reduction, which is best performed by the explicit construction of a transforming matrix, with valuable by-products of identities useful for the computation of averages pertaining to the crystal. . . . It so happens that the representations of maximal dimension, which

contain the two largest roots, are identified with ease from simple general properties of the operators and their representative matrices. The largest roots whose eigenvectors satisfy certain special conditions can be found by a moderate elaboration of the procedure; these results will suffice for a qualitative investigation of the spectrum. To determine the thermodynamic properties of the model it suffices to compute the largest root of the secular equation as a function of temperature.

The passage to the limiting case of an infinite base involves merely the substitution of integrals for sums. The integrals are simplified by elliptic substitutions, whereby the symmetrical parameter of Kramers and Wannier appears in the modulus of the elliptic functions. The modulus equals unity at the "Curie point"; the consequent logarithmic infinity of the specific heat confirms a conjecture made by Kramers and Wannier.

Onsager's partition function for the two-dimensional Lenz-Ising problem deserves to be recorded here, since it is the first *exact* solution of a nontrivial problem in statistical mechanics, in which interparticle forces are taken into account without approximation:

$$\lim_{n,m \to \infty} \frac{\ln Z}{nm} = \ln 2 + (1/2\pi^2) \int_0^\pi d\omega \int_0^\pi d\omega' \ln (\cosh 2K \cosh 2K'$$
$$- \sinh 2K \cos \omega - \sinh 2K' \cos \omega'),$$

where $K = U/kT$, $K' = U'/kT$. The lattice is assumed to be rectangular with $n \times m$ points, and may have different interactions (U and U') in the horizontal and vertical directions.

Onsager became famous not only for his major contributions to theoretical physics (he received the Nobel Prize in chemistry in 1968 for his "reciprocal relations" of irreversible thermodynamics, in my opinion a less significant achievement than his solution of the Lenz-Ising problem) but for the casual way he announced them (Montroll 1977). Thus his important formula for the spontaneous magnetization of the square lattice was, according to Montroll et al. (1963),

first exposed to the public on 23 August 1948 on a blackboard at Cornell University on the occasion of a conference on phase transitions. . . . To tease a wider audience, the formula was again exhibited during the discussion which followed a paper by Rushbrooke at the first postwar IUPAP statistical mechanics meeting in Florence in 1948; it finally appeared in print as a discussion remark. However,

Onsager never published his derivation. The puzzle was finally solved by C. N. Yang.

Longuet-Higgins and Fisher (1978) provide many other fascinating glimpses of Onsager's personality and career.

The relevance of Onsager's solution to critical phenomena in general was not evident for another two decades. According to Domb (1971), "This two dimensional Ising model was highly artificial and did not represent any known experimental situation; also the solution applied only in zero magnetic field, it could not yield a complete equation of state and deduced only certain aspects of critical behavior." Moreover, "The exact method which gave such beautiful results for the two dimensional Ising model failed completely in three dimensions. Several different lines were pursued, and the most successful proved to be the use of series expansions at high and low temperatures for the various quantities of interest." Series expansions were first used to estimate properties at the critical point by C. Domb in his Ph.D. thesis at Cambridge University (1949), and the method was further developed by M. F. Sykes, M. E. Fisher, J. W. Essam, D. S. Gaunt, and others. By the 1960s the mathematical techniques of computing and extrapolating these series expansions had been so highly refined that it became feasible to make numerical computations of any property of the model to any desired degree of accuracy. What was needed was some evidence that the model had enough physical content to make such computations worthwhile.

6.5 Theory of Gas Condensation

The success of van der Waals' equation gave some hope for explaining the gas-liquid transition and critical point by starting with a theory valid for dilute gases and taking account of intermolecular forces at higher densities. But to develop such a theory would apparently require the calculation of a large number of virial coefficients, and even then one could not be sure that the virial series actually converges up to the point of condensation. Little was done on this problem until 1936, presumably because most of the theoreticians who might have been interested in it were occupied with working out the new quantum mechanics. There was also a feeling that it might be impossible in principle to derive discontinuous thermodynamic functions from a partition function, which is a sum of continuous functions; this feeling was expressed at the van der Waals centennial conference in 1937, and largely dispelled by H. A. Kramers, who pointed out that such functions may well become discontinuous in the limit when the number of particles becomes infinite though N/V remains finite (Uhlenbeck 1978).

The modern theory of gas condensation was initiated by Joseph E. Mayer in 1936; at that time he was in the chemistry department at Johns Hopkins University, and much of the work in developing the theory was done by a graduate student, Sally F. Harrison.[5] In a series of papers in 1937–38, Mayer and Harrison showed how the statistical-mechanical expressions for the virial coefficients in terms of interatomic forces could be systematically arranged and enumerated as "cluster integrals" representing interactions of two, three, and higher numbers of molecules. They credited a British mathematician, H. D. Ursell, with the initial formulation of the problem; and in estimating the number of terms involving N molecules they relied on a number-theoretic result first proved in 1917 by another British mathematician, G. H. Hardy, and his Indian colleague Ramanujan (number of ways N can be expressed as a sum of numbers). One of the crucial theorems (on the reduction of cluster integrals) was proved by Mayer's wife, Maria Goeppert Mayer (later to become more famous than her husband since she won the Nobel Prize for her work on the nuclear shell model).

Mayer and Harrison concluded that the existence of condensation phenomena followed as a logical consequence of their theory, even though precise evaluation of all the integrals was not practical. Moreover, they predicted from the theory not only a critical point at a temperature T_c but another special temperature T_m below T_c, at which the surface tension of the condensed phase disappears. Between T_m and T_c there would be no visible sharp meniscus between liquid and gas phases, and a range of densities would be equally stable in the "derby hat" region. They found experimental evidence for this "derby hat" theory in the work of O. Maass and others during the previous few years. (Maass had followed up earlier work which suggested the persistence of the liquid phase and density variations above the critical point, although these phenomena were not reproducible and had usually been attributed to the effects of gravity or insufficient mixing; see Levelt Sengers 1979.)

In 1945 another student of Mayer, W. G. McMillan, Jr., wrote a Ph.D. dissertation at Columbia University (where Mayer was then located) applying similar techniques to the statistical-mechanical theory of mixtures. McMillan and Mayer suggested that the curve of osmotic pressure vs. concentration should be horizontal over a finite range of density at the temperature where separation into two phases first occurs. They

---

[5] The name given on the original publications was "S. F. Harrison." Harrison, now Sally H. Dieke, later became an astronomer.

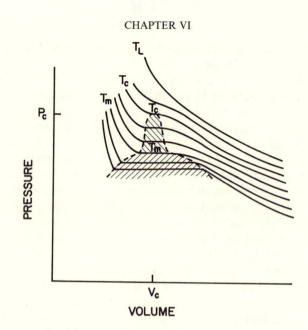

FIG. 6.5.1. Derby-hat theory of Mayer & Harrison.

noted that experimenters as early as 1900 had remarked on the flatness of such curves.

A physical chemist at the University of North Carolina, O. K. Rice, examined the Mayer-Harrison theory and proposed a modification of it in 1947. Agreeing with Mayer and Harrison that there is a "critical region" above T_m, he concluded that the isotherm would be flat at T_m but not at higher temperatures. His conclusion was based on a physical interpretation of Mayer's "cluster integrals" in terms of actual clusters of molecules in statistical equilibrium; at condensation there would be a sudden increase in the number of very large clusters. In a review of critical phenomena in binary liquid systems (1949), Rice stated that there was considerable experimental evidence for horizontal critical isotherms or coexistence curves.[6]

As a direct result of the theoretical predictions by Mayer and Harrison, and Rice's predictions based on his modification of their theory, there was considerable interest in performing very accurate experiments near the critical point after World War II. E. A. Guggenheim in 1945 published a review of older data, and claimed that the difference between gas and

[6] See obituary of Rice by Widom (1979).

liquid density near the critical temperature for many substances followed a cube-root law,

$$\rho_L - \rho_G \propto \left(\frac{T_c - T}{T}\right)^{1/3}, \tag{1}$$

so that the exponent β would be $1/3$ rather than $1/2$, as predicted from the van der Waals theory—though the cube-root law goes back to the beginning of the century, as Levelt Sengers (1976) has shown. New very accurate experiments near T_c by W. G. Schneider and his collaborators in Ottawa, Canada, confirmed this exponent, and failed to find any evidence for a flat portion of the critical isotherm (Atack & Schneider 1951, Weinberger & Schneider 1952, Habgood & Schneider 1954). Hans Lorentzen, in Oslo, Norway, found (1953) that the critical isotherm of carbon dioxide has a rounded top and that the density difference could be represented by the above formula with an exponent $\beta = 0.357$ to within $.0014°$ of T_c; the reason for this choice of exponent was that it had been found to represent the results of a prewar experiment in Holland (Michels, Blaisse, and Michels 1937), and in later work (1965) Lorentzen supported the value $1/3$.

In another experiment specifically designed to test the McMillan-Mayer theory, Bruno Zimm (1950) at Berkeley found that the shape of the coexistence curve for a binary mixture of CCl_4 and C_7F_{14} is rounded on top, not flat, and that it obeys a $1/3$ power law near T_c. Analysis of solubility experiments on several other binary systems showed that a similar value of $1/3$ can be derived from the shape of the coexistence curve (or curve of partial pressure of one component vs. concentration) near T_c.

In 1951 Zimm suggested that the two critical temperatures (T_c and T_m) proposed by Mayer and Harrison might not really be a valid consequence of their theory, and that if one adopted some reasonable assumptions about the statistical distribution functions for two and three particles (equivalent to the "superposition principle" proposed by J. G. Kirkwood in 1935), there should be only one critical temperature. It would have the mathematical properties of an "essential singularity"—derivatives of every order of the pressure with respect to the density would be zero. Zimm noted that his own experimental data on mixtures did not suggest any second critical temperature, nor did Onsager's exact solution of the two-dimensional Ising model.

Mayer, in a brief reply published immediately after Zimm's paper of 1951, admitted that Zimm's argument "weakens the case for the existence of an anomalous critical region" and that his conclusions "are at least as plausible as this author's earlier hypotheses." Mayer abandoned the derby-hat theory in his later publications.[7] A theorem of Yang and Lee (1952) apparently makes it untenable (Dorfman 1978).

The treatment of the Mayer-Harrison derby-hat theory of the critical region by later writers on critical phenomena is typical of the "Whig interpretation" of the history of science: they either ignore it or dismiss it as a dead end. Yet for a decade it represented the best, or at least a very respectable, theoretical view which seemed to be supported by experiments. In the light of later developments its main significance was the strong impetus it gave to experimental work on the critical region. The outcome of this work (on gas-liquid systems, binary liquid mixtures, and other systems to be mentioned below) was that the top of the coexistence curve is not flat nor does it follow the square-root law predicted by the van der Waals equation; but it is fairly well represented by an exponent β approximately equal to 1/3.

The Mayer cluster-integral theory (without the derby hat) has been developed to a high degree of mathematical sophistication and may eventually provide a good explanation of gas condensation. But the main line of development was elsewhere.

6.6 THE HARD-SPHERE PHASE TRANSITION

At the opposite extreme from the highly mathematical approach of Mayer is the method of "molecular dynamics" which led to the discovery of the hard-sphere phase transition by Berni Alder and Thomas Wainwright in 1957. Since the time of Clausius and Maxwell it had been taken for granted that in dealing with the large number of particles present in any macroscopic sample of matter it would be necessary to use statistical methods. The assumption was that even if one knew (or wanted to specify) the initial positions and velocities of all the particles in a system, the labor of computing their trajectories, taking account of mutual interactions, would be prohibitive. When the Clausius-Maxwell theory itself made possible a reasonably accurate estimate of Avogadro's number (in the late 1860s), this assumption was further reinforced; the number of molecules in a cubic centimeter of gas under ordinary conditions is about 10^{19}. But with the development of fast large-memory electronic computers after World War II, partly stimulated by the need to solve complicated scientific and technical problems, it became possible to reconsider the situation. Alder and Wainwright used an IBM-704 computer at the Livermore Radiation Laboratory (its main purpose was to calculate the

---

[7] In particular the discussion of condensation and critical phenomena, which occupied an entire chapter in the first edition of *Statistical Mechanics* (1940) by Joseph and Maria Goeppert Mayer, was almost entirely dropped from the second edition (1977).

predicted behavior of various nuclear-weapon designs). The computer kept track of the positions and velocities of about 100 particles with short-range forces, initially assumed to be hard spheres that interact only on collision. After each short interval of time the new positions of the particles were computed, and if any came within the collision distance the new velocities were computed using simple Newtonian mechanics. There was no element of chance except for the inevitable roundoff error in doing any calculation with a finite number of digits. The main drawback of the method is that surface effects are expected to be disproportionately large for such a small number of particles; this difficulty was mitigated to some extent by imposing "periodic boundary conditions" so that a particle which goes out one side of the system comes in the opposite side rather than being reflected from a wall (Alder & Wainwright 1959, 1960).

Alder had studied under J. G. Kirkwood, who had predicted in a paper with Elizabeth Monroe in 1941 that a system of hard spheres with no attractive forces would crystallize at sufficiently small volumes. (The hard-sphere phase transition had also been suggested by the Russian physicist I. Z. Fisher in 1955.) Alder and Wainwright found that in a certain range of densities, from about 60 to 65% of the maximum (close-packed) density, the system oscillates between two different pressures; these seemed to correspond to fluid and solid phases. At higher densities there is a single phase (continuous with the higher of the two pressures in the transition region), with the spheres moving around the equilibrium positions in a crystalline lattice. At lower densities there is a single fluidlike phase, and the pressure is given fairly accurately by the first few terms in the virial equation of state.

The Alder-Wainwright transition appears to be a liquid-solid rather than a gas-liquid transition; the hard-sphere system does not seem to have anything resembling a condensation from gas to liquid, unless some kind of attractive force is added to the model. So one might infer that the phenomenon of freezing is primarily due to repulsive forces that cause the molecules to fall into an ordered pattern at high densities and low temperatures, whereas condensation is primarily due to attractive forces that hold the molecules together. This qualitative explanation of phase transitions is not universally accepted, but the evidence for it has been increasing during the past two decades.

6.7 Toward a Universal Theory of Critical Phenomena

It was recognized by F. Cernuschi and H. Eyring in 1937, and subsequently discussed by several Japanese physicists, that formulae derived for the Lenz-Ising model could apply equally well to systems of atoms and holes in a lattice; that is, the values $\sigma = 1$ and $\sigma = -1$ could be interpreted as "lattice site occupied" or "unoccupied" respectively; thus the theory might apply to a gas-liquid system with the peculiar property that the atoms are restricted to the positions of a regular lattice (Ono 1947, Muto 1948, Tanaka et al. 1951). The modern term "lattice gas" was first applied to this system in 1952 by T. D. Lee and C. N. Yang, the Chinese-American physicists better known for their work in elementary particle physics (for example parity violation). Lee and Yang gave a systematic account of the transcription of variables and formulae from the magnet to the gas-liquid system as well as deriving some important new results for the Lenz-Ising model.

Whereas the set of all possible spin values (σ_i) for N lattice sites corresponds to a canonical ensemble for a magnet, it corresponds to a grand canonical ensemble for a lattice gas. The total volume is fixed but the total number of atoms can vary. To calculate the partition function one sums over all possible values of the number of atoms, from zero up to the total number of lattice sites. Thus the *magnetization* of the magnet (difference between number of up and down spins) is directly related to the *density* of the lattice gas (fraction of occupied sites), as suggested by Pierre Curie in 1895. The role of the external magnetic field, which is a controllable parameter for the magnetic model, is now played by the fugacity (which is related to the chemical potential, see §1.13) for the lattice gas. The case of zero external field—the only one that can be treated exactly by Onsager's method—now corresponds to the case in which half of the lattice sites are filled on the average. Below the transition temperature the system may split into two phases of different densities; these correspond to the two possible states of spontaneous magnetization of a magnet in zero field (whether the magnetization is $+M$ or $-M$ depends on the history of the system, or on random fluctuations).

The lattice gas model can represent a system of atoms with either attractive or repulsive forces, depending on whether the parameter U in Eq. 6.4-1 is negative or positive. If the forces are attractive, one has a possible model for the gas-liquid transition; if repulsive, configurations with alternating filled and vacant sites are favored, and one can interpret this ordered phase as a solid. (In the magnetic system, this would correspond to *antiferromagnetic* ordering.) Following the Alder-Wainwright discovery of the hard-sphere transition, H.N.V. Temperley in England and others proposed the lattice gas model with repulsive forces as a possible interpretation of the liquid-solid transition.

As mentioned above, the Lenz-Ising model was not taken seriously for several decades because it grossly oversimplified the nature of interatomic forces and did not seem to correspond closely to any real physical system, with the possible exception of alloys and binary mixtures. The situation changed dramatically in 1961,[8] when George A. Baker, Jr., at the Los Alamos Scientific Laboratory,[9] showed that a mathematical technique invented by the French mathematician Henri Padé in 1891 could be used to determine the singularities at the critical point from known terms in the series expansions for the Lenz-Ising model. For example, he found that the intensity of spontaneous magnetization near the critical temperature T_c should vary as

$$M \propto (T_c - T)^\beta, \tag{1}$$

where $\beta = 0.30 \pm 0.01$, in a three-dimensional lattice. (It was known from Onsager's work that $\beta = 1/8$ for the two-dimensional lattice.)

Knowledge of terms in the series expansions for the Lenz-Ising model was concentrated in Domb's group at King's College, London, and Baker's new technique was quickly taken up by two theorists there, John Essam and Michael Fisher. In a calculation announced at a conference at Brown University in June 1962 and published in full early in 1963, they used several additional terms beyond those available to Baker and estimated that the exponent β lies between 0.303 and 0.318, conjecturing that its exact value is $5/16 = 0.3125$.[10] They noted that P. Heller (Harvard) and G. B. Benedek (MIT) had found (1962) by nuclear magnetic resonance that the spontaneous magnetization in an antiferromagnet, MnF_2, varies according to a similar law with an exponent $\beta = 0.335 \pm 0.01$. Moreover, it was "well known" that the density difference near the gas-liquid critical point obeys a 1/3 power law. Their conclusion was that

the closeness of the values provides support for the view that the dimensionality and the statistics are the main factors determining critical behavior, the details of the interactions (which are grossly oversimplified by the lattice model) being of secondary importance.

Baker also estimated that the exponent γ' in the theoretical formula for the low-temperature susceptibility,

$$\chi'_T = \left(\frac{\partial M}{\partial H}\right)_T \propto (T_c - T)^{-\gamma'}, \tag{2}$$

[8] My interpretation differs slightly from that of Domb (1971), who says that the next major phase in the history of critical phenomena began in 1965 with the papers by Domb and Hunter and by Widom.

[9] His calculations, like those of Alder and Wainwright, were done on an IBM 704 computer and financed by the Atomic Energy Commission.

[10] For current theoretical value see table at the end of this section.

has the value $\gamma' \approx \frac{5}{4} = 1.25$. For the lattice gas the corresponding quantity is the compressibility $K_T = -(1/V)(\partial V/\partial P)_T$. Both quantities become infinite at the critical point. Experimental values for fluids and magnets around 1960 ranged from 1.1 to 1.4, so the Lenz-Ising model seemed to have a small but significant advantage over the "classical" (van der Waals or Weiss) theories, which predicted $\gamma = 1$. Subsequent calculations suggested that the exponent below T_c is slightly higher than the exponent above T_c (these are denoted by γ' and γ respectively.)

Finally, Essam and Fisher proposed on the basis of an approximate theory that three critical-point exponents are related by an equation,

$$\alpha' + 2\beta + \gamma = 2, \tag{3}$$

where α' is the exponent for the specific-heat singularity on the low-temperature side,

$$c_v \propto (T_c - T)^{\alpha'}. \tag{4}$$

This relation is satisfied for the two-dimensional Ising model, where it is known that $\alpha' = 0$ (logarithmic singularity in the specific heat), $\beta = 1/8$, and $\gamma' = 7/4$. For the three-dimensional model, accepting the estimated values $\beta = 5/16$ and $\gamma = 5/4$, it would predict $\alpha = 1/8$. (The prediction was confirmed by theoretical calculations of Sykes, Martin, and Hunter in 1967.)

The British theorist G. S. Rushbrooke then proved (1963) that for a magnetic system it follows directly from thermodynamics that Eq. (3) is valid at least as an inequality,

$$\alpha + 2\beta + \gamma \geq 2, \tag{5}$$

and since then a large number of such inequalities have been proved.

In Russia, a measurement of the specific heat of argon near the critical point was published by M. I. Bagatskii, A. V. Voronel, and V. G. Gusak in 1962; an English translation was available the following year. They pointed out that the classical theory, as formulated by L. D. Landau (including the van der Waals and Langevin-Weiss models as special cases), predicted a finite discontinuity in the specific heat. However, their measurements suggested that the specific heat becomes infinite at the critical point, and probably has a logarithmic singularity. They noted that a similar logarithmic singularity had been observed by Fairbank, Buckingham, and Kellers (1957) at the λ-transition of liquid helium.

Voronel and his Russian collaborators reported similar results for oxygen in 1963. Fisher, assuming that the results of the three-dimensional Lenz-Ising calculations should be applicable, suggested that an equation of the form (4) be used to analyze the data. The Russians tried this (Voronel, Snigirev, & Chashkin 1965) and conceded that such an equation would give an equally good fit to their data, with an exponent of 1/4, but said that they preferred to keep their original logarithmic form because it fits the most reliable points and they wanted to preserve some kind of discontinuity.[11]

Although there still remained some discrepancies between experimental and theoretical values of critical-point exponents, it was clear by 1965 that the lattice theory based on the Lenz-Ising model was far more successful in treating the gas-liquid critical phenomena than the supposedly more "realistic" continuum theories such as that of Mayer. This was a rather surprising development which called for some explanation. To quote Domb (1971:93–96):

Why has so much progress been achieved with lattice models, whereas the cluster integral theory provided no detailed information regarding critical behaviour? To answer this question, let me refer in more detail to the researches of Uhlenbeck and his collaborators in the 1950's, the general message of which was "if graphs play a major part in theory, one must treat them seriously and study them in their own right". Hence they provided us with a systematic enumeration of connected graphs and stars, and tried to obtain information regarding the asymptotic behaviour of the numbers of these graphs. The key to the whole development lay in Polya's theorem, and for *labeled* graphs and stars they were able to obtain asymptotic formulae which showed that the elimination of nonstar graphs in the Mayer theory gave little numerical advantage ... the number of stars increased with extraordinary rapidity with the number of points....

When we remember that for each star we must evaluate a multidimensional integral, it is not surprising that even with the best computers available not many terms of the virial expansion have been evaluated ... for the simplest hard sphere potential all the exact results were already obtained by Boltzmann in 1899. Approximation and Monte Carlo methods have allowed us to add a total of three new terms....

---

[11] Presumably what they mean by this is that the specific heat at $T = T_c - t$ is always greater than at $T = T_c + t$, so the singularity is not symmetrical or "continuous at infinity." The implication is that some residuum of their compatriot Landau's ideas will be preserved if the data are represented by a logarithmic rather than a power-law singularity.

The specific gains of a lattice model can be listed as follows:—

a) Instead of topological graphs we consider graphs embedded in the lattice constructed from bonds which correspond to interactions; instead of evaluating cluster integrals we have to count the number of possible embeddings and computers can be used in a very positive way.

b) Many of the stars correspond to zero embeddings, particularly for lattices with low coordination like the simple quadratic and diamond. Hence one can contemplate the possibility of 15 or 20 terms of a Mayer series.

c) For lattice models the most important results have been obtained from high temperature expansions of the partition function. . . . The grouping of the graphs is according to lines l rather than points p, and . . . (i) the number of connected graphs $C(l)$ rises much more slowly [than the number of stars $S(p)$] and (ii) there is a much greater gain in eliminating non-star graphs.

d) The symmetry of the partition function to reversal of magnetic field gives rise to many simplifying features.

e) We can look at many different lattices [and] search for "long-range properties which are lattice independent.

In spite of its quantitative success there seemed to be something lacking in the theory of the critical point based on the Lenz-Ising model. This something was hinted at by G. E. Uhlenbeck in his opening address to a conference on critical phenomena held at the National Bureau of Standards (Washington, D.C.) in April 1965. He regretted the fact that the "classical" theories of van der Waals and Weiss had been replaced by numerical computations based on series expansions, since the latter did not supply any "general point of view" about the behavior of the system. Such a point of view had been offered by the classical theories, although they obviously fail "close to the critical point where the substance remembers so to say Onsager." What was needed, according to Uhlenbeck, was "the reconciliation of Onsager with van der Waals" (1966:6).

According to Domb (1971) the theoretical developments of the next five years in fact went "a long way towards achieving such a reconciliation." The first step was taken in 1965 by Benjamin Widom (Chemistry Department, Cornell University), who developed a theory of surface tension intended to incorporate modern knowledge of the critical point (based on the Lenz-Ising model) into the classical van der Waals theory. He proposed that in a subvolume $v = L^s$ of a homogeneous fluid in equilibrium with its conjugate phase (for example liquid and gas, or two saturated phases of a binary mixture) there will be density fluctuations near the critical point which produce the density corresponding to the conjugate phase (L = correlation length, s = dimensionality). If the free energy associated with this fluctuation is assumed proportional to kT, then one of the relations between critical-point exponents can be derived.

Widom's ideas were taken up by Leo Kadanoff (Physics Department, Brown University), who used them as a basis for his "scaling laws" (1966). A cell of L lattice sites on a side, where L is now assumed to be *much less* than the correlation length ξ corresponding to the temperature T (close to the critical temperature T_c), has most of its spins aligned. This cell may be considered a single atom with fixed total spin, in a new system composed of cells of the same size and spin, oriented in various directions. Since for the new system ξ will be relatively smaller by a factor $1/L$, the new system will be further from its critical point. By analyzing the effects of such changes from the original system to new systems of cells, Kadanoff could show that the relations between magnetization and magnetic field and temperature must have definite functional relations. (These relations were consistent with equations derived slightly earlier by C. Domb and D. L. Hunter in London, 1965.) Moreover, he could derive equations between critical-point exponents.

In 1971, Kenneth Wilson, an elementary-particle theorist at Cornell, presented a mathematical analysis of Kadanoff's scaling laws using techniques similar to those developed earlier by Gell-Mann and Low, and by Stueckelberg and Petermann, in quantum electrodynamics; Wilson called his technique the "renormalization group."[12] The following year, in collaboration with Michel Fisher (Chemistry Department, Cornell) he showed that the Lenz-Ising model could be generalized to d dimensions in such a way as to show a continuous transition between classical (van der Waals) and nonclassical (Onsager) behavior. The nonclassical behavior depends on the fact that the correlation length becomes infinite at the critical point so that the critical phenomena are dominated by very long-range correlations. However, for four or more dimensions this effect is absent. Thus to see how nonclassical behavior appears, Wilson and Fisher introduced a continuous dimensionality

$$d = 4 - \epsilon$$

and derived perturbation expansions for the critical-point exponents as power series in ϵ. As it happened, the perturbation series converges fairly

[12] As noted by Fisher (1974), the application of the renormalization group to phase transitions was anticipated to some extent by Larkin and Khmel'nitskii (1969). A summary of the work that led to the award of the 1982 Nobel Prize in Physics to Wilson is given by Anderson (1982).

TABLE 6.7.1. Values of the Critical Exponents
(Levelt Sengers & Sengers 1981)

	Ising model series expansions	Renormalization group theory	Adopted by Levelt Sengers & Sengers
α	0.125 ± 0.010	0.109 ± 0.004	0.1085
β	0.312 ± 0.005	0.325 ± 0.001	0.325
γ	1.250 ± 0.003	1.241 ± 0.002	1.2415

well for ϵ as large as one. Then it became possible to derive the values of the exponents from a general theory.

A further generalization included in the renormalization-group theory is based on the number of components of the spin. In the Ising model the spin has only one component ($n = 1$), whereas in the Heisenberg model it has three ($n = 3$). H. E. Stanley (MIT) and others, beginning around 1968, discussed models in which n may have other values including infinity. It is now believed that all phase transitions can be assigned to universality classes according to the value of n; the critical exponents would then be functions of n and d. The case $n = \infty$ happens to be exactly soluble (it corresponds to the "spherical model" introduced in 1952 by T. H. Berlin and M. Kac); one method for treating models with finite n is thus to start at $n = \infty$ and expand in powers of $1/n$ (see Fisher 1974 for a review of this approach). As noted at the end of §4.9, the λ-transition of liquid helium-4 is described fairly accurately by a lattice model with $n = 2$; the theoretical values of α (critical exponent for the specific heat) are now in fairly good agreement wih those determined by experiment.

Table 6.7.1 shows some values of the exponents α, β, and γ for the three-dimensional Lenz-Ising model ($n = 1$); there is still some disagreement in the second or third significant figure. (For further discussion of this topic see Wilson 1979, Bruce 1980.)

VII. STATISTICAL MECHANICS AND THE PHILOSOPHY OF SCIENCE

7.1 REDUCTION, STATISTICS, AND IRREVERSIBILITY

The historical development of statistical physics has generated several problems, some solutions, and a few singular events that are or should be of interest to philosophers of science. I will begin by mentioning the issues that have most frequently been discussed by philosophers,[1] and then try to put them in a broader historical context. This will provide a review, from a different perspective, of some of the topics covered in earlier chapters of this book.

To nonspecialists, "statistical mechanics" often seems to mean the kinetic theory of ideal gases, based on the "billiard-ball model" (elastic spheres with no long-range forces or internal structure). This theory has frequently been cited as a good example of "scientific explanation." That fact in itself may have some philosophical significance, for the appeal of the theory seems to reside in features that are *not* typical of 20th-century physical theories: a visualizable model with contact interactions governed by Newtonian mechanics, and derivations that employ only simple algebra.

But even if we agree that most scientific theories cannot be expected to have those features, the kinetic theory remains as the best available example of the *reduction* of a theory formulated in terms of observable macroscopic quantities to one formulated in terms of unobservable microscopic entities. Thus it is often used to illustrate the procedure of theory reduction in general, or to argue whether reduction is actually possible at all.

Reduction as a mode of scientific progress is associated with positivistic and neopositivistic philosophies, and thus the attack on the possibility of reduction is part of the attack on such philosophies, although one can defend reduction without being a positivist. Reduction also threatens the autonomy of some scientific disciplines (chemistry may be reduced to physics, see §5.10) and the survival of some viewpoints such as holism and vitalism, which postulate nonreducible explanatory concepts; thus it

---

[1] For additional references see the original version of this chapter (Brush 1977).

impinges on the boundaries of research fields. It aroused antagonism in the neoromantic counterculture of the late 1960s and 1970s yet is somehow essential at least in the physical sciences. Roth (1976) blames the prevalence of reductionism on the male ego and suggests, "It is likely that a more synthetic and holistic approach, which seems to be more of a feminine characteristic, will be necessary for science in the future."

One often encounters the claim that "thermodynamics has been reduced to statistical mechanics." I consider this quite misleading. Thermodynamics is the science that deals with the general relations between the thermal and mechanical energy of substances whose special constitutive properties are assumed to be known. By a constitutive property I mean for examaple the "equation of state" relating pressure, volume, and temperature for the substance when it is in thermal equilibrium. Given the equation of state, one may use thermodynamics to compute such quantities as the work done when a fluid is heated from one temperature to another, at constant pressure. But a theoretical calculation of the equation of state from statistical mechanics is not a derivation or reduction of thermodynamics itself. Yet that is what many philosophers present as the only concrete justification for the claim that thermodynamics has been reduced to statistical mechanics: the derivation of the ideal gas law from kinetic theory, using a model of infinitesimal billiard balls.

One reason for this confusion, I think, is the use of the term "law" to describe the ideal gas equation of state (as in "Boyle's law" or "Charles's law"). The "ideal gas law" is often used to illustrate the application of thermodynamics, because of its simplicity, but it is not one of the "laws of thermodynamics." Indeed by restricting oneself to ideal gases one misses some of the most important applications of thermodynamics to modern science, including almost all cases in which intermolecular forces play a significant role.[2]

Looking at the other side of this alleged reduction, I would also point out that the reducing theory in the typical textbook example is not really

---

[2] I think most physicists would agree with the following assessment of Boyle's law: "Although exactly true, by definition, for an ideal gas, it is obeyed only approximately by real gases and is not a fundamental law like Newton's laws or the law of conservation of energy" (Sears & Zemansky 1970:247). One may agree with the statement of Andrews (1971:26), that "in some branches of science very minor empirical correlations are called laws, but in thermodynamics the term is kept for the most important generalizations on which the whole subject is based"—though the author also refers to the "ideal-gas law" (p. 117). Margenau is one of the few physicists who specifically includes the ideal gas law among the laws of thermodynamics—but he clearly distinguishes such laws from the more general "principles" or "Laws of Thermodynamics" (1950:208–212).

statistical mechanics but rather a crude kinetic theory of gases in which statistical considerations do not play any essential role. One gets the same formula for the pressure of a gas in a cubical container whether one assumes that all the molecules move at constant speed at right angles to the walls of the container or that there is a statistical distribution of the directions and magnitudes of the molecular velocity vectors. There is nothing wrong with using this as a textbook example, but it is too trivial to support inferences about the nature of reduction in general and too restricted to give any insight into the relation between thermodynamics and statistical mechanics. Indeed, philosophers have been hard pressed to find in it anything worth arguing about, other than the problem of whether the word "temperature" has changed its meaning during the reduction.

So far I have not found any philosophical discussion of the reduction of *thermodynamics* to statistical mechanics except for the problem of irreversibility, where the time asymmetry implied by the second law of thermodynamics is in question. Perhaps this is because this reduction is so far only a program, not a fact. But philosophers of science generally do not seem to be familiar with the work that has actually been done in this area, in the 19th century and more recently.[3]

Turning now to the second issue that seems to be of philosophical interest, we find that the kinetic theory of gases is usually considered to have represented the first significant application of probabilistic reasoning in physics. Thus it provides historical background for discussions of the general notion of "statistical law" in science. Usually it is asserted that there is a fundamental difference between the use of statistics by 19th-century physicists, who allegedly believed as an article of faith that all molecular motions are deterministic, and the probabilistic character of quantum physics, which assumes an inherent randomness or indeterminism in nature.

The third philosophical problem related to statistical mechanics is irreversibility, often expressed somewhat dramatically as the "direction of time." Ernest Nagel wrote in 1939,

> perhaps the greatest triumph of probability theory within the framework of nineteenth-century physics was Boltzmann's interpretation of the irreversibility of thermal processes. (1955:355)

---

[3] On the 19th-century reduction attempts see Daub (1969) and Klein (1972). Some examples of noteworthy recent efforts are Tisza and Quay (1963), Dyson and Lenard (1967), Ruelle (1969). For the current state of ergodic theory see the lectures published in Moser (1975). I think Sklar (1973) goes too far in downgrading the importance of ergodic theory, and recommend the article by Quay (1978) as a corrective. See also Friedman (1976).

Hans Reichenbach echoed this sentiment (1951:159) and devoted a major part of his last book (1956) to a discussion and extension of Boltzmann's concept of alternating time directions, originally an attempt to reconcile the second law of thermodynamics with reversible Newtonian mechanics through statistical mechanics. Other philosophers concerned with the nature of time have taken up the question of entropy increase and the H-theorem. The related problem of whether Brownian movement constitutes a violation of the second law of thermodynamics has attracted some attention.

In addition to these three major issues, philosophers of science have mentioned the kinetic theory of gases and statistical mechanics in several other connections. The billiard-ball model of kinetic theory has been invoked in arguments about the use of *models* in science. Suppes mentioned that Gibb's work in statistical mechanics could easily be formulated so as to show the distinction between theories and models; it gives a "particularly good example for applying the exact notion of model used by logicians, for there is not a direct and immediate tendency to think of Gibb statistical mechanical theories as being the theories of the one physical universe" (1961:168).

More frequently the kinetic theory has been invoked for just the opposite reason: to illustrate the triumph of an atomistic/mechanistic view of nature. At the same time it is recognized that the debate between Boltzmann and his opponents (Mach, Ostwald, and others) about the role of atomism in science has not been adequately studied by philosophers. There is also a certain amount of embarrassment among the followers of Mach on this point, and Philipp Frank has tried to minimize the extent of disagreement between Mach and Boltzmann by claiming that the latter was not really a strict mechanist either (1949:142–44).

There have been two attempts to describe Boltzmann's work on kinetic theory in the framework of Imre Lakatos' "methodology of scientific research programmes" (1971). Peter Clark (1976) argued that Boltzmann's original research programme was "degenerating" after 1870 and therefore that it was reasonable (by Lakatosian criteria) for scientists to abandon it in favor of the rival programme based on macroscopic thermodynamics, even though Boltzmann's programme had to be revived after 1900 when the thermodynamic programme proved to be inadequate.[4] Yehuda Elkana (1974), also writing from the Lakatos viewpoint, claimed that Boltzmann himself changed his views in the 1890s and no longer considered meaningful the question of whether atoms really exist. This revisionist interpretation of Boltzmann has now gone so far that a selection of his

---

[4] For a detailed critique of Clark's interpretation see Nyhof (1981).

philosophical writings has recently been published in a series titled "Vienna Circle Collection"! (McGuinness 1974; cf. S. R. De Groot's remarks on p. xi of this book).

Does the history of statistical mechanics play any role in the current debate about paradigms and scientific revolutions? Very little, as far as I can see; Kuhn mentioned it only in passing as an example of the discussion of the nature and standards of physics that accompanied the transition from Newtonian to modern physics (1970:48). Karl Popper mentioned Boltzmann as an example of a great scientist who doesn't fit into Kuhn's normal/revolutionary dichotomy (1970:54). Perhaps the impression that scientific revolutions occur as a response to a "crisis" produced by "anomalies" in the accepted theory has been fostered by the story of the "ultraviolet catastrophe" of classical statistical mechanics— the idea that Planck invented his quantum theory in order to avoid a divergence predicted by Rayleigh's theory of black-body radiation. The story continues to be repeated in the literature (Post 1971:222; Salmon 1971:321; Blackwell 1969:56–57), and the phase has even been used as the title of a children's book (Mahy 1975).

7.2 Beginnings of Statistical Mechanics

The early history of statistical mechanics can provide much useful material for those philosophers who investigate the process by which scientific theories are proposed and established. The most interesting period here is the half-century from 1815 to 1865. At the beginning of that period the caloric theory of heat and gases was generally accepted, and the kinetic theory received little serious consideration when it was proposed by John Herapath. At the end of the period the caloric theory had been replaced by the kinetic theory, and Maxwell and Boltzmann were able to start developing statistical mechanics in its modern form without having to worry too much about whether all its predictions agreed with the latest experiments.

What factors were mainly responsible for this success?

First, it is important to realize that the "billiard ball" kinetic-theory derivation of the ideal gas law, which I mentioned earlier as the textbook example of theory reduction, was not a major factor. Other theories of gases, based on a caloric fluid or rotating molecular vortices, could account for the ideal gas law and also for many other phenomena not yet explained by the billiard-ball model. The "domain" which defined the problems to be solved by the theory (Shapere 1974) included specific heats of liquids and solids as well as gases, latent heat of phase changes, and the properties of radiant heat. The caloric theory of gases offered plausible explanations of all these phenomena, though its proponents were still debating the merits of two different versions (see Fox 1971).

Second, experiments and theories in fields other than gas physics prepared the way for a revival of the kinetic theory after 1845. The major event here was the establishment of the principle of conservation and convertibility of energy by Mayer, Joule, Helmholtz, and Colding. This made it plausible that heat, being a form of energy interconvertible with mechanical energy, is in reality just mechanical energy on the molecular level. Even before the general principle of energy conservation was adopted, the caloric (substance) theory of heat had been transmuted into a wave (motion) theory as a result of experiments on radiant heat and light (§1.8). Both phenomena were thought to be qualitatively similar, and when the particle (substance) theory of light was replaced by a wave (motion) theory as a result of Fresnel's work around 1820, it seemed logical to assume that radiant heat is also a form of wave motion in the ether. At that time all heat was assumed to be similar to radiant heat.

Third, physicists took an interest in certain gas properties that had not been systematically investigated before: free expansion (Joule-Thomson effect), diffusion and effusion, viscosity, and heat conduction. In some cases experiments were stimulated by predictions of the kinetic theory itself. The result was to open up a new domain which turned out to be more congenial to the kinetic theory than to its rivals.

Fourth, the phenomena that could be explained better by caloric theory than by kinetic theory were simply excluded from the redefined domain, at least temporarily. The most striking example of this exclusion was radiant heat, which began to be considered a phenomenon separate from other kinds of heat; thus kinetic theory ignored the "ether" and assumed that molecules move through "empty" space. This redefinition of the domain failed to get rid of one awkward problem, the specific heats of gases, where there was an obstinate discrepancy between experimental values and the kinetic-theory predictions.

Fifth, the kinetic theory offered the theoretical advantage of a simple, clearly defined molecular model which could be treated mathematically; and the fact that reasonably consistent numerical values for parameters such as the molecular diameter could be inferred from different properties made the model quite plausible. Here we have favorable conditions for the elaboration of a more sophisticated theory.

These five factors could serve as the basis for a "rational reconstruction" of the early history of statistical mechanics, but something essential to history is missing: the personalities and social environments of the scientists who actually made the history, and the accidental events that helped or hindered them. So I would like to say something about two pioneers:

J. J. Waterston (1811–83), who never received proper recognition during his own lifetime, and James Clerk Maxwell (1831–79), who was well known for his work in several areas of physics and held the prestigious Cavendish professorship at Cambridge University.

One might claim that the history of statistical mechanics really began in 1845 when Waterston completed his paper "On the Physics of Media that are composed of free and perfectly elastic molecules in a State of Motion" and submitted it to the Royal Society of London. (He was at that time living in Bombay but later returned home to Scotland.) Unlike most scientists (before and after him) who worked on the kinetic theory of gases, Waterston wanted only a consistent theory based on a simple model whose properties could be worked out mathematically from a few plausible postulates, rather than a comprehensive flexible theory that could explain all the data. "Whether gases do consist of such minute elastic projectiles or not, it seems worthwhile to inquire into the physical attributes of media so constituted, and to see what analogy they bear to the elegant and symmetrical laws of aeriform bodies," he wrote (1928:214).

His first postulate was that heat is some kind of motion of the smallest parts of bodies. To justify this he did not cite the principle of energy conservation, still in its infancy in 1845, but the experiments on radiant heat which had suggested the wave theory of heat mentioned earlier.

Waterston's second postulate was that the molecules move so rapidly and in so many directions, with frequent "fortuitous encounters" (1928:214), that one can compute the properties of the system at any time by simply averaging over all possible molecular states. A more sophisticated version of this postulate later became known as the "ergodic hypothesis."

The third postulate, which seemed to be a consequence of the first and second, was that in a mixture of elastic particles of different masses in thermal equilibrium, each kind of particle will have the same average kinetic energy. This postulate is now called the "equipartition theorem," though Waterston did not explicitly extend it to the case where some of the particles are bound together by forces to form molecules.

Waterston applied his theory to gas properties with moderate success. That is, he explained some known facts, such as the ratio of specific heats and the decrease of temperature with height in the atmsophere, by arguments that were later found to be somewhat defective. In other cases his calculations were sound. But the Royal Society refused to publish his paper, on the recommendation of two referees who called it nonsense. Even worse, they refused to give it back to him, so he was prevented from publishing it elsewhere. As a result Waterston's work remained generally unknown until a decade after his death.

The axiomatic style that makes Waterston's paper attractive to a modern reader was probably responsible in part for its unfavorable reception by his contemporaries. He did not attempt to show that the billiard ball is a reasonable model for a molecule, and the amount of mathematics may have seemed excessive. Add to this the circumstance that Waterston had no reputation in the scientific community and few friends to defend his work, and one may perhaps be able to understand why his theory was ignored—though it still appears to me that he was treated rather harshly by the Royal Society.

Rudolf Clausius (1857) gave the kinetic theory what Waterston had failed to supply: a comprehensive justification in terms of the latest knowledge of the physical and chemical properties of gases, a realistic molecular model, and the endorsement of an acknowledged authority on heat. While Clausius was thus primarily responsible for reviving the kinetic theory and persuading other scientists to take it seriously, I suppose we must criticize him for adjusting the postulates of the theory in somewhat arbitrary ways in order to escape experimental refutation. Indeed his most famous contribution to kinetic theory, the "mean free path" concept, was devised for just this purpose. He had originally assumed that the gas molecules have effectively zero size and simply bounce back and forth between the solid sides of their container. But a Dutch meteorologist, C.H.D. Buys-Ballot, pointed out that such molecules, moving (as Clausius estimated) at speeds of several hundred meters per second, would diffuse quickly through a large room, contrary to experience. To extricate himself from this difficulty, Clausius (1858) had to attribute to his molecules a diameter large enough so they could not diffuse too rapidly, but small enough to leave unaffected the other deductions from the theory. He had no independent means of estimating the diameter, so we must conclude that this was a purely ad hoc hypothesis as of 1858. Yet the way in which the hypothesis was introduced— by assuming that the molecule travels a certain distance, on the average, before hitting another one (mean free path)—turned out to be useful in the next stage of development of the theory.

James Clerk Maxwell read Clausius' paper in English translation in 1859 and figured out a clever way to refute the kinetic theory by deducing from it a falsifiable consequence: the viscosity of a gas of tiny billiard balls must be independent of density and must increase with temperature. Everyone knows that a fluid flows more slowly as it gets thicker and colder, but since common experience does not necessarily extend to gases, Maxwell asked G. G. Stokes, the expert on hydrodynamics, if any experiments had been done on this point. He was presumably assured by Stokes that experiments on the damping of pendulum-swings in air proved that the viscosity of a gas goes to zero as its density goes to zero; in any case he wrote that "the only experiment I have met with on the subject does not seem to confirm" the prediction that viscosity is independent of density (1860a; Brush 1965:166).

To clinch his argument Maxwell also worked out the ratio of specific heats for a gas of diatomic molecules and found that it should be $c_p/c_v = 1.333$, whereas experimental data clearly indicated $c_p/c_v = 1.40$ or 1.41. Whereas Clausius had evaded this problem by allowing the proportion of internal energy to be determined empirically by the specific heats, Maxwell insisted that in a consistent theory each mechanical degree of freedom must have the same average energy.

Armed with these calculations Maxwell went off to the British Association meeting at Oxford (1860), where he announced (referring specifically to the specific heats discrepancy): "This result of the dynamical theory, being at variance with experiment, overturns the whole hypothesis, however satisfactory the other results may be" (1860b). (For some reason this conclusion, published in the report of the British Association meeting, was not included in Maxwell's collected papers.)

7.3 VICTORY

Fortunately for the progress of science Maxwell did not really believe in the hypothetico-deductive method; the kinetic theory had become such an intriguing mathematical game that he couldn't bear to abandon it just because its predictions disagreed with experiment. Moreover, in spite of Stokes, he suspected that the density dependence of the viscosity of air was not reliably established. So he decided to measure it himself.

The year 1865 saw the triumph of the kinetic theory. Maxwell found that the viscosity of air is indeed constant over a wide range of densities. This result was fully confirmed by other physicists. It turned out that what Stokes had claimed to be an experimental fact was not a fact at all but a theory-laden observation: in reducing the data it had been *assumed* that the viscosity of air goes to zero as the density goes to zero. It seems obvious that if there is no air it cannot exert any viscous resisting force. Common sense, or a principle of continuity in nature, would suggest that if there is a very small amount of air the viscosity must be very small. But the actual pressure range in which this behavior occurs happens to be below that attainable in the first half of the 19th century, so the presumed continuous decrease of viscosity to zero was not directly observed. This episode might provide a good example for analyzing the relation between theory and experiment—only with the help of the theory was it possible to arrive at the correct interpretation of an experiment which was supposed to test the theory.

The other major event of 1865 was Josef Loschmidt's determination of the size of a molecule, using the mean-free-path formula from the kinetic theory of gases together with the ratio of gas-to-liquid densities. Other estimates, giving somewhat similar results (diameter 10^{-8} to 10^{-7} cm), were published soon afterward by G. J. Stoney, L. Lorenz, and William Thomson. Suddenly the atom was real—for now atoms could be measured, weighed, and counted. But not everyone was convinced, and it was another 40 years before the "real existence of the atom" could be completely established (with the help of statistical mechanics). Even then scientists could not claim that they could *see* an atom; by the time that was possible (Mueller 1956), the question of the existence of atoms was a dead issue.

7.4 ENTROPY AND PROBABILITY

During the 1860s and 1870s a few physicists were also interested in the problem I mentioned at the beginning, the "reduction of thermodynamics." Since the first law of thermodynamics was recognized to be only a special case of the general principle of conservation of energy, it seemed reasonable to look for another general principle corresponding to the second law. The most popular candidate was Hamilton's version of the principle of least action. But these attempts to reduce the second law to mechanics dealt only with the problem of identifying a quantity that could serve as an integrating factor for the heat transferred in a reversible change. In this way one could obtain a mechanical analogue of *entropy* (defined in thermodynamics as heat transferred, divided by absolute temperature—Clausius 1865), but the irreversible increase of entropy with time implied by the generalized second law was still unexplained.

Ludwig Boltzmann and James Clerk Maxwell realized around 1870 that the reduction of the second law must somehow involve statistical considerations. Boltzmann took an important step toward the goal with his "H-theorem" for low-density gases (1872). According to this theorem, a quantity H depending on the molecular velocity distribution always decreases with time unless the gas is in thermal equilibrium; H is then proportional to $(-)$ the thermodynamic entropy. Later Boltzmann proposed a more general but very simple interpretation: the entropy of a system is proportional to its probability. In modern notation,

$$S = k \log W,$$

here W is the number of molecular states, or "microstates," corresponding to a state of the macroscopic system—a "macrostate," defined by thermodynamic variables such as pressure and temperature. If one samples at random from the microstates, the probability of finding a given macrostate is proportional to its W. Conversely, if one is told that a system has a certain macrostate, one is left in the dark about its molecular nature to the extent that it might be in any one of W microstates. If W is large, one therefore says the system is "disordered." According to Boltzmann, the statement "entropy tends to increase" can be interpreted as "a system tends to become more disordered."

But there is a fundamental objection to any such reduction of *irreversibility* to mechanics. The objection was first pointed out explicitly by William Thomson (later Lord Kelvin), the same person who gave the first general statement of the principle of irreversibility in 1852. In 1874 Thomson called attention to the apparent contradiction between the irreversibility of thermodynamics (and of macroscopic processes in general) and the reversibility of abstract Newtonian dynamics. This was the "reversibility paradox," later made famous by the debate between Loschmidt and Boltzmann, just as the "recurrence paradox" was made famous by the debate between Zermelo and Boltzmann in the 1890s.

The controversy about the origin of irreversibility and the direction of time, which I have discussed in Chapter Two, still offers interesting material for philosophical analysis. I might mention in particular Boltzmann's semiserious suggestion that the direction of time as we experience it *depends* on the direction of entropy change in the environment, and may itself be reversed during some epochs of the history of the universe.

In the present context it seems appropriate to point out that the attempt to reduce thermodynamics to mechanics led to the introduction of statistical ideas which changed the absolute character of thermodynamics. The "reduced theory" thereby was modified but at the same time strengthened, since it could now be applied with more confidence to a wider range of phenomena, in particular those involving fluctuation phenomena such as Brownian movement. It was just this kind of application of a generalized thermodynamics that led to the final establishment of the existence of atoms at the beginning of the 20th century.[5]

---

[5] Einstein (1905b) showed that there is an intermediate situation—small particles suspended in a fluid—where the fluctuations and reversals predicted by molecular theory are observable. The experimental confirmation of Einstein's theory by Jean Perrin was regarded by many skeptical scientists (for example Wilhelm Ostwald) as convincing evidence for the real existence of atoms (Brush 1976a:698–99; Nye 1972). The claim that matter consists of discrete atoms had previously been regarded as a metaphysical proposition by many scientists; surely the historical events that led to this proposition being accepted as a scientific truth deserve the attention of philosophers of science!

But the "reducing theory" was also affected: the introduction of statistical thermodynamics and in particular the recognition that irreversibility involves some kind of randomness at the molecular level[6] helped to undermine the rigid determinism of Newtonian mechanics. The result was that physicists in the first part of the 20th century were prepared to accept a new mechanics based on indeterminism.[7] In this way one can see a close connection between the three philosophical aspects of the history of statistical mechanics—reduction, statistical law, and irreversibility.

7.5 PHASE TRANSITIONS

Until now the most important developments in statistical mechanics, aside from those associated with quantum theory, have been in the theory of phase transitions. The origin of this theory, or at least the idea that such a theory is possible, may be found in the 1873 dissertation of J. D. van der Waals. Van der Waals developed an "equation of state" which took account of intermolecular forces, thereby improving on the "ideal gas law." He found that this equation of state provided a remarkably simple explanation for phenomena near the gas-liquid critical point.

If a theory such as that of van der Waals can successfully account for the transition from gaseous to liquid states of matter on a molecular basis, it should have some relevance to philosophical discussions of "levels of reality." One of the arguments of antireductionists is that new concepts must be invoked in going from a lower to a higher level of complexity; the laws for the higher level, being formulated in terms of those concepts, are therefore nonreducible. Now it would seem that one of the simplest examples of this is the set of concepts "gas," "liquid," and "solid," which apparently have no meaning when applied to a single atom[8] but emerge

---

[6] "... *the fortuitous concourse of atoms is the sole foundation in Philosophy on which can be founded the doctrine that it is impossible to derive mechanical effect from heat otherwise than by taking heat from a body at a higher temperature, converting at most a definite proportion of it into mechanical effect, and giving out the whole residue to matter at a lower temperature*" (Kelvin 1894:464), italics in original).

[7] Just so the reader will not assume that these questions have been permanently settled, I wish to call attention to the recent papers of Goldstein et al. (1981) and of Misra & Prigogine (1982), in which randomness is attributed to instabilities in *classical* dynamical systems.

[8] One assumes implicitly that the individual atom or molecule is the same entity whether it is in a gas, liquid, or solid; but that is not really obvious. In the 1880s some scientists proposed that when a gas changes to a liquid the molecules combine chemically to form larger complexes, so that "liquid molecules" are different from "gas molecules." This possibility was later rejected on the basis of further investigation of the properties of gases and liquids near the critical point. A detailed study of this episode has been undertaken by Levelt Sengers (1975).

gradually as one puts together enough atoms to form a macroscopic system. But if statistical mechanics can show exactly how this happens, and derive the qualitative differences between states of matter from quantitative properties of atomic systems, we would have to conclude that the distinction between macroscopic and microscopic levels is not absolute but only a matter of convenience. So I am suggesting that the success of the statistical-mechanical theory of phase transitions might be taken as an important example of reduction, much more significant than the derivation of the ideal gas law and more clear-cut than the explanation of irreversibility in terms of randomness.

Although van der Waals' theory suggested the possibility of explaining the gas-liquid transition in terms of intermolecular forces, it was not really an application of statistical mechanics. The first and simplest example of a phase transition derivable from statistical mechanics was discovered by Einstein in 1924. This is the famous condensation of a Bose-Einstein gas at very low temperatures, to a peculiar substance in which a finite fraction of the particles are in the lowest quantum state. The physical significance of this mathematical result was not appreciated for another decade, but it was eventually established that the Bose-Einstein condensation is a first approximation to the "λ-transition" between the normal and "superfluid" states of liquid helium at 2.2°K, and that the anomalous flow properties below the transition can be understood in this way (§4.9).

The λ-transition is the gateway to a spectacular realm of quantum phenomena—the unusual behavior of superfluid helium is easily visible on the macroscopic level. There is also a macroscopic quantum theory of superfluidity, developed by L. D. Landau (1941), which succeeded in explaining and predicting some (but not all) of the properties of helium below the λ-temperature, without using Bose-Einstein statistical mechanics. Thus one can even discuss the reduction of a macroscopic quantum theory to quantum statistical mechanics. Yet those philosophers of science who have written extensively on the meaning of quantum mechanics seem to have little interest in what seems to me to be one of its most striking manifestations.

Shortly after Einstein announced his discovery of the condensation of a Bose-Einstein gas, Ernst Ising published a short paper (1925) on an idealized model that has become the basis for a large amount of contemporary research on the theory of phase transitions (§6.4, 6.7). The so-called Ising model was actually proposed in 1920 by Ising's mentor, Wilhelm Lenz. It consists of a system of atoms arranged on a regular crystalline lattice. Each atom is fixed in position and has only one property: a magnetic moment of "spin" which can have one of two possible values, $+1$ or -1. There is a force between nearest neighbors only, tending to make the atomic spins align in the same direction; in addition there may be an externally imposed magnetic field. The energy of a microstate of the system depends on the number of nearest-neighbor pairs of atoms having the same or opposite spins, and on the total number of spins having each value. The basic question about the Ising model is: Will the system display spontaneous magnetization; that is, will more than half of the atomic spins have the same value in the absence of an external field, when the temperature goes below some critical temperature T_c (the "Curie temperature")?

Ising himself calculated the properties of his model only in the one-dimensional case, and found that there is no phase transition. From this result he jumped to the erroneous conclusion that there would be no transition in a two- or three-dimensional model either.

In the 1930s there was considerable interest among solid-state physicists and physical chemists in models similar to Ising's, applied to alloys and liquid mixtures. By a simple change of algebraic sign in the energy formula one can make the force such that nearest-neighbor spins tend to be unlike rather than like, and one can translate "spin $+1$ or -1" into "lattice site occupied by atom of type A or type B." Thus at low temperatures micro-states with an alternating ABABAB . . . arrangement would be favored, the so-called superlattice ordering. On the other hand if the force is such as to favor like pairs rather than unlike pairs (as in the ferromagnetic case), the system may split into two phases at low temperatures, one phase containing more A atoms and the other more B atoms. Thus the Ising model might be able to explain order-disorder transitions in alloys and the partial insolubility of liquids in each other as well as ferromagnetism.

The concept of the "Ising model" as a mathematical object existing independently of any particular physical application seems to have been developed by the Cambridge University group led by R. H. Fowler in the 1930s. Rudolph Peierls (1936) was apparently the first to point out explicitly the equivalence of the Ising theory of ferromagnetism to the theories of other physical systems mentioned above.

The major breakthrough in the theory of phase transitions was Lars Onsager's exact solution of the zero-field two-dimensional Ising model, announced in 1942 and published in 1944. I consider this one of the most important events in theoretical physics of the 20th century. For the first time it was possible to determine precisely the physical properties of a many-particle system in which interparticle forces play a significant role. Previously the only such results available were based on approximations such as series expansions valid only in temperature-density regions far away from suspected phase transitions. Although such approximations did predict that there would be a transition to an ordered state at low temperatures, they failed to describe the nature of the phase transition itself. For example, the so-called critical exponents, which characterize the rate at which certain properties go to zero or infinity at the transition point, can be computed accurately only from mathematically rigorous theories like that of Onsager (or approximations whose error bounds are precisely established).

Although it has not yet been possible to solve the three-dimensional Ising model by finding a closed analytical formula for the partition function, as Onsager did for the two-dimensional case, various methods have been refined to the point where it may be said that all the properties of the model can be computed to any desired degree of accuracy. The Ising model is thus the first nontrivial case in which a phase transition has been completely explained in terms of interatomic forces. It is a much more interesting example of the reduction of thermodynamics to statistical mechanics than the textbook derivation of the ideal gas law from billiard-ball kinetic theory. In both cases one is talking about the computation of thermodynamic properties as predicted for a particular system from an atomic model rather than the derivation of *laws* of thermodynamics from *laws* of atomic physics. But a phase transition is a qualitative change in macroscopic properties that suddenly occurs at a definite temperature and pressure—spontaneous magnetization, or phase separation in a mixture—and therefore the ability to explain it seems a much more spectacular achievement.

During the last 30 years it has been shown that the Ising model applies to an even wider class of phase transitions. Around 1950 it was pointed out that the model can represent ordinary gases, liquids, and solids as mixtures of atoms and "holes"—the spin variable is defined as $+1$ if an atom occupies the lattice site and as -1 if it is empty. This is known as the "lattice gas" model. With attractive forces between neighboring atoms the system may condense into a two-phase arrangement with the atoms clustered together in one region and the holes in another. With repulsive forces the system may crystallize into an ordered state with atoms and holes alternating on adjacent sites. Thus the gas-liquid transition would be analogous to the onset of spontaneous magnetization at the Curie temperature, while the liquid-solid transition would be analogous to super-lattice formation in alloys.

The lattice gas model might seem to be very unrealistic, especially for gases. But M. E. Fisher (1965) pointed out that it gives a fairly good estimate of the critical exponents for the gas-liquid transition—better than the van der Waals equation—and the further development of this model has been stimulated by the intense interest in critical-point experiments during the past 15 years.

Although it has long been thought that the gas-liquid transition is in some way related to attractive forces, the idea that the liquid-solid transition is associated with repulsive forces is relatively recent. The possibility of crystallization of the billiard-ball gas (with *no* attractive forces) was suggested by Kirkwood and Monroe (1941) and by I. Z. Fisher (1955); it was definitely established by B. J. Alder and T. E. Wainwright (1957). Systems with "softer" repulsive forces such as the inverse square also crystallize at low temperatures and high densities.[9]

The fact that the Ising model gives a reasonably good description of several different kinds of phase transitions shows that the precise nature of the interatomic force law is less important than the long-range correlations produced by these forces. Theorists now speak of a "universal" theory of phase transitions. Near the transition there is a delicate balance between the ordering effects of the forces and the disordering effects of thermal agitation. As the temperature drops there is a sudden "landslide" into an ordered state; the correlations that previously influenced only nearest neighbors now extend over macroscopic distances. (This point is completely missed in the only philosophical discussion I have seen on this subject, by Girill 1976.)

The success of the Ising model vindicates the approach advocated by Waterston and Maxwell. An atomic model should not incorporate all the physical factors that are believed to influence the system; instead it must be kept simple enough to permit the accurate mathematical calculation of its properties.

(Additional references for this section: Domb 1971, Stanley 1971, Fisher 1972, Wightman 1979.)

7.6 WHAT IS REDUCTION?

Throughout this chapter I have used the term *reduction* without giving it an explicit definition. This omission is probably quite annoying to a philosophical audience, but I will now try to make amends in my concluding remarks. In fact none of the definitions known to me when I started to write the chapter now seem satisfactory. The difficulty is that one cannot simply say "A is reduced to B" because in all the nontrivial historical examples with which I am familiar, A and B both undergo significant changes as a result of the reduction. The attempt to reduce the second law of thermodynamics changed it from an absolute to a statistical

---

[9] Kirzhnits (1960), Abrikosov (1960), Salpeter (1961), Brush et al. (1966). The situation is not completely symmetrical, since a system with only attractive forces is not thermodynamically stable (Lieb 1979).

law; and a significant part of the "reducing theory," statistical mechanics, came into existence because of this attempt. Similarly the phenomeno-logical theory of phase transitions was substantially modified as the statistical-mechanical theory developed, and the current theory contains surprising new features on both macroscopic and microscopic levels.

In my view "reduction" refers to what scientists do whenever they try to replace an arbitrary or complicated empirical description by a simpler, more plausible theory, and in this sense the logical positivists were right in seeing reduction as a central feature of progress in science. But their use of "reduction" to describe other kinds of theory change (for example the replacement of the phlogiston theory by the oxygen theory) is unsatis-factory to me; it should be restricted to vertical motion along a hierarchy of sciences or levels, so that the reducing theory is in some way more fundamental than the reduced one.[10]

At the same time the antagonism toward reduction voiced by many scientists and nonscientists is also justified insofar as they have been led to accept the literal connotation of the word: an entity *loses* something if it is "reduced." But the history of science shows that the "reduced" theory usually gains something else to replace the arbitrary or incomprehensible aspects it has lost. Often this gain will include a more precise knowledge of the limits of validity of the original theory (so that it can be used with more confidence within those limits) and indications of what has to be done to extend it beyond those limits.[11]

---

[10] But one must also allow the usage "theory A reduces to theory B when a parameter in A takes on a singular value"—thus quantum statistical mechanics reduces to classical statistical mechanics when Planck's constant goes to zero. (Yaglom has given a perspicuous way of demonstrating this reduction by means of Feynman's path integral formulation of quantum mechanics—see Brush 1961.)

[11] Charles Misner (colloquium at the University of Maryland, May 1976) has argued that our most reliable knowledge of the world comes from a theory (such as Newtonian mechanics) that has been refuted, that is, shown to be wrong outside certain limits, since the theory that replaced it (relativity) usually confirms its accuracy *within* those limits. The more recent theory, whose limits of validity have not yet been determined, is *less* reliable. In the same way one could say that thermodynamics, *after* its reduction by statistical mechanics (insofar as that has occurred), is more reliable than statistical mechanics. Failure to recognize this point has led some philosophers of science to claim that "we cannot retain both classical and statistical thermodynamics" because, in the case of Brownian movement, the former has been falsified and "replaced" by the latter.

VIII. OUTSTANDING PROBLEMS IN STATISTICAL PHYSICS

When this book was finished and accepted for publication, the series editor suggested that readers would be interested to know about some of the outstanding problems which remain to be solved or are likely to attract special attention in the next few years. Intrigued by this suggestion, I wrote to a number of colleagues and people who had recently published research papers in statistical physics, asking them to tell me about "outstanding problems remaining to be solved, and topics on which research is likely to focus in the next couple of decades." This should not be considered a random or representative sample, but simply indicates the views of the 39 people who were kind enough to respond on rather short notice in October 1981.[1] All but one gave permission to quote from their letters; I have listed their names and affiliations at the end of this chapter.

In the area of equilibrium statistical mechanics, the most frequently mentioned problem was the need for an exact solution of some nontrivial three-dimensional model of cooperative phenomena such as the Lenz-Ising model. No one suggested a specific strategy for finding such a solution, but two people pointed out the need for a better justification of the currently popular technique for estimating the critical behavior of the model:

> Does the renomalization group theory of critical phenomena based as it is on the weak coupling expansion in the continuum limit for a boson field theory correctly describe the critical behavior of the three-dimensional, spin-$\frac{1}{2}$, ferromagnetic Ising model? (Baker)

> Approximate calculations based on this [RNG] theory are successful in explaining the main features of the critical phenomena, as the universality of the critical exponents and the scaling laws. However, up to date, there is no satisfactory rigorous approach to this theory. (Cammarota)

---

[1] Only one person declined to reply for a specific reason: "I want to leave the direction of research activities up to the individual."

Cammarota lists several conjectured properties of the renormalization transformation of the Hamiltonian, convergence of the block-spin variable to a Gaussian distribution, etc., which could justify the use of RNG theory, but which have so far been established only for special cases.

Of the various phase transitions that occur in nature, the solid-liquid transition was thought to be the most in need of a satisfactory explanation. Here is one description of what the problem involves:

No one has succeeded in proving rigorously the existence of the solid-liquid transition or the appearance of crystalline long range order in any continuum system. Moreover, it is not precisely known what property of the interaction potential leads to the solid-liquid transition. At present, we can roughly say that the short range repulsive part of the interaction is essential in the melting transition. But the precise condition is not known yet.

Another and related fundamental problem is to clarify the nature of the solid-liquid transition, especially the nature of its broken symmetry. . . . Recently, the nature of second-order transitions and associated critical phenomena have been well understood by the success of the scaling theory and the renormalization group method. However, there seems to be no such understanding in the case of the first-order transition. . . . We cannot tell why the solid-liquid transition has . . . properties which are seemingly quite different from those of, say, gas-liquid transition. Another important problem is to construct a reasonable analytic theory of melting which can predict the solid-liquid transition quantitatively, or at least semi-quantitatively. Such a theory should be formulated both for classical and quantum systems. (Kawamura)

Several respondents also mentioned problems relating to the "glassy" state of matter or to a lattice model known as the "spin glass."

Almost all materials can be prepared as glasses by suitable means. At the present time, although it is clear that these are not first order transitions, it is not known whether the transitions are a purely dynamic phenomenon or whether the transition has a static limit. The problem is of immense technological importance, because many synthetic materials solidify by vitrification. For these materials it is important to know, not only the apparent behavior of variables such as the density and entropy, but the behavior of fluxes of transported mass, momentum, and energy. (Frisch)

From the theoretical viewpoint the problem may be posed:

> Show that the following system does or does not undergo a phase transition: An Ising model in which the nearest-neighbor exchange interactions have a random distribution such that any particular interaction can be positive or negative with probability one half. ... So far as I know, this problem is unsolved for dimension two or more, despite a lot of recent work on "spin glass," of which this is an example. (Griffiths)

The problem of spin-glass behavior exhibited by some dilute magnetic alloys like CuMn is attracting the attention of a large number of workers due to some novel statistical mechanical questions. These alloys show various physical characteristics, e.g., susceptibility, which suggest that with the lowering of temperature the magnetic moments in the system freeze in random directions. A burning question is whether the freezing process sets in as a result of a phase transition, as in an ordinary ferromagnet, or the freezing process is gradual. The question is enormously complicated due to the randomness of magnetic interactions which do not permit the existence of any kind of long range order. This seems to be a prototype problem in which randomness and critical thermodynamic fluctuations which are normally responsible for phase transitions interfere in an essential way to cause a very complex behavior. (Kumar)

A more ambitious goal for statistical mechanics would be to develop a "unified theory of substances containing an explanation of the entire phase diagram, the gas-liquid phase transitions and a representation of the shifts due to quantum effects" (Nordholm). Presumably one would start with the simplest real substance, hydrogen. A quantum treatment would require among other things "a definite (ab initio) treatment of the problem of the finite electronic partition function for bound electronic states of a monatomic atom (ion) at finite densities" (Rouse). It would also involve the metallic transition of hydrogen, a problem currently of interest in high-pressure physics (Alder).

As for phase transitions due *essentially* to quantum effects, the explanation of the λ-transition of helium in terms of a Bose condensation is still regarded as unfinished business:

> I am not convinced that our current ideas concerning the mechanisms of the lambda transition (superfluidity) of liquid ^{4}He are completely correct, even qualitatively. A truly convincing theory of the lambda transition in a system of real ^{4}He atoms (not "Bose billiard balls") is, in my opinion, still lacking. (Girardeau)

It has also been suggested that the relation between statistical mechanics and quantum mechanics could be more than just inserting the latter as a known set of postulates into the former in order to find the "quantum effect."

There are already some hints of this in "stochastic electrodynamic approaches to quantum mechanics.... Although these approaches are still crude and speculative, it seems likely that a complicated and chaotic microstructure of the "vacuum" is an essential feature of the real world. If so, current approaches to quantum mechanics and high energy theory may well be only semiphenomenological approximations to some future microscopic theory embodying statistical ideas in a more fundamental way. (Girardeau)

Another example along similar lines is the analogy between certain two-dimensional statistical-mechanical models and quantum field theory models. If one dimension of a lattice model is called "space" and the other direction "time," then one can associate the "time-evolution operator" with a transfer matrix defined for neighboring rows.

How can statistical physics help quantum field theory? In statistical physics there is a lot of experience in solving the thermodynamics of lattice systems. A powerful tool is the above mentioned transfer matrix method, which reduces the evaluation of the two-dimensional partition function to the determination of the eigenvalue of a one-dimensional quantum lattice-Hamiltonian. Suppose you have studied such a lattice Hamiltonian and can express it in suitable Fermion operators. Then you can investigate its continuum limit. Under certain conditions ... a relativistic quantum field theory may result. The properties of the lattice theory can be taken over for the field theory so constructed. The continuum limit involves mathematical problems not completely solved up to now. To be specific, it is the weak limit of an operator sequence $H(s)$ on the infinite direct product and it depends on the countable subspace you choose.... There is another approach to construct a field theory out of the thermodynamics of a lattice system.

Of course, the described difficulties do not exist for the continuum limit of expectation values of the lattice theory, such as energy values and correlation functions. Now a very famous theorem (Osterwalder, Schrader 1975) allows one to reconstruct a field theory from the continuum correlation functions (Schwinger functions) of a statistical model.... To calculate the correlation function for various statistical models, however, is an unsolved problem as well.... (Wolf)

Turning to nonequilibrium statistical physics, I find two outstanding problems mentioned about equally often. One, turbulence, is generally considered of great importance, but it is not clear whether statistical mechanics will be able to contribute to its solution. The other, the density dependence of transport coefficients, seems less fundamental, but more likely to depend on a statistical-atomistic approach.

Probably the biggest unsolved problem of classical physics is turbulence. How much will be contributed by statistical physics is unclear to me. Presumably it is all in the Navier-Stokes equations, and we are just too dumb to be able to see the behavior of all the possible solutions in a mathematical sense. (Mason)

A problem which appears to be related to turbulence is the transition of a nonergodic system to chaotic behavior, as it is subjected to a continuously increasing perturbation.

A significant feature of the behavior of the perturbed system is that as the strength of the perturbation, and thus the total energy of the system, is increased so does the set of initial conditions that gives chaotic behavior become a larger fraction of the set of all possible initial conditions. This result, which has been established by way of an existence theorem has been supplemented by a variety of numerical computations. Not only have these computations verified the existence of ordered behavior at low energies, but they have also indicated the existence of a critical energy, the value of which depends on the nature of the particular system under investigation, at which ordered behavior at lower energies changes rather sharply to chaotic behavior at higher energies.

It is the nature of this transition that is of extreme interest. The numerical computations are performed necessarily on a system having a finite number N of particles. As $N \to \infty$ does the transition become completely sharp, as happens for a phase transition? And, if it does do so, is the transition of the same nature as a phase transition? What then is the significance of the parameters which parallel those of a phase transition?

But, if indeed there is a transition point separating ordered and chaotic behavior in an infinitely large system, how then is the success of the methods of equilibrium statistical mechanics to be explained? (Farquhar)

The other major nonequilibrium problem arises from the discovery in the mid-1960s that the coefficients in the density expansion of transport coefficients show a logarithmic divergence. Thus there is a need for a different approach to calculate the transport coefficients of dense gases and liquids.

Kubo's theory of correlation functions gives a very beautiful and elegant theory. But we can't compute correlation functions in dense media very well, if at all. The theory of Prigogine and coworkers is another example of a development which is not particularly useful in the liquid range. The theory of Kirkwood (late 40's–early 50's) extended by Rice, et al. (mid 50's–early 60's), based on Brownian motion ideas, is perhaps the best theory available, from the computational point of view, but agreement with experiment is only fair, as I recall. (Mazo)

The correlation function representation is elegant, but has yet to yield formulae in which the role of the intermolecular potential is transparently displayed. The usual schemes for calculation based upon the correlation function formalism involve either parametric representations of some memory function, or other function, or involve assumptions which are not physical, such as using a wave vector dependent molecular diameter in order to fit experimental data. I do not regard any of this as satisfactory. On the other hand, the kinetic equation approach remains essentially undeveloped, though the recent work of Dahler, which makes contact with older work of mine, looks promising. I believe it to be desirable to develop a kinetic equation approach to the description of both steady state transport phenomena and high frequency transport phenomena. (Rice)

While the calculation of transport coefficients of dense fluids has reached a fairly high level of sophistication and the outstanding problems, though quite difficult, appear to be a rather technical nature, progress in dealing with other nonequilibrium phenomena seems to be held up at a more fundamental level.

At present a satisfactory theory of macroscopic systems not at equilibrium does not exist. One is able to:
1) calculate expectation values of macroscopic observables,
2) define correlation functions at two different time points.
One is not able to:
a) construct a joint probability distribution for measurements at two different time points.

b) deduce correlation functions from such a probability distribution,
c) obtain from quantum mechanics the concept of trajectory, in the space of the macroscopic state-parameters.

As a consequence: i) the theory of measurement in quantum mechanics cannot be founded on a satisfactory theory of macrosystems, ii) the theory is lacking in the case of experiments in which macroscopic variables are measured with extremely high precision; such experiments will have an important role in testing general relativity. (Loinger and Lanz)

According to Griffiths, a basic difficulty is the lack of anything comparable to Gibbs' "canonical" distribution for nonequilibrium situations. Even if we were able to solve the mechanical equations of motions, this lack would prevent us from finding the nonequilibrium flow of a system, except by going down to the level of individual atoms, in which case the classical problem would become completely deterministic and we would no longer be doing *statistical* mechanics.

Our failure to bridge the gap satisfactorily between a deterministic theory of individual atoms and a statistical theory of many-particle systems shows up in the fact that the classical ergodic hypothesis has not been proved except for rather special systems. It would be desirable to show "that some slightly complicated system actually is ergodic" (Griffiths). According to ter Haar the problem is "why systems with non-integrable Hamiltonians behave like integrable systems." In particular:

What is the "closest" integrable Hamiltonian to a given non-integrable Hamiltonian, and how "close" must the two Hamiltonians be, and in what sense, before the lack of ergodicity—which after all is shown by some actually existing systems—will show up. (ter Haar)

Dorfman is optimistic about the chances of solving such problems with the help of new mathematical techniques from dynamical systems theory, which are finally filtering into statistical-mechanics courses.

Most of the problems mentioned in both equilibrium and nonequilibrium statistical physics involve very simple models of molecules; systems of hard spheres, Lennard-Jones particles, and three-dimensional spin lattices are not yet completely understood. But several researchers suggest that the most important problems lie in the area of molecular physics, where the three-dimensional structure of the molecule plays a significant role. It will be necessary to calculate the effect of "a realistic number of degrees of freedom on the relaxation time in polyatomic gases" (Liley). The role of molecular orientation in dielectric effects needs more study:

... the amazing anomalies of nonlinear dielectric phenomena and electro- and magneto-optical effects appearing in the broad critical region strengthen our conviction that, owing to a flattening of the Lennard-Jones potential curve, a soft mode and amplitude enhancement appear, resulting in the 'knocking-out' of molecules from imperfect lattices (of a liquid) and the initiation of random walks. Thus a drastic lowering of the diffusion in a given direction occurs, creating an anomaly in the region of the critical point. Similar anomalies occur in the behavior of other transport phenomena. (Piekara)

The current state may be summarized as follows:

I do not believe that any of the current attempts to understand the relationship between the coupling of internal motions and translation in a liquid of polyatomic molecules, e.g. vibrational relaxation, is completely satisfactory. Each of the treatments is carefully tailored to a particular phenomenon and even when successful is not suggestive of how a general formalism can be written down. I think it is desirable to develop a general formalism for the kinetic theory of dense fluids of polyatomic molecules, including the quantized internal motions of the molecules. I am aware that some very formal kinetic theories of this type exist, but they appear to me to be impenetrable and unuseable. What is needed is something which uses the kinds of concepts and terms used by the experimentalists in interpreting observations. (Rice)

Finally, several respondents mentioned the need for a satisfactory theory of nonuniform continuous systems such as those including a liquid-vapor interface where boundary effects are important.

The above listing of outstanding problems is limited to those mentioned by at least three people; I must apologize to several respondents for omitting their description of a problem simply on the grounds that no more than one other person mentioned that problem.

Another kind of sample of the currently outstanding problems is provided by the lists of topics dicussed at international meetings. Two recent examples are the XIVth International Conference on Thermodynamics and Statistical Mechanics held at the University of Alberta in August 1980 (Table 8.1) and the "Conferencias sobre Fisica Estadistica" held at Cocoyoc, Mexico, in January 1981 (Table 8.2). I am indebted to the editor of the proceedings of the latter conference, Jorge Barojas, for sending me a copy.

TABLE 8.1. Contents of the Proceedings of the XIVth International
Conference on Thermodynamics and Statistical Mechanics,
University of Alberta, Edmonton, Canada, August 17–22, 1980.
(*Physics* 106A/1–2, 1981. Reprinted as *Statphys* 14, ed. J. Stephenson.
Amsterdam: North-Holland Pub. Co., 1981.)

"Not by discoveries alone: The Centennial of Paul Ehrenfest." *M. J. Klein*

"Corner transfer matrices." *R. Baxter*

"Simple Ising Models still thrive! A review of some recent progress."
M. E . Fisher

"Hyperscaling and universality in 3 dimensions." *B. Nickel*

"Mathematical properties of renormalization-group transformations."
R. B. Griffiths

"The hard sphere gas in the Boltzmann-Grad limit." *O. E. Lanford III*

"Advances and challenges in the kinetic theory of gases." *J. R. Dorfman*

"Melting of a two-dimensional crystal of electrons." *C. C. Grimes*

"Two-dimensional melting." *D. R. Nelson*

"The Potts and Cubic Models in Two Dimensions: A renormalization-
group description." *E. K. Reidel*

"Connections among different phase transition problems in two dimensions." *L. P. Kadanoff* and *M.P.M. Don Nijs*

"Recent results on instabilities and turbulence in Couette flow." *M.
Gorman* and *H. L. Swinney*

"Rhythms and turbulence in populations of chemical oscillators." *Y.
Kuramoto*

"Physisorption kinetics." *H. J. Kreuzer*

"An introduction and review of lattice gauge theories." *R. B. Pearson*

"Disordered one-dimensional conductors at finite temperature." *S.
Kirkpatrick*

"Theory and experiment at the sol-gel phase transition." *D. Stauffer*

"Simple Coulomb liquids." *M. Baus*

"Thermodynamics of relativistic systems" *W. Israel*

"A photographic record of the solidification and melting of ^{4}He at
pressures between 4 and 7 kbar." *J. P. Franck*

"Phase diagram of the two-dimensional Lennard-Jones system; evidence
for first-order transitions." *J. A. Barker, D. Henderson,* and *F. F.
Abraham*

"Molecular orientation and interfacial properties of liquid water." *C. A.
Croxton*

"Interpretation of the unusual behavior of H_2O and D_2O at low temperature: Are concepts of percolation relevant to the 'puzzle of liquid
water'?" *H. E. Stanley, J. Teixeira, A. Geiger,* and *R. L. Blumberg*

"Bound-state contributions to the triple-collision operator." *J. A.
McLennan*

"Equilibrium statistical mechanics of a nearly classical one-component
plasma in a magnetic field, in three or two dimensions." *A. Alastuey* and
B. Jancovici

"The Critical isotherm of the modified F model." *X. Sun* and *F. Y. Wu*

"Exact Differential real-space renormalization: Ising, Gaussian and Ashkin-Teller models."*H. J. Hilhorst* and *J. M. van Leeuwen*

"Monte Carlo realization of Kadanoff block transformation in the 2d plane-rotor model." *H. Betsuyaki*

"Numerical studies on the Anderson localization problem." *J. Stein* and *U. Krey*

"Order parameters and non-linear susceptibilities in spin glasses." *M. Suzuki* and *S. Miyashita*

TABLE 8.2.   Titles of Talks at the *X Reunion de Fisica Estadistica*, Cocoyoc, Mexico, January 1981.
(Published in *Kinam: Revista de Fisica*, vol. 3, ser. A, pp. 1–174, ed. Jorge Barojas, 1981.)

"Recent developments in non equilibrium statistical mechanics."[2] *R. Zwanzig*

"On the contraction problem in stochastic processes." *H. Hasegawa*

"Statistical mechanics of a two-dimensional electron fluid." *A. Isihara* and *S. C. Godoy*

"Light scattering from fluids not in thermal equilibrium." *E.G.D. Cohen*

"A fruitful application of generalized hydrodynamics: depolarized light scattering." *D. Kivelson, M. P. Allen,* and *P. J. Chapell*

"Hopping models of ultrasonic absorption in dilute polymer solutions II." *M. López de Haro*

"The electric double layer." *D. Henderson*

"Theory of the dielectric constant of ionic solutions." *H. L. Friedman*

"Molecular dynamics simulation of vibrating hard-dumbells." *G. A. Chapela* and *S. Martinez*

"Molecular orientation effects near surfaces." *K. E. Gubbins*

"Three-phase equilibrium and the tricritical point." *B. Widom*

"Structural aspects of the melting transitions." *F. H. Stillinger* and *T. A. Weber*

"Dissipation in driven oscillatory chemical systems." *P. Rehmus, Y. Termonia,* and *J. Ross*

---

[2] This talk includes answers to the question asked in my survey. The author mentions the following goals of current research: (1) Understanding the approach to equilibrium— "we still cannot characterize classes of Hamiltonians which lead to equilibrium distributions." (2) "Understanding the properties of nonequilibrium steady states, for example a fluid in a uniform time-independent temperature gradient." (3) Understanding dynamic metastability, for example the supercooling of a liquid. (4) Understanding the structure of transport equations. "For example, what are the correct hydrodynamic equations of motion for strongly nonlinear flows?" (Zwanzig 1981:6).

LIST OF RESPONDENTS

Berni J. Alder, Lawrence Livermore National Laboratory, USA.

George A. Baker, Jr., Los Alamos Scientific Laboratory, USA.

J. P. Boon, Université Libre de Bruxelles, Belgium.

Camillo Cammarota, Università di Roma, Italy.

Tony Clifford, transmitted by P. Gray, University of Leeds, UK.

C. Domb, Bar-Ilan University, Ramat-Gan, Israel.

J. Robert Dorfman, University of Maryland, USA.

I. E. Farquhar, University of St. Andrews, Scotland, UK.

Harold L. Friedman, State University of New York at Stony Brook, USA.

H. L. Frisch, State University of New York at Albany, USA.

Marvin D. Girardeau, University of Oregon, USA.

M.A.F. Gomes, Università degli Studi, Rome, Italy.

Robert B. Griffiths, Carnegie-Mellon University, USA.

Dirk ter Haar, University of Oxford, UK.

William Hoover, Lawrence Livermore National Laboratory, USA.

Shigetoshi Katsura, Tohuku University, Sendai, Japan.

Hikaru Kawamura, University of Tokyo, Japan.

Deepak Kumar, University of Roorkee, India.

P. T. Landsberg, University of Southampton, UK.

P. E. Liley, Purdue University, USA.

A. Loinger & A. Lanz, Università degli studi di Milano, Italy.

Benoit Mandelbrot, IBM, Yorktown Heights, New York, USA.

Edward A. Mason, Brown University, USA.

Robert M. Mazo, University of Oregon, USA.

Petter Minhagen, Indiana University, USA

L. Mistura, Università di Roma, Italy.

John F. Nagle, Carnegie-Mellon University, USA.

S. Nordholm, University of Sydney, Australia.

Pierre Pfeuty, Université de Paris-Sud, Orsay, France.

A. H. Piekara, Polish Academy of Sciences, Warsaw, Poland.

Stuart A. Rice, University of Chicago, USA.

Marc Robert, Cornell University, USA.

Carl A. Rouse, General Atomic Co., San Diego, California, USA.

J. S. Rowlinson, Oxford University, UK.

M. Suda, Österreichisches Forschungszentrum Seibersdorf, Austria.

D. I. Uzunov, Bulgarian Academy of Sciences, Sofia, Bulgaria.

Renzo Vallauri, Istituo di Elettronica Quantistica, Florence, Italy.

Gregory H. Wannier, University of Oregon, USA.

Dietrich Wolf, Universität zu Köln, Germany.

BIBLIOGRAPHY

AAPT Committee on Resource Letters. [n.d.]. *Liquid Helium, Selected Reprints.* New York: American Institute of Physics.

ABRIKOSOV, A. A. 1960. "Some Properties of Strongly Compressed Matter, I." *Soviet Physics JETP* 2:1254–59. Translated from *Zhurnal Eksperimentalnoi i Teoreticheskoi Fiziki*, 39:1797–1805.

———. 1973. "My years with Landau." *Physics Today*, 26 (1):56–60.

ADAMS, Henry. 1909. "The Rule of Phase applied to History." Reprinted in *The Degradation of the Democratic Dogma*, pp. 261–305. New York: Capricorn Books, G. P. Putnam's Sons, 1958.

AHLERS, Guenter. 1973. "Thermodynamics and experimental tests of static scaling and universality near the superfluid transition in He^4 under pressure." *Physical Review*, ser. 3, A8:530–68.

———. 1976. "Experiments near the superfluid transition in ^4He and ^3He-^4He mixtures." In Bennemann & Ketterson (1976), pp. 85–206.

———. [1978]. "Critical phenomena, Rayleigh-Benard instability, and turbulence at low temperatures." Tenth Fritz London Memorial Award Lecture, preprint. "Critical Phenomena at Low Temperatures," *Reviews of Modern Physics*, 52:489–503 (1980).

ALDER, B. J., & WAINWRIGHT, T. E. 1957. "Phase transition for a hard sphere system." *Journal of Chemical Physics*, 27:1208–9.

———. 1959. "Studies in Molecular Dynamics. I. General method." *Journal of Chemical Physics*, 31:459–66.

———. 1960. "Studies in molecular dynamics. II. Behavior of a small number of elastic spheres." *Journal of Chemical Physics*, 33:1439–51.

ALEXANDER, H. G. 1956. *The Leibniz-Clarke Correspondence.* Manchester, Eng.: Manchester University Press.

ALLEN, J. F. 1952. "Liquid Helium." In *Low Temperature Physics, Four Lectures*, by F. Simon et al., pp. 66–94. New York: Academic Press.

ALLEN, J. F., & MISENER, A. D. 1938. "Flow of Liquid Helium II." *Nature*, 141:75.

AMPERE, A. M. 1832. "Idées de M. Ampère sur la chaleur et sur la lumière." *Bibliothèque Universelle de Genève, Sciences et Arts*, 49:225–35.

————. 1835. "Note sur la chaleur et sur la lumière considerées comme resultant de mouvement vibratoires." *Annales de Chimie*, 58:432–44. English translation, "Note by M. Ampère on Heat and Light considered as the Results of Vibratory Motion," *Philosophical Magazine*, ser. 3, 7:342–49.

ANDERSON, P. W. 1982. "The 1982 Nobel Prize in Physics." *Science*, 218:763–64.

ANDREWS, Frank C. 1971. *Thermodynamics*. New York: Wiley-Interscience.

ANDREWS, Thomas. 1863. Announced in Miller, W. A., *Chemical Physics*, 3d ed., vol. 1, p. 328. London: Parker.

ANDRONIKASHVILLI, E. 1946. "A direct observation of two kinds of motion in Helium II." *Journal of Physics (USSR)*, 10:201–6. Reprinted in Galasiewicz (1971).

ANDRONIKASHVILLI, E. L., & MAMALDZE, Y. G. 1966. "Quantization of macroscopic Motions and Hydrodynamics of rotating Helium II." *Reviews of Modern Physics*, 38:567–625.

ARMITAGE, Jonathan G. M., & FARQUHAR, Ian E., eds. 1975. *The Helium Liquids*. Proceedings of the Fifteenth Scottish Universities Summer School in Physics, 1974. New York: Academic Press.

ARRHENIUS, S. 1900. "Zur Physik des Volcanismus." *Geologiska Föreningens i Stockholm Förhandlingar*, 22:395–420.

ATACK, D., & SCHNEIDER, W. G. 1951. "The coexistence curve of sulfur hexafluoride in the critical region." *Journal of Physical Chemistry*, 55:532–39.

ATKINS, K. R. 1951. "Summary of work in Liquid Helium." *Proceedings of the International Conference on Low Temperature Physics, Oxford, August 22–28, 1951*, pp. 60–61.

————. 1952. "Wave Propagation and Flow in Liquid Helium II." *Advances in Physics*, 1:169–208.

ATKINS, K. R., & HART, K. H. 1954. "The Attenuation of Second Sound in Liquid Helium II above 1°K." *Canadian Journal of Physics*, 32:381–92 (1954).

AVOGADRO, Amedeo. 1811. "Essai d'une manière de déterminer les masses relatives des molecules élémentaires des corps, et les proportions selon lesquelles elles entrent dans ces combinations." *Journal de Physique*, 73:58–76. English translation in *Foundations of the Molecular Theory*, Alembic Club Reprints No. 4, reissue ed. Edinburgh: Livingstone, 1961.

BAADE, W., & ZWICKY, F. 1934. "Cosmic rays from Super-Novae." *Proceedings of the National Academy of Sciences of the U.S.A.*, 20:259–63.

BAGATSKII, M. I.; VORONEL, A. V.; & GUSAK, V. G. 1962. "Measurement of the specific heat c_v of argon in the immediate vicinity of the critical point." *Soviet Physics JETP*, 16:517–18. Translated from *Zhurnal eksperimentalnoi i teoreticheski fiziki*, 43:728–29. Reprinted in Stanley (1973).

BAILLY, J. S. 1777. *Lettres sur l'Origine des Sciences*. London & Paris.

BAKER, George A., Jr. 1961. "Application of the Padé approximant method to the investigation of some magnetic properties of the Ising model." *Physical Review*, ser. 2, 124:768–74.

BAKER, George A., Jr.; NICKEL, Bernie G.; & MEIRON, Daniel I. 1978. "Critical Indices from Perturbation Analysis of the Callan-Symanzik Equation." *Physical Review*, B17:1365–74.

BALMER, J. J. 1885. "Notiz über die Spectrallinien des Wasserstoffs." *Verhandlungen der Naturforschenden Gesellschaft zu Basel*, 7:548–60, 750–52. English translation in Hindmarsh (1967).

BARDEEN, John. 1963. "Development of concepts in superconductivity." *Physics Today*, 16 (1):19—28.

———. 1973. "Electron-phonon interactions and superconductivity." *Physics Today*, 26 (7):41–47.

BARDEEN, J.; COOPER, L. N.; & SCHRIEFFER, J. R. 1957. "Theory of Superconductivity." *Physical Review*, ser. 2, 108:1175–1204.

BARKER, J. A. 1976. "Gibbs' contribution to statistical mechanics." *Proceedings of the Royal Australian Chemical Society*, 131–36.

BENNEMANN, K. H., & KETTERSON, J. B., eds. 1976. *The Physics of Liquid and Solid Helium*, pt. 1. New York: Wiley-Interscience.

BENNEWITZ, Kurt, & SIMON, Franz. 1923. "Zur Frage de Nullpunktsenergie." *Zeitscrhift für Physik*, 16:183–99.

BERLIN, T. H., & KAC, M. 1952. "The spherical model of a ferromagnet." *Physical Review*, ser. 2, 86:821–35.

BERNOULLI, Daniel. 1738. *Hydrodynamica*. Argentorati [Strassburg]: Dulsecker. English translation by T. Carmody & H. Kobus, *Hydrodynamics*. New York: Dover Pubs., 1968. A translation of the section on kinetic theory is in Brush (1965:57–65).

BETHE, H. A. 1935. "Statistical theory of superlattices." *Proceedings of the Royal Society of London*, A150:552–75.

———. 1939. "Energy production in stars." *Physical Review*, ser. 2, 55:434–56.

BETHE, H. A., & CRITCHFIELD, C. L. 1938. "The formation of deuterons by proton combination." *Physical Review*, ser. 2, 54:248–54, 860.

BETTS, D. D., & CUTHIELL, D. 1980. "A lattice model of a quantum fluid." *Theoretical Chemistry*, 5:141–83.

BIJL, A. 1940. "The lowest Wave Function of the Symmetrical Many Particles System." *Physica*, 7:869–86.

BIJL, A.; DE BOER, J.; & MICHELS, A. 1941. "Properties of Liquid Helium II." *Physica*, 8:655–75.

BJERRUM, Niels. 1949. *Selected Papers.* Copenhagen: Munksgaard. Includes an English translation of "On the specific heat of gases" (1911).

BLACKWELL, Richard J. 1969. *Discoveries in the Physical Sciences.* Notre Dame, Ind.: University of Notre Dame Press, 1969.

BLANPIED, William A. 1972. "Satyendranath Bose: Co-Founder of quantum statistics." *American Journal of Physics*, 40:1212–20.

BLATT, J. M. 1964. *Theory of Superconductivity.* New York: Academic Press.

BLEANEY, B., & SIMON, F. 1939. "Vapor pressure of liquid helium below the λ point." *Transactions of the Faraday Society*, 35:1205–14.

BLOCH, Felix. 1928. "Über die Quantenmechanik der Elektronen in Kristallgittern." *Zeitschrift für Physik*, 52:555–600.

BOAS, Marie. 1952. "The establishment of the Mechanical Philosophy." *Osiris*, 10:412–541.

BOGOLIUBOV, N. N. 1947. "On the Theory of Superfluidity." *Journal of Physics* (USSR), 11:23–32. Reprinted in Galasiewicz (1971).

———. 1967. *Lectures on Quantum Statistics*, vol. 1. *Quantum Statistics.* Translated from Ukranian. New York: Gordon & Breach.

BOHR, Niels. 1911. *Studier over Metallernes Elektrontheori.* Copenhagen, dissertation. Reprinted with English translation in Bohr (1972).

———. 1913. "On the Constitution of Atoms and Molecules." *Philosophical Magazine*, ser. 6, 26:1–25, 476–502, 857–75.

———. 1924. "On the Application of the Quantum Theory to Atomic Structure. I. The Fundamental Postulates." *Proceedings of the Cambridge Philosophical Society*, supplement. Translated from *Zeitschrift für Physik*, 13:117–65 (1923).

———. 1927. "The Quantum Postulate and the Recent Development of Atomic Theory." Lecture at Como, Italy, published in 1928 and reprinted in *Atomic Theory and the Description of Nature.* Cambridge, Eng.: Cambridge University Press, 1934.

———. 1972. *Collected Works*, vol. 1. J. R. Nielsen, ed. Amsterdam: North-Holland.

BOHR, N.; KRAMERS, H. A.; & SLATER, J. C. 1924. "The quantum theory of radiation." *Philosophical Magazine*, ser. 6, 47:785–802.

BOLTZMANN, Ludwig. 1868. "Studien über das Gleichgewicht der lebendige Kraft zwischen bewegten materiellen Punkten." *Sitzungsberichte, K. Akademie der Wissenschaften, Wien, Mathematisch-Naturwissenschaftliche Klasse*, 58:517–60.

————. 1871. "Einige allgemeine Sätze über Wärmegleichgewicht." *Sitzungsberichte, K. Akademie der Wissenschaften, Wien, Mathematisch-Naturwissenschaftliche Klasse*, 63:679–711.

————. 1872. "Weitere Studien über das Wärmegleichgewicht unter Gasmolekülen." *Sitzungsberichte, K. Akademie der Wissenschaften in Wien, Math.-Naturwiss. Kl.*, 66:275—370. English translation in Brush (1966).

————. 1876. "Über die Natur der Gasmolecüle." *Sitzungsberichte, K. Akademie der Wissenschaften in Wien, Math.-Naturwiss. Kl.*, 74:553–60.

————. 1877a. "Über die Beziehung eines allgemeine mechanischen Satzes zum zweiten Hauptsatze der Wärmetheorie." *Sitzungsberichte, K. Akademie der Wissenschaften in Wien, Math.-Naturwiss. Kl.*, 75:67–73. "On the Relation of a General Mechanical Theorem to the Second Law of Thermodynamics," in Brush (1966).

————. 1877b. "Über die Beziehung zwischen des zweiten Hauptsatze der mechanischen Wärmetheorie und der Wahrscheinlichkeitsrechnung, respective den Satzen über das Wärmegleichgewicht." *Sitzungsberichte. K. Akademie der Wissenschaften in Wien, Math.-Naturwiss. Kl.*, 76:373–435.

————. 1884. "Über die Eigenschaften monocyclischer und anderer damit verwandter Systeme." *Sitzungsberichte, K. Akademie der Wissenschaften, Wien, Mathematisch-Naturwissenschaftliche Klasse*, 90:231–45.

————. 1896. "Entgegnung auf die wärmetheoretischen Betrachtungen des Hrn. E. Zermelo." *Annalen der Physik*, ser. 3, 57:773–84. "Reply to Zermelo's remarks on the theory of heat," in Brush (1966).

————. 1897. "Zu Hrn. Zermelo's Abhandlung Über die mechanische Erklärung irreversibler Vorgänge," *Annalen der Physik*, ser. 3, 60:392–98. "On Zermelo's Paper 'On the mechanical explanation of irreversible processes,'" in Brush (1966).

————. 1964. *Lectures on Gas Theory*. Translated by S. G. Brush from *Vorlesungen über Gastheorie* (1896–98). Berkeley: University of California Press.

BOORSE, Henry A., & MOTZ, Lloyd, eds. 1966. *The World of the Atom*. New York: Basic Books.

BORN, Max. 1926a. *Problems of Atomic Dynamics*. Cambridge, Mass.: MIT Press.

————. 1926b. "Quantenmechanik der Stossvorgänge." *Zeitschrift für Physik* 38:803–27. "Quantum Mechanics of Collision Processes," in Ludwig (1968).

————. 1948. *Natural Philosophy of Cause and Chance*. Oxford: Clarendon Press.

————. 1964. "The Statistical Interpretation of Quantum Mechanics." In *Nobel Lectures in Physics, 1942–62*, pp. 256–67. New York: Elsevier.

————, (ed.) 1971. *The Born-Einstein Letters*. New York: Walker.

BORN, Max, & HEISENBERG, Werner. 1924. "Über den Einfluss der Deformierbarkeit der Ionen auf optische und chemische Konstanten." *Zeitschrift für Physik*, 23:388–410.

BORN, Max, & KARMAN, Theodore von. 1912. "Über Schwingungen in Raumgittern." *Physikalische Zeitschrift*, 13:297–309.

————. 1913a. "Zur Theorie der spezifischen Wärme." *Physikalische Zeitschrift*, 14:15–19.

————. 1913b. "Über die Verteilung der Eigenschwingungen von Punktgittern." *Physikalische Zeitschrift*, 14:65–71.

BORN, Max, & OPPENHEIMER, [J.] Robert. 1927. "Zur Quantentheorie der Molekeln." *Annalen der Physik*, ser. 4, 84:457–84.

BOSANQUET, R.H.M. 1877. "Notes on the theory of sound." *Philosophical Magazine*, ser. 5, 3:271–78, 343–49, 418–24; 4:25–39, 125–36, 216–22.

BOSCOVICH, Roger [Boskovic, Rudjer]. 1758. *Theoria Philosophiae Naturalis redacta ad unicam legem virium in natura existentium*. Vienna. Rev. ed. Venice, 1763. English translation by J. M. Child, *A Theory of Natural Philosophy*. Chicago: Open Court Pub. Co., 1922. Reprint, Cambridge, Mass.: MIT Press, 1966.

BOSE, S. N. 1924. "Plancks Gesetz und Lichtquantenhypothese." *Zeitschrift für Physik*, 26:178–81. English translation in Boorse & Motz (1966).

BOYLE, Robert. 1660. *New Experiments Physico-Mechanical, touching the Spring of the Air, and its Effects*. Oxford.

————. 1662. *A Defence of the Doctrine touching the Spring and Weight of the Air, proposed by Mr. Boyle in his New Physico-Mechanical Experiments; Against the objections of Franciscus Linus, wherewith the Objector's Funicular Hypothesis is also Examined*. Oxford.

BRAGG, W. L., & WILLIAMS, E. J. 1934. "The effect of thermal agitation on atomic arrangement in alloys." *Proceedings of the Royal Society of London*, A145:699–730.

————. 1935. "The effect of thermal agitation on atomic arrangement in alloys. Part II." *Proceedings of the Royal Society of London*, A151:540–.

BRIDGMAN, P. W. 1931. *The Physics of High Pressure*. Bell. Reprint, New York: Dover, 1970.

BROGLIE, Louis de 1924. *Recherches sur la théorie des quanta*. Paris: Masson. Partial English translation in Ludwig (1968).

BRUCE, Alastair D. 1980. "Structural phase transitions. II. Static critical behaviour." *Advances in Physics*, 29:111–217.

BRUECKNER, K. A., & SAWADA, K. 1957. "Bose-Einstein Gas with Repulsive Interactions—General Theory" & "Bose-Einstein Gas with Repulsive Interactions—Hard Spheres at High Density." *Physical Review*, ser. 2, 106:1117–27, 1128–35.

BRUNHES, Bernard. 1900. (Discussion remark). *Travaux du Congres International de Physique, Paris, 1900.* vol. 4, p. 29. Paris.

BRUSH, Stephen G. 1957, 1958. "The Transition Temperature in Liquid Helium." *Proceedings of the Royal Society of London*, A242:544–57; A247:225–36.

———. 1961. "Functional Integrals and Statistical Physics." *Reviews of Modern Physics*, 33:79–92.

———. 1965, 1966. *Kinetic Theory.* 2 vols. New York: Pergamon Press.

———. 1970. "Francis Bitter and 'Landau Diamagnetism.'" *Journal of Statistical Physics*, 2:195–97.

———. 1971. "Proof of the impossibility of ergodic systems: The 1913 papers of Rosenthal and Plancherel." *Transport Theory and Statistical Physics*, 1:287–311.

———. 1972. *Kinetic Theory*, vol. 3. New York: Pergamon Press.

———. 1976a. *The Kind of Motion We Call Heat: A History of the Kinetic Theory of Gases in the 19th Century.* Amsterdam: North-Holland.

———. 1976b. "Irreversibility and Indeterminism: Fourier to Heisenberg." *Journal of the History of Ideas*, 37:603–30.

———. 1977. "Statistical Mechanics and the Philosophy of Science: Some Historical Notes." In *PSA 1976* (Proceedings of the Philosophy of Science Association meeting, Chicago, October 1976), ed. F. Suppe and P. D. Asquith, pp. 551–584. East Lansing, Mich.: Philosophy of Science Association.

———. 1978 [published 1979]. *The Temperature of History: Phases of Science and Culture in the Nineteenth Century.* New York: Burt Franklin.

———. 1979. "Nineteenth-century debates about the inside of the Earth: Solid, Liquid or Gas?" *Annals of Science*, 36:225–54.

———. 1980. "The Chimerical Cat: Philosophy of Quantum Mechanics in Historical Perspective." *Social Studies of Science*, 10:393–447.

———; SAHLIN, H. L.; & TELLER, E. 1966. "Monte Carlo Study of a One-Component Plasma. I." *Journal of Chemical Physics*, 45: 2102–18.

BUCHANAN, James. 1857. *Modern Atheism.* Boston: Gould & Lincoln.

BUCKINGHAM, R. A. 1958. "The repulsive interaction of atoms in s states." *Transactions of the Faraday Society*, 54:453–59.

BUFFON, George-Louis Leclerc, Comte de. 1774. *Introduction à l'Histoire des Minéraux*. Paris. *Oeuvres Complètes de Buffon*, ed. M. Flourens, nouv. ed., t. 9. Paris: Garnier Freres, n.d.

BURBURY, S. H. 1890. "On some problems in the kinetic theory of gases." *Philosophical Magazine*, ser. 5, 30:298–317.

——. 1894. "Boltzmann's Minimum Function." *Nature*, 51:78.

BURKE, J. G. 1966. "Descartes on the refraction and the velocity of light." *American Journal of Physics*, 34:390–400.

BURRAU, Oyvind. 1927. "Berechnung des Energiewertes des Wasserstoffmolekel-Ions (H_2^+) im Normalzustand." *Det Kgl. Danske Videnskabernes Selskab, Matematisk-Fysiske Meddelelser*, vol. 7, no. 14.

BUTLER, S. T.; BLATT, J. M.; & SCHAFROTH, M. R. 1956. "Nature of the λ-transition in Liquid Helium." *Nuovo Cimento*, ser. 10, 4:674–75.

BUYS-BALLOT, C.H.D. 1858. "Ueber die Art von Bewegung, welche wir Wärme und Electricitat nennen." *Annalen der Physik*, ser. 2, 103:240–59.

BYCKLING, Eero. 1965. "Vortex Lines and the λ-transition." *Annals of Physics*, 32:367–76.

BYKOV, G. V. 1965. "Historical sketch of the electron theories of organic chemistry." *Chymia*, 10:199–253.

BYRNE, Patrick H. 1980. "Statistical and causal concepts in Einstein's early thought." *Annals of Science*, 37:215–28.

CAGNIARD DE LA TOUR, Charles. 1822. "Exposé de quelques résultats obtenus par l'action combinée de la chaleur et de la compression sur certaines liquides, tels que l'eau, l'alcool, l'éther sulfurique et l'essence de pétrole rectifiée." *Annales de Chimie*, 21:127–32. "Supplément au Mémoire. . . ." *Ibid*, 21:178–82.

CAMPBELL, Lewis, & GARNETT, William. 1969. *The Life of James Clerk Maxwell*. Reprint of 1882 ed. with additions, ed. R. H. Kargon. New York: Johnson Reprint.

——. 1970. *The Life of James Clerk Maxwell*. Reprint of the 1st ed., London, 1882, with a selection of letters from the 2d ed., 1884, and a new preface by R. H. Kargon. New York: Johnson Reprint Corp.

CARDWELL, D.S.L. 1971. *From Watt to Clausius: The Rise of Thermodynamics in The Early Industrial Age*. Ithaca: Cornell University Press.

CARNOT, N. L. Sadi. 1824. *Réflexions sur la puissance motrice du feu et sur les machines propres a développer cette puissance*. Paris. *Reflections on the Motive Power of Fire by Sadi Carnot and Other*

Papers on the Second Law of Thermodynamics by E. Clapeyron and R. Clausius, ed. E. Mendoza. New York: Dover Pubs.

CASIMIR, H.B.G., & POLDER, D. 1946. "Influence of retardation on the London-van der Waals forces." *Nature*, 158:787–88. *See also Physical Review*, ser. 2, 73:360–72 (1948).

CERNUSCHI, F., & EYRING, H. 1939. "An elementary theory of condensation." *Journal of Chemical Physics*, 7:547–51.

CHANDRASEKHAR, S. 1931. "The Maximum Mass of Ideal White Dwarfs." *Astrophysical Journal*, 74:81–82.

———. 1935a. "The highly collapsed configurations of a stellar mass (second paper)." *Monthly Notices of the Royal Astronomical Society*, 95:207–25.

———. 1935b. "Stellar configurations with degenerate cores." *Monthly Notices of the Royal Astronomical Society*, 95:226–60.

———. 1939. *An Introduction to the Study of Stellar Structure*. Chicago: University of Chicago Press. Reprint, New York: Dover Pubs., 1957.

———. 1945. "Ralph Howard Fowler 1889–1944." *Astrophysical Journal*, 101:1–5.

———. 1964. "The Case for Astronomy." *Proceedings of the American Philosophical Society*, 108:1–6.

CHAPMAN, Sydney. 1916. "On the Law of Distribution of Molecular Velocities, and on the Theory of Viscosity and Thermal Conduction, in a Non-Uniform Simple Monatomic Gas." *Philosophical Transactions of the Royal Society of London*, A216:279–348.

———. 1917. "On the Kinetic Theory of a Gas; Part II, A Composite Monatomic Gas, Diffusion, Viscosity, and Thermal Conduction." *Philosophical Transactions of the Royal Society of London*, A217: 115–97.

CHAPMAN, Sydney, & COWLING, T. G. 1939. *The Mathematical Theory of Non-Uniform Gases*. Cambridge, Eng.: Cambridge University Press. 2d ed. 1952, 3d ed. 1970.

CHASE, C. E. 1953. "Ultrasonic Measurements in Liquid Helium." *Proceedings of The Royal Society of London*, A220:116–32.

CHESTER, G. V. 1955. "λ Transition in Liquid Helium." *Physical Review*, ser. 2, 100:455–62.

———. 1966. Discussion remark in *Quantum Fluids*, D. F. Brewer, ed., p. 254. New York: Wiley.

———. 1975. "Introductory Lectures on Liquid Helium Four and Three." In Armitage & Farquhar (1975), pp. 1–52.

CLARK, Peter. 1976. "Atomism versus Thermodynamics." In *Method and Appraisal in the Physical Sciences*, Colin Howson, ed., pp. 41–105. New York: Cambridge University Press.

CLAUSIUS, Rudolf. 1850a. "Ueber die bewegende Kraft der Wärme und die Gesetze welche sich daraus für die Wärmelehre selbst ableiten lassen." *Annalen der Physik*, ser. 2, 79:368–97, 500–24. For English translation see above under Carnot (1824), book edited by E. Mendoza.

———. 1850b. "Notiz über den Einfluss des Drucks auf das Gefrieren der Flüssigkeiten." Annalen der Physik, ser. 2, 81:168–72. English translation in *Philosophical Magazine*, ser. 4, 2:548–50 (1851).

———. 1854. "Ueber eine veränderte Form des zweiten Hauptsatzes der mechanischen Wärmetheorie." *Annalen der Physik*, ser. 2, 93:481–506. English translation, "On a modified form of the second fundamental theorem in the Mechanical theory of heat," *Philosophical Magazine*, ser. 4, 12:81–98 (1856).

———. 1857. "Ueber die Art der Bewegung, welche wir Wärme nennen." *Annalen der Physik*, ser. 2, 100:353–80. English translation in Brush (1965).

———. 1858. "Ueber die mittlere Länge der Wege, welche bei der Molecularbewegung gasförmigen Körper von den einzelnen Molecülen zuruckgelegt werden; nebst einigen anderen Bemerkungen über die mechanische Wärmetheorie." *Annalen der Physik*, ser. 2, 105:239–58. English translation in Brush (1965).

———. 1865. "Ueber verschiedene für die Anwendung bequeme Formen der Hauptgleichung der mechanischen Wärmetheorie." *Annalen der Physik*, ser. 2, 125:353–400. "On several convenient forms of the fundamental equations of the mechanical theory of heat," in *The Mechanical Theory of Heat*, T. A. Hirst, ed. London: Van Voorst, 1867. Also under same title, W. R. Browne, ed. London: Macmillan, 1879.

———. 1868. "On the Second Fundamental Theorem of the Mechanical Theory of Heat." *Philosophical Magazine*, ser. 4, 35:405–19.

COHEN, M., & FEYNMAN, R. P. 1957. "Theory of Inelastic Scattering of Cold Neutrons from Liquid Helium." *Physical Review*, ser. 2, 107:13–24.

COMPTON, Arthur Holly. 1923. "A quantum theory of the scattering of X-rays by light elements." *Physical Review*, ser. 2, 21:483–502.

COMPTON, K. T., & RUSSELL, H. N. 1924. "A possible explanation of the behaviour of the hydrogen lines in giant stars." *Nature*, 114:86–87.

CONANT, James Bryant. 1950. "Robert Boyle's Experiments in Pneumatics." In *Harvard Case Histories in Experimental Science*, J. B. Conant, ed., pp. 1–63. Cambridge, Mass.: Harvard University Press.

CONIGLIO, A., & MARINARO, M. 1967. "On Condensation for an Interacting Bose System." *Nuovo Cimento*, ser. 10, 48:249–61.

COOPER, L. N. 1956. "Bound electron pairs in a degenerate Fermi gas." *Physical Review*, ser. 2, 104:1189–90.

———. 1973. "Microscopic quantum interference in the theory of superconductivity" Nobel Lecture. *Science*, 181:909–16.

COPELAND, D. A., & KESTNER, N. R. 1968. "Accurate 'effective' intermolecular pair potentials in gaseous argon." *Journal of Chemical Physics*, 49:5214–22.

CULVERWELL, E. P. 1890. "Note on Boltzmann's Kinetic Theory of Gases, and on Sir W. Thomson's Address to Section A, British Association, 1884." *Philosophical Magazine*, ser. 5, 30:95–99.

CUMMINGS, F. W. 1971. "The Condensate in ^{4}He II as a Pilot Wave." *Physics Letters*, 34A:196–97.

CURIE, Pierre. 1895. "Propriétés magnetique des corps a diverses températures." *Annales de Chimie*, ser. 7, 5:289–405.

DALTON, John. 1793. *Meteorological Observations and Essays*. London.

———. 1805. "On the Absorption of Gases by Water and other liquids." *Memoirs of the Literary and Philosophical Society of Manchester*, ser. 2, 1:271–87.

DARWIN, C. G., & FOWLER, R. H. 1922. "On the partition of energy." *Philosophical Magazine*, ser. 6, 44:450–79, 823–42.

DAUB, Edward E. 1969. "Probability and Thermodynamics: The Reduction of the Second Law." *Isis*, 60:318–30.

———. 1971. "Waterston's influence on Kronig's kinetic theory of gases." *Isis*, 62:512–15.

———. 1976. "Gibbs phase rule: a centenary retrospect." *Journal of Chemical Education*, 53:747–51.

DAVIES, Gordon. 1969. *The Earth in Decay*. London: Macdonald.

DE BOER, J., & VAN KRANENDONK, J. 1948. "The Viscosity and Heat Conductivity of Gases with Central Intermolecular Forces." *Physica*, 14:442–52.

DEBYE, Peter J. W. 1912. "Zur Theorie der spezifischen Wärmen." *Annalen der Physik*, ser. 4, 39:789–838. English translation in Debye (1954).

———. 1920. "Die van der Waalsschen Köhasionskräfte." *Physikalische Zeitschrift*, 21:178–87. English translation in Debye (1954).

———. 1954. *Collected Papers*. New York: Interscience.

DE KLERK, D.; HUDSON, R. P.; & PELLAM, J. R. 1954. "Second Sound Propagation below 1°K." *Physical Review*, ser. 2, 93:28–37.

DELBRÜCK, M. 1980. "Was Bose-Einstein statistics arrived at by serendipity?" *Journal of Chemical Education*, 57:467–70.

DEWAR, M.J.S. 1949. *Electronic Theory of Organic Chemistry*. Oxford: Clarendon Press.

DIRAC, P.A.M. 1926. "On the Theory of Quantum Mechanics." *Proceedings of the Royal Society of London*, A112:661–77.

———. 1929. "Quantum mechanics of many-electron systems." *Proceedings of the Royal Society of London*, A123:714–33.

DOMB, C. 1949. "Order-disorder statistics." *Proceedings of the Royal Society of London*, A196:36–50; A199:199–221.

———. 1971. "The Curie Point." In *Statistical Mechanics at the Turn of the Decade*, E.G.D. Cohen, ed., pp. 81–128. New York: Dekker.

DOMB, C., & HUNTER, D. L. 1965. "On the critical behaviour of ferromagnets." *Proceedings of the Physical Society of London*, 86: 1147–51.

DORFMAN, J. R. 1978. "Book Review. Statistical Mechanics, 2nd ed. By J. E. Mayer and M. Goeppert Mayer." *Journal of Statistical Physics*, 18:415–20.

DORLING, Jon. 1971. "Einstein's introduction of photons: argument by analogy or deduction from the phenomena?" *British Journal for the Philosophy of Science*, 22:1–8.

DOROZYNSKI, Alexander. 1965. *The Man They Wouldn't Let Die*. New York: Macmillan.

DÖRRE, P.; HAUG, H.; & TRAN THOAI, D. B. 1979. "Condensate Theory versus Pairing Theory for Degenerate Bose Systems." *Journal of Low Temperature Physics*, 35:456–85.

DORTOUS DE MAIRAN, J. J. 1719. "Sur la cause générale du Froid en Hyver, & de la chaleur en Eté." *Mémoires de Mathématique et de Physique tirés des Registres de l'Académie Royale des Sciences de l'Année*, M.DCCXIX, 104–35.

DRUDE, Paul. 1900. "Zur Elektronentheorie der Metalle." *Annalen der Physik*, ser. 4, 1:566–613; 3:369–402.

DYSON, Freeman J., & LENARD, A. 1967, "Stability of Matter." *Journal of Mathematical Physics*, 8:423–34. *See also ibid.*, 9:698–711 (1968).

EDDINGTON, Arthur S. 1917. "Further notes on the Radiative Equilibrium of the Stars." *Monthly Notices of the Royal Astronomical Society*, 77:596–612.

———. 1932. "The Decline of Determinism." *Mathematical Gazette*, 16: 66–80.

———. 1935. "On 'Relativistic Degeneracy.'" *Monthly Notices of the Royal Astronomical Society*, 95:194–206.

EGGERT, John. 1919. "Über den Dissoziationszustand der Fixsterngase." *Physikalische Zeitschrift*, 20:570–74.

EHRENFEST, Paul. 1913. "Bemerkung betreffs der spezifischen Wärme zweiatomiger Gase." *Verhandlungen der Deutsche Physikalische Gesellschaft*, 15:451–57.

———. 1933. "Phasenumwandlungen in ueblichen und erweitert Sinn, classifiziert nach den entsprechenden Singularitäten des thermodynamischen Potentials." *Proceedings of the Academy of Sciences, Amsterdam*, 36:153–57.

EHRENFEST, Paul & EHRENFEST, Tatiana. 1911. "Begriffliche Grundlagen der statistischen Auffassung in der Mechanik." *Encyklopäsdie der mathematische Wissenschaften*, vol. 4, pt. 32. *The Conceptual Foundations of the Statistical Approach in Mechanics*. Translated by M. J. Moravcsik. Ithaca: Cornell University Press, 1959.

EHRENFEST, P., & TRKAL, V. 1920. "Deduction of the dissociation-equilibrium from the theory of quanta and a calculation of the chemical constant based on this." *Proceedings of the Section of Sciences, Akademie van Wetenschappen, Amsterdam*, 23:162–83.

EINSTEIN, Albert. 1905a. "Über einen die Erzeugung und Verwandlung des lichtes betreffenden heuristischen Gesichtspunkt." *Annalen der Physik*, ser. 4, 17:132–48. English translation in ter Haar (1967).

———. 1905b. "Ueber die von der molekular-kinetischen Theorie der Wärme geforderte Bewegung von in ruhenden Flüssigkeiten suspenierten Teilchen." *Annalen der Physik*, ser. 4, 17:549–60. English translation in *Investigations on the Theory of the Brownian Movement*. New York: Dover Pubs., 1956 (reprint of 1926 ed.).

———. 1907. "Plancksche Theorie der Strahlung und die Theorie der spezifischen Wärme." *Annalen der Physik*, ser. 4, 22:180–90.

———. 1916. "Zur Quantentheorie der Strahlung." *Mitteilungen der Physikalische Gesellschaft, Zürich*, 18:47–62. "On the Quantum Theory of Radiation," in B. L. Van der Waerden (1967).

———. 1924. "Quantentheorie des einatomigen idealen Gases." In *Sitzungsberichte, Akademie der Wissenschaften, Berlin*, pp. 261–267. *See also ibid*, pp. 3–14 (1925).

———. 1949. "Autobiographical Notes." *Albert Einstein Philosopher-Scientist*, P. Schilpp, ed., pp. 1–95. New York: Library of Living Philosophers.

EINSTEIN, Albert, & STERN, O. 1913. "Einige Argumente für die Annahme einer molekular Agitation beim absoluten Nullpunkt." *Annalen der Physik*, ser. 4, 40:551–60.

EISENSCHITZ, R., & LONDON, F. 1930. "Über das Verhältnis der van der Waalsschen Kräfte zu den homöopolaren Bindungskraften." *Zeitschrift für Physik*, 60:491–527.

ELKANA, Yehuda. 1974a. "Boltzmann's Research Programme and its Alternatives." In *The Interaction between Science and Philosophy*, Y. Elkana, ed., pp. 243–79.

———. 1974b. *The discovery of the Conservation of Energy*. London: Hutchinson Educational.

ENSKOG, David. 1717. *Kinetische Theorie der Vorgänge in mässig verdünnten Gasen*. Uppsala: Almqvist & Wiksell. "Kinetic Theory of Processes in Dilute Gases," in Brush (1972).

———. 1922. "Kinetische Theorie der Wärmeleitung, Reibung und Selbstdiffusion in gewissen verdichteten Gasen und Flüssigkeiten." *Kungliga Svenska Vetenskapsakademiens Handlingar*, Ny Följd, 63 (4). English translation in Brush (1972).

ESSAM, JOHN W., & FISHER, M. E. 1963. "Padé approximant studies of the lattice gas and Ising ferromagnet below the critical point." *Journal of Chemical Physics*, 38:802–12.

EULER, Leonhard. 1727. "Tentamen explicationis phaenomenorvm aeris." *Commentarii Academiae Scientiarum Imperialis Petropolitanae*, 2:347–68.

EVANS, E. J. 1913. "The spectra of helium and hydrogen." *Nature*, 92:5.

EVANS, W.A.B. 1975. "A Theoretical Argument against a Bose-Einstein Condensate in Superfluid ^{4}He." *Nuovo Cimento*, ser. 11, 30B: 145–52.

———. 1978. "The Pairing Theory of Superfluid ^{4}Helium." In *Acta Universitatis Wratislaviensis*, no. 436, xiv-th Winter School of Theoretical Physics in Karpacz, March 7–20, 1977, pp. 39–60. Wroclaw.

EVANS, W.A.B., & HARRIS, C. G. 1978. "Potential Dependence of Pairing Solution for the Bose Superfluid." *Journal of Physique*, 39: Supplement C6-237–239.

———. [1979]. "New Condensate-less pair solution for the Bose Superfluid." Preprint.

EXNER, Franz. 1919. *Vorlesungen über die physikalischen Grundlagen der Naturwissenschaften. Wien: Deuticke.*

FAIRBANK, W. M. 1963. "The Nature of the λ-Transition in Liquid Helium." In *Liquid Helium*, Proceedings of the International School of Physics "Enrico Fermi" Course xxi, Varenna, July 1961, G. Careri, ed., pp. 293–304. New York: Academic Press.

FAIRBANK, W. M.; BUCKINGHAM, M. J.; & KELLERS, C. F. 1957. "Specific Heat of Liquid He^4 near the λ Point." *Bulletin of the American Physical Society*, ser. II, 2:183.

FAIRBANK, W. M., & KELLERS, C. F. 1966. "The Lambda Transition in Liquid Helium." In *Critical Phenomena*, M. S. Green & J. V.

Sengers, eds., pp. 71–78. Washington, D.C.: National Bureau of Standards, Misc. Pub. 273.

FERMI, E. 1923. "Sulla probabilita degli stati quantici." *Rendiconti, Academia dei Lincei, Classe di Scienze Fisiche, Matematiche e Naturali*, ser. 5, 32 (2):493–95. Reprinted in *The Collected Papers of Enrico Fermi*, vol. I, pp. 118–20. Chicago: University of Chicago Press, 1962.

———. 1926. "Über die Wahrscheinlichkeit der Quantenzustände." *Zeitschrift für Physik*, 26:54–56.

FEYNMAN, R. P. 1948. "Space-Time Approach to Non-Relativistic Quantum Mechanics." *Reviews of Modern Physics*, 20:367–87.

———. 1953a. "The λ-Transition in Liquid Helium." *Physical Review*, ser. 2, 90:1116–17.

———. 1953b. "Atomic Theory of the λ-Transition in Helium." *Physical Review*, ser. 2, 91:1291–1301.

———. 1953c. "Atomic Theory of Liquid Helium near Absolute Zero." *Physical Review*, ser. 2, 91:1301–8.

———. 1955. "Application of Quantum Mechanics to Liquid Helium." In *Progress in Low Temperature Physics*, vol. 1, C. J. Gorter, ed., chap. II. New York: Interscience. Reprinted in Galasiewicz (1971) and in AAPT.

FEYNMAN, R. P., & COHEN, M. 1955. "The Character of the Roton State in Liquid Helium." *Progress of Theoretical Physics*, 14:261–63.

———, 1956. "Energy Spectrum of the Excitations in Liquid Helium." *Physical Review*, ser. 2, 102:1189–1204.

FINN, Bernard S. 1964. "Laplace and the Speed of Sound." *Isis*, 55:7–19.

FISHER, I. Z. 1955. "On the Stability of a Homogeneous Phase." *Soviet Physics JETP*, 1:154–60, 273–79, 280–83. Translated from *Zhurnal Eksperimentalnoi i Teoreticheskoi Fiziki*, 28:171–80, 437–46, 447–51.

FISHER, Michael E. 1967. "The theory of equilibrium critical phenomena." *Reports on Progress in Physics*, 30:615–730. Extensive list of corrigenda & addenda available from author.

———. 1972. "Phase transitions, symmetry, and dimensionality." In *Essays in Physics*, G.K.T. Conn & G. N. Fowler, eds., vol. 4, pp. 43–89. London: Academic Press.

———. 1974. "The renormalization group in the theory of critical behaviour." *Reviews of Modern Physics*, 46:597–616.

———. 1979. "The states of matter—A theoretical perspective." *Proceedings of the Robert A. Welch Conferences on Chemical Research, XXIII, Modern Structural Methods*, Houston, Texas, November 12–14, pp. 73–145.

FORD, Kenneth W. 1968. *Basic Physics*. Waltham, Mass.: Blaisdell.

FORMAN, Paul. 1971. "Weimar Culture, Causality, and Quantum Theory, 1918–1927; Adaptation by German Physicists and Mathematicians to a Hostile Intellectual Environment." *Historical Studies in the Physical Sciences*, 3:1–115.

FOURIER, J.B.J. 1807. "Théorie de la propagation de la chaleur dans les solides." Manuscript presented to the Institut de France. First published in *Joseph Fourier 1768–1830*, by I. Grattan-Guinness with J. R. Ravetz. Cambridge, Mass.: MIT Press, 1972.

————. 1819. "Mémoire sur le refroidissement séculaire du globe terrestre." *Annales de Chimie*, 13:418–37.

————. 1822. *Théorie Analytique de la Chaleur*. Paris: Didot. English translation by A. Freeman, *The Analytical Theory of Heat*, London, 1978. Reprinted, New York: Dover Pubs., 1955.

————. 1890. *Oeuvres de Fourier*, t. 2. Paris: Gauthier-Villars.

FOWLER, Ralph Howard. 1923. "Dissociation-equilibria by the method of partitions." *Philosophical Magazine*, ser. 6, 45:1–33.

————. 1925. "Notes on the theory of absorption lines in stellar spectra." *Monthly Notices of the Royal Astronomical Society*, 85:970–77.

————. 1926. "On dense stars." *Monthly Notices of the Royal Astronomical Society*, 87:114–22. Reprinted in *A Source Book in Astronomy and Astrophysics, 1900–1975*, K. R. Lang & O. Gingerich, eds., pp. 435–39. Cambridge, Mass.: Harvard University Press, 1979.

————. 1935. "A theory of the rotations of molecules in solids and of the dielectric constant of solids and liquids." *Proceedings of the Royal Society of London*, A149:1–28.

FOWLER, R. H., & JONES, H. 1938. "The Properties of a Perfect Bose-Einstein Gas at low temperatures." *Proceedings of the Cambridge Philosophical Society*, 34:573–76.

FOWLER, R. H., & GUGGENHEIM, E. A. 1925. "Applications of statistical mechanics to determine the properties of matter in stellar interiors." *Monthly Notices of the Royal Astronomical Society*, 85:939–60, 961–70.

————. 1940. "Statistical thermodynamics of superlattices." *Proceedings of the Royal Society of London*, A174:189–206.

FOWLER, R. H., & MILNE, E. A. 1923. "The intensities of absorption lines in stellar spectra, and the temperatures and pressures in the reversing layers of stars." *Monthly Notices of the Royal Astronomical Society*, 83:403–24.

————. 1924. "The maximum of absorption lines in stellar spectra

(Second Paper)." *Monthly Notices of the Royal Astronomical Society*, 84:499–515.

FOX, Robert. 1971. *The Caloric Theory of Gases from Lavoisier to Regnault*. Oxford: Clarendon Press.

FRANCK, J., & HERTZ, G. 1914. "Über Zusammenstösse zwischen Elektronen und den Molekülen des Quecksilberdampfes und die Ionisierungsspannung desselben," and "Über die Erregung der Quecksilberresonanzlinie 253,6 $\mu\mu$ durch Elektronenstösse." *Verhandlungen der Duetschen Physikalischen Gesellschaft*, 16:457–67, 512–17. English translation of the first paper in Boorse & Motz (1966), and of the second in ter Haar (1967).

FRASER, A. R. 1951. "The Condensation of a Perfect Bose-Einstein Gas." *Philosophical Magazine*, ser. 7, 42:156–64, 165–75.

FRENKEL, J. 1936. *Wave Mechanics, Elementary Theory*. Oxford: Oxford University Press.

FRENKEL, V. J. 1974. "Yakov (James) Il'ich Frenkel (1894-1952): Materials for his Scientific Biography." *Archive for History of Exact Sciences*, 13 :1–26 (1974).

FRIEDMAN, K. S. 1976. "A partial vindication of ergodic theory." *Philosophy of Science*, 43:151–62.

FRÖHLICH, H. 1937. "Zur Theorie des λ-Punktes des Heliums." *Physica*, 4:639–44.

———. 1950. "Theory of the superconducting state. I. The ground state at the absolute zero of temperature." *Physical Review*, ser. 2, 79:845–56.

FRÖHLICH, H.; PELZER, H; & ZIENAU, S. 1950. "Properties of slow electrons in polar materials." *Philosophical Magazine*, ser. 7, 41:221–42.

FROST, A. A., & WOODSON, J. H. 1958. "Semi-empirical potential energy functions. III. Generalization for Ionic Molecules and the Includion of London Forces." *Journal of the American Chemical Society*, 80:2615–18.

GALASIEWICZ, Zygmunt. 1970. *Superconductivity and Quantum Fluids*. Oxford: Pergamon Press.

———. 1971. *Helium 4*. Oxford: Pergamon Press.

GALILEI, Galileo. 1638. *Discorsi e dimostrazioni matematiche intorno a due nuoue scienze, attenenti alla Mecanica & i Movimenti Locali*. Leiden. English translation by S. Drake, *Two New Sciences*. Madison: University of Wisconsin Press, 1974.

GALISON, Peter. 1981. "Kuhn and the Quantum Controversy." *British Journal for the Philosophy of Science*, 32:71–84.

GARBER, Elizabeth. 1970. "Clausius and Maxwell's Kinetic Theory of Gases." *Historical Studies in the Physical Sciences,* 2:299–319.

———. 1972. "Aspects of the introduction of probability into Physics." *Centaurus,* 17:11–39.

———. 1976. "Some reactions to Planck's law, 1900–1914." *Historical Studies in the Physical Sciences,* 7:89–126.

———. 1978. "Molecular science in Late-Nineteenth-Century Britain." *Historical Studies in the Physical Sciences,* 9:265–97.

GARDNER, Michael R. 1979. "Realism and instrumentalism in 19th century Atomism." *Philosophy of Science,* 46:1–34.

GASPARINI, F. M., & GAETA, A. 1978. "Universality of the Specific Heat of ^{3}He-^{4}He Mixtures at the λ-transition." *Physical Review,* ser. 3, B17:1466–71.

GASPARINI, F. M., & MOLDOVER, M. R. 1974. "Renormalization of the ^{4}He λ-Transition in ^{3}He-^{4}He Mixtures. In *Low Temperature Physics—LT 13,* Proceedings of the 13th International Conference on Low Temperature Physics, 1972, K. D. Timmerhaus et al., eds., vol. 1, pp. 618–622. New York: Plenum.

———. 1975. "Specific Heat of ^{4}He and ^{3}He-^{4}He Mixtures at their λ-transition." *Physical Review,* ser. 3, B12:93–113.

GAY-LUSSAC, Louis Joseph. 1807. "Premier essai pour déterminer les variations de température qu'épreuvent les gaz en changeant de densité, et considérations sur leur capacité pour le calorique." *Mémories de Physique et de Chimie de la Société d'Arcueil,* 1:180–204.

GERBER, J. 1969. "Geschichte der Wellenmechanik." *Archive for History of exact sciences,* 5:349–416.

GIBBS, J. Willard. 1902. *Elementary Principles in Statistical Mechanics, developed with especial reference to the rational foundation of Thermodynamics.* New York: Scribner. Reprinted in Gibbs (1948).

———. 1948. *Collected Works.* New Haven: Yale University Press.

GILBERT, William. 1600. *De Magnete.* London. English translation by P. Fleury Mottelay, *On the Lodestone and Magnetic Bodies, and on the Great Magnet the Earth.* New York: Wiley, 1893. Reprint, New York: Dover, 1958.

GILLISPIE, C. C. 1963. "Intellectual factors in the background of analysis by probabilities." In *Scientific Change,* A. C. Crombie, ed., pp. 431–53. New York: Basic Books.

———. 1972. "Probability and politics: Laplace, Condorcet and Turgot." *Proceedings of the American Philosophical Society,* 116:1–20.

GINZBURG, V. L. 1943. "Scattering of Light in Helium II." *Journal of Physics* (USSR), 7:305–6.

GIRARDEAU, M. 1962. "Simple and Generalized Condensation in Many-Boson Systems." *Physics of Fluids*, 5:1468–78.

———. 1969. "Fermi versus Bose Condensation in Superconductors and Liquid ^{4}He." *Physics Letters*, 29A:64–65.

———. 1971. "Off-Diagonal Long-Range Order and the Momentum Distribution of Electron Pairs in Superconductors, and Helium Atoms in Liquid ^{4}He." *Physical Review*, ser. 3, A4:777–88.

———. 1978. "Equilibrium Superradiance in a Bose Gas." *Journal of Statistical Physics*, 18:207–15.

GIRILL, T. R. 1976. "The Problem of Micro-Explanation." In *PSA 1976*, F. Suppe & P. D. Asquith, eds., vol. 1, pp. 47–55. East Lansing, Mich.: Philosophy of Science Association.

GOBLE, D. F. 1971. "Liquid Helium and the Properties of a Bose-Einstein Gas, IV. The Effect of Various Approximations." *Canadian Journal of Physics*, 49:3099–3114.

GOLDBERG, Stanley. 1976. "Max Planck's philosophy of nature and his elaboration of the special theory of relativity." *Historical Studies in the Physical Sciences*, 7:125–60.

GOLDSTEIN, Louis. 1964. "Lambda transformation of liquid helium four." *Physical Review*, ser. 2, 135:A1471–80.

GOLDSTEIN, S.; MISRA, B.; & COURBAGE, M. 1981. "On intrinsic randomness of dynamical systems." *Journal of Statistical Physics*, 25:111–26.

GORSKY, W. 1928. "Röntgenographische Untersuchungen von Umwandlungen in der Legierung CuAu." *Zeitschrift für Physik*, 50: 64–81.

GORTER, C. J. 1952. "Heat pulses in He II below 1°K." *Physical Review*, ser. 2, 88:681.

GORTER, C. J., & CASIMIR, H. 1934. "On Supraconductivity I." *Physica*, 1:306–20.

GOUY, Leon. 1888. "Note sur le mouvement brownien." *Journal de Physique*, ser. 2, 7:561–64.

GRAHAM, Loren R. 1964. "A Soviet Marxist View of Structural Chemistry: the Theory of Resonance Controversy." *Isis*, 55:20–31.

GRATTAN-GUINNESS, I. 1972. *Joseph Fourier 1768–1830*. Cambridge, Mass.: MIT Press.

GRIFFITHS, R. B. 1964. "Peierls proof of spontaneous magnetization in a two-dimensional Ising ferromagnet." *Physical Review*, ser. 2, 136:A437–39.

GUGGENHEIM, E. A. 1935. "Statistical mechanics of regular solutions." *Proceedings of the Royal Society of London*, 148A:304–12.

————. 1938. "Statistical mechanics of cooperative assemblies." *Proceedings of the Royal Society of London,* 169A:134–48.

————. 1952. *Mixtures.* Oxford: Clarendon Press.

HAAR, D. ter. 1967. *The Old Quantum Theory.* Oxford: Pergamon Press.

HAAS, Arthur, ed. 1936. *A Commentary on the Scientific Writings of J. Willard Gibbs.* New Haven: Yale University Press.

HABGOOD, H. W., & SCHNEIDER, W. G. 1954. "PVT measurements in the critical region of zenon." "Thermodynamic properties of xenon in the critical region." *Canadian Journal of Chemistry,* 32:98–112, 164–73.

HALDANE, J. S. 1928. "Memoir of J. J. Waterston." In *The Collected Scientific Papers of John James Waterston,* J. S. Haldane, ed., pp. xiii–xviii. Edinburgh: Oliver and Boyd.

HALICIOGLU, Timur, & SINANOGLU, Oktay. 1968. "Medium-independent intermolecular potential for liquids and its use in obtaining free energy and entrophy." *Journal of Chemical Physics,* 49:996–1000.

HALL, Marie Boas, ed. 1970. *Nature and Nature's Laws: Documents of the Scientific Revolution.* New York: Harper & Row.

HALLOCK, Robert B. 1982. "Resource Letter SH-1: Superfluid Helium." *American Journal of Physics,* 50:202–12.

HANLE, Paul A. 1975. *Erwin Schrödinger's Statistical Mechanics, 1912–1925.* Ph.D. Dissertation, Yale University.

————. 1977a. "Erwin Schrödinger's reaction to Louis de Broglie's thesis on the quantum theory." *Isis,* 68:606–9.

————. 1977b. "The coming of age of Erwin Schrödinger: His quantum statistics of ideal gases." *Archive for History of Exact Sciences,* 17:165–92.

————. 1979a. "The Schrödinger-Einstein correspondence and the sources of wave mechanics." *American Journal of Physics,* 47:644–48.

————. 1979b. "Indeterminacy before Heisenberg: the case of Franz Exner and Erwin Schrödinger." *Historical Studies in the Physical Sciences,* 10:225–69.

HANSON, W. B., & PELLAM, J. R. 1954. "Second Sound Attenuation in Liquid Helium II." *Physical Review,* ser. 2, 95:321–27.

HARRISON, S. F., & MAYER, J. E. 1938. "Statistical mechanics of condensing systems IV." *Jouurnal of Chemical Physics,* 6:101–4.

HASSE, H. R., & COOK, W. R. 1929. "The Determination of Molecular Forces from the Viscosity of Gases." *Proceedings of the Royal Society of London,* A125:195–221.

HAYASHI, C.; HOSHI, R.; & SUGIMOTO, D. 1962. "Evolution of the stars." *Progress of Theoretical Physics*, supplement, 22:1–183.

HEILBRON, John L. 1977. "Lectures on the history of atomic physics 1900–1922." In *History of Twentieth Century Physics* (Varenna International School of Physics "Enrico Fermi" Course LVII, 1972), C. Weiner, ed., pp. 40–108. New York: Academic Press.

———. 1981. "Rutherford-Bohr atom." *American Journal of Physics*, 49:223–31.

HEILBRON, John L., & KUHN, Thomas S. 1969. "The genesis of the Bohr atom." *Historical Studies in the Physical Sciences*, 1:211–90.

HEISENBERG, Werner. 1925. "Über quantentheoretische Umdeutung kinematischer und mechanischer Beziehungen." *Zeitschrift für Physik*, 33:879–93. English translations in Ludwig (1968) and in Van der Waerden (1967).

———. 1926a. "Mehrkörperproblem und Resonanz in der Quantenmechanik." *Zeitschrift für Physik*, 39:411–26.

———. 1926b. "Über die Spektra von Atomsystemen mit zwei Elektronen." *Zeitschrift für Physik*, 39:499–518.

———. 1927. "Über den anschaulichen Inhalt der quantentheoretischen Kinematik und Mechanik." *Zeitschrift für Physik*, 43:172–98.

———. 1928. "Zur Theorie des Ferromagnetismus." *Zeitschrift für Physik*, 49:619–36.

———. 1971. *Physics and Beyond*. New York: Harper & Row.

———. 1973. "Tradition in Science." *Science and Public Affairs* (*Bulletin of the Atomic Scientists*), 29 (10):4–10.

———. 1974. *Across the Frontiers*. New York: Harper & Row.

HEITLER, W. 1967. "Quantum chemistry: the early period." *International Journal of quantum chemistry*, 1:13–36.

HEITLER, W., & LONDON, F. 1927. "Wechselwirkung neturaler Atome und homopolare Bindung nach der Quantenmechanik." *Zeitschrift für Physik*, 44:455–72.

HELLER, P., & BENEDEK, G. B. 1962. "Nuclear magnetic resonance in MnF_2 near the critical point." *Physical Review Letters*, 8:428–32.

HERAPATH, John. 1821. "A mathematical inquiry into the causes, laws and principal phenomenae of heat, gases, gravitation, etc." *Annals of Philosophy*, ser. 2, 1:273–93, 340–51, 401–6. Reprinted in 1972 with Herapath (1847).

———. 1836. "Exact calculation of the velocity of sound." *Railway Magazine*, new ser. 1:22–28.

———. 1847. *Mathematical Physics*. London: Whittaker, 1847. Reprinted with Selected Papers by John Herapath, ed. S. G. Brush. New York: Johnson Reprint Corp., 1972.

————. 1860. "On the Dynamical Theory of Airs." *Atheneum*, 722.

HERIVEL, John. 1965. *The Background to Newton's Principia*. Oxford: Clarendon Press.

HERMANN, Armin. 1971. *The Genesis of Quantum Theory* (1899–1913). Cambridge, Mass.: MIT Press. Translation of *Frühgeschichte der Quantentheorie* (1899–1913). *Mosbach in Baden: Physik*, 1969.

HERSCHEL, John. 1850. "Quetelet on probabilities." *Edinburgh Review*, 92:1–57.

HERZFELD, K. F. 1916. "Zur Statistik des Bohrschen Wasserstoffatommodells." *Annalen der Physik*, ser. 4, 51:261–84.

HESTENES, David. 1970. "Entropy and indistinguishability." *American Journal of Physics*, 38:840–45.

HIEBERT, E. N. 1978. "Nernst, Hermann Walther." *Dictionary of Scientific Biography*, 15:432–53.

HINDMARSH, W. R. 1967. *Atomic Spectra*. Oxford: Pergamon Press.

HIROSIGE, T., & NISIO, S., 1970. "The genesis of the Bohr atom model and Planck's theory of radiation." *Japanese Studies in the History of Science*, 9:35–47.

HIRSCHFELDER, J. O., ed. 1967. *Intermolecular Forces* (*Advances in Chemical Physics*, vol. 12). New York: Interscience.

HIRSCHFELDER, J. O.; BIRD, R. B.; & SPOTZ, E. L. 1949. "The Transport Properties of Gases and Gaseous Mixtures. II." *Chemical Reviews*, 44:205–31.

HIRSCHFELDER, J. O.; CURTISS, C. F.; & BIRD, R. B. 1954. *Molecular Theory of Gases and Liquids*. New York: Wiley.

HOHENBERG, P. C. 1978. "The state of low temperature physics." *Science*, 201:336–37.

HOHENBERG, P. C., & PLATZMAN, P. M. 1966. "High-energy Neutron Scattering from Liquid He^4." *Physical Review*, ser. 2, 152: 198–200.

HOLM, E. 1913. "Anwendung der neueren Plackschen Quantenhypothese zur Berechnung der rotatorischen Energie der zweiatomigen Gases." *Annalen der Physik*, ser. 4, 42:1311–20.

HOOYKAAS, R. 1948. "The First Kinetic Theory of Gases (1727)." *Archives Internationales d'Histoire des Sciences*, 2:180–84.

HOPKINSON, John. 1889. "Magnetic and other physical properties of iron at a high temperature." *Philosophical Transactions of the Royal Society of London*, A180:443–65.

HOYER, Ulrich. 1973. "Über die Rolle der Stabilitätsbetrachtungen in der Entwicklung der Bohrschen Atomtheorie." *Archive for History of Exact Sciences*, 10:177–206.

————. 1974. *Die Geschichte der Bohrschen Atomtheorie.* Weinheim: Physik-Verlag.

————. 1980. "Von Boltzmann zu Planck." *Archive for History of Exact Sciences,* 23:47–86.

HOYLE, F., & SCHWARZSCHILD, M. 1955. "On the evolution of type II stars." *Astrophysical Journal Supplement,* 2:1–40.

HÜCKEL, Erich. 1931. "Quantentheoretische Beiträge zum Benzol-problem." *Zeitschrift für Physik,* 70:204–86; 72:310–37.

HUFBAUER, Karl. 1981. "Astronomers take up the stellar-energy problem, 1917–1920." *Historical Studies in the Physical Sciences,* 11:277–303.

HUND, Friedrich. 1926. "Zur Deutung einiger Erscheinungen in den Molekelspectren." *Zeitschrift für Physik,* 36:657–74.

HUTCHISON, Keith. 1973. "Der Ursprung der Entropiefunktion bei Rankine und Clausius." *Annals of Science,* 30:341–64.

————. 1981a. "W.J.M. Rankine and the rise of thermodynamics." *British Journal for the History of Science,* 14:1–26.

————. 1981b. "Rankine, atomic vortices, and the entropy function." *Archives Internationales d'Histoire des Sciences,* 31:72–134.

HYLAND, G. J., & ROWLANDS, G. 1972. "On the microscopic deriva-tion of the two-fluid Thermohydrodynamic Equations for ^{4}He II." *Journal of Low Temperature Physics,* 7:271–89.

IBERALL, A., & SOODAK, H. 1978. "Physical basis for complex systems—Some Propositions relating Levels of Organization." *Collective Phenomena,* 3:9–24.

IHDE, Aaron. 1964. *The Development of Modern Chemistry.* New York: Harper & Row.

IMRY, Y. 1970. "Self-consistent pairing theory of the Bose superfluid." In *Quantum Fluids,* N. Miser & D. J. Amit, eds., pp. 603–14. New York: Gordon & Breach.

ISING, Ernst. 1925. "Beitrag zur Theorie des Ferromagnetismus." *Zeitschrift für Physik,* 31:253–58.

JACKSON, H. W. 1974. "Reexamination of evidence for a Bose-Einstein Condensate in Superfluid He^4." *Physical Review,* ser. 3, A10: 278–94.

JAMMER, Max. 1966. *The Conceptual Development of Quantum Me-chanics.* New York: McGraw-Hill.

JAUCH, J. M. 1973. "Determinism in classical and quantal physics." *Dialectica,* 27 (1):13–24.

JEVONS, W. Stanley. 1958. *The Principles of Science.* Reprint of 2d ed. (1877). New York: Dover.

JOHNSON, M. C. 1930. "A Method of Calculating the Numerical

Equation of State for Helium below 6° Absolute, and of estimating the relative importance of Gas Degeneracy and Interatomic Forces." *Proceedings of the Physical Society of London*, 42:170–79. Discussion by J. E. Lennard-Jones et al., 179–80.

JOULE, James Prescott. 1843. "On the calorific effects of magneto-electricity and the mechanical value of heat." *Philosophical Magazine*, ser. 3, 23:263–76, 347–55, 435–43.

———. 1848. "Some remarks on heat, and the constitution of elastic fluids." *Memoirs of the Manchester Literary and Philosophical Society*, 9:107–14 (published 1851).

JOULE, James Prescott, & THOMSON, William. 1854. "On the thermal effects of fluids in motion." *Philosophical Transactions of the Royal Society of London*, 321–64.

———. 1862. "On the thermal effects of fluids in motion." *Philosophical Transactions of the Royal Society of London*, 579–89.

KADANOFF, L. 1966. "Scaling laws for Ising models near T_c." *Physics* 2:263–72.

KAHN, B., & UHLENBECK, G. E. 1938. "On the Theory of Condensation." *Physica* 5:399–416.

KAMERLINGH ONNES, Heike. 1911. "Verdere proeven met vloeibaar helium. D. Over de verandering van den galvanischen weerstand van zuivere metalen bij zeer lage temperaturen, enz. V. Het verdwijnen van den weerstand van kwik." *Verslag van de gewone vergaderingen der wis-en natuurkundige Afdeeling*, 20:81–83. "Further experiments with liquid helium. D. On the change of the electrical resistance of pure metals at very low temperatures, etc. V. The disappearance of the resistance of Mercury." *Communications from the Physical Laboratory of the University of Leiden*, no. 122b (1911).

———. 1967. "Investigations into the Properties of Substances at Low Temperatures, which have led, amongst other things, to the preparation of Liquid Helium." Nobel Lecture, December 11, 1913. In *Nobel Lectures, including Presentation Speeches and Laureates' Biographies, Physics 1901–1921*, pp. 306–336. New York: Elsevier.

KANGRO, Hans. 1970. *Vorgeschichte der Planckschen Strahlungsgesetzes.* Weisbaden: Steiner. English translation, *Early History of Planck's Radiation Law.* New York: Crane, Russak, 1976.

———. 1972. *Planck's Original Papers in Quantum Theory.* London: Taylor & Francis.

KAPITZA, P. 1938. "Viscosity of Liquid Helium below the λ-Point." *Nature*, 141:74. Reprinted in AAPT, *Liquid Helium.*

KEESOM, W. H. 1912. "On the Deduction from Boltzmann's entropy principle of the 2nd virial coefficient for material particles (in the limit rigid spheres of central symmetry) which exert central forces upon each other and for rigid spheres of central symmetry containing at electric doublet at their centre." *Communications of the Physical Laboratory, Leiden*, supplement 24b. Translated from *Verslagen van de Afd. Nat. K. Akad. Wetenschappen*, Amsterdam, 1912.

————. 1942. *Helium*. Amsterdam: Elsevier.

KEESOM, W. H., & CLUSIUS, K. 1932. "Specific Heat of Liquid Helium." *Proceedings of the Section of Sciences, Koninklijke Akademie van Wetenschappen te Amsterdam*, 35:307–20. Reprinted in Galasiewicz (1971).

KEESOM, W. H., & KEESOM, A. P. 1932. "On the Anomaly in the Specific Heat of Liquid Helium." *Proceedings of the Section of Sciences, Koninklijke Akademie van Amsterdam*, 35:736–42.

KEESOM, W. H., & WOLFKE, M. 1928. "Two Different Liquid States of Helium." *Proceedings of the Section of Sciences, Koninklijke Akademie van Wetenschappen te Amsterdam*, 31:90–94. Reprinted in Galasiewicz (1971).

KEKULÉ, August. 1865. "Sur la constitution des substances aromatiques." *Bulletin de la Société Chimique de Paris*, 3:98–110.

[Kelvin, Lord] THOMSON, William. 1848. "On an absolute thermometric scale, founded on Carnot's theory of the motive power of heat, and calculated from Regnault's observations." *Philosophical Magazine*, ser. 3, 33:313–17.

————. 1852. "On a Universal Tendency in Nature to the Dissipation of Mechanical Energy." *Proceedings of the Royal Society of Edinburgh*, 3:139–42. *Philosophical Magazine*, ser. 4, 4:304–6.

————. 1874. "The Kinetic Theory of the Dissipation of Energy." *Proceedings of the Royal Society of Edinburgh*, 8:325–34.

KELVIN, Lord. 1894. *Popular Lectures and Addresses*, vol. 2. London: Macmillan.

————. 1901. "Nineteenth Century Clouds over the Dynamical Theory of Heat and Light." *Philosophical Magazine*, ser. 6, 2:1–40.

KHALATNIKOV, I. M. 1950. "Pogloshchenie zvuka v gelii II." *Zhurnal eksperimentalnoi i teoreticheskoi fiziki*, 20:243–66. German translation in *Abhandlungen aus der Sowjetischen Physk*, 1:191–220 (1951). For English translation see *Technical Translations*, 3:811 (1960).

————. 1952. "Kineticheskie koeffitsienty v gelii II" & "Teploprovodnost' i pogloshchenie zvuka v gelii II." *Zhurnal eksperimentalnoi i teoreticheskoi fiziki*, 23:8–20, 21–34. English translations, RT-2672 and RT-2905, available from Reilly Translations, 3690 Jasmine Avenue, Los Angeles, CA 90034.

————. 1976. "Phenomenological Theory of Superfluid ^{4}He." In Bennemann & Ketterson (1976), pp. 1–84.

KIHARA, Taro, & KOTANI, Maso. 1943. "Determination of Intermolecular Forces from Transport Phenomena in Gases. II." *Proceedings of the Physico-Mathematical Society of Japan*, ser. 3,
25:602–14.

KIKUCHI, Ryoichi. 1954. "λ Transition of Liquid Helium." *Physical
Review*, ser. 2, 96:563–68.

KIKUCHI, R.; DENMAN, H. H.; & SCHRIEBER, C. L. 1960. "Statistical mechanics of Liquid He^4." *Physical Review*, ser. 2, 119:1823–31.

KINGSTON, A. E. 1965. "Derivation of Interatomic Potentials for Inert-
Gas Atoms from the Second Virial Coefficient." *Journal of Chemical
Physics*, 42:719–22.

KIRKWOOD, J. G. 1935. "Statistical mechanics of fluid mixtures."
Journal of Chemical Physics, 3:300–312.

————. 1938. "Order and disorder in binary solid solutions." *Journal of
Chemical Physics*, 6:70–75.

KIRKWOOD, J. G., & MONROE, Elizabeth. 1941. "Statistical Mechanics of Fusion." *Journal of Chemical Physics*, 9:514–26.

KIRZHNITS, D. A. 1960. "On the Internal Structure of Super-Dense
Stars." *Soviet Physics JETP*, 11 : 365–68. Translated from *Zhurnal
Eksperimentalnoi i Teoreticheskoi Fiziki*, 38:503–8.

KLEIN, M. L., & MUNN, R. J. 1967, "Interaction potential of the inert
gases." *Journal of Chemical Physics*, 47:1035–37.

KLEIN, Martin J. 1962. "Max Planck and the Beginnings of the Quantum Theory." *Archive for History of Exact Sciences*, 1:459–79.

————. 1963a. "Planck, entropy, and quanta 1901–1906." *The Natural
Philosopher*, 1:83–108.

————. 1963b. "Einstein's first paper on quanta." *The Natural Philosopher*, 2:57–86.

————. 1964. "Einstein and the wave-particle duality." *The Natural
Philosopher*, 3:1–49.

————. 1965. "Einstein, specific heats, and the early quantum theory."
Science, 148:173–80.

————. 1966. "Thermodynamics and quanta in Planck's work." *Physics
Today*, 19 (11):23–32.

————. 1967. "Thermodynamics in Einstein's thought." *Science*, 157:
509–16.

————. 1970. *Paul Ehrenfest*. Vol. 1. *The Making of a Theoretical Physicist*. New York: American Elsevier.

————. 1972. "Mechanical Explanation at the End of the Nineteenth
Century." *Centaurus*, 17:58–82.

————. 1975. "Einstein, Boltzmann's principle, and the mechanical world view." *Proceedings of the XIV International Congress of History of Science, 1974,* 1:183–94.

KNOTT, Gargill Gilston. 1911. *Life and Scientific Work of Peter Guthrie Tait.* Cambridge: At the University Press.

KOHLER, Robert E., Jr. 1974. "Irving Langmuir and the 'Octet' Theory of Valence." *Historical Studies in the Physical Sciences,* 4:39–87.

————. 1975a. "G. N. Lewis's views on Bond Theory 1900–1916." *British Journal for the History of Science,* 8:233–39.

————. 1975b. "The Lewis-Langmuir Theory of Valence and the Chemical Community, 1920–1928." *Historical Studies in the Physical Sciences,* 6: 431–68.

KRAGH, Helge. 1977. "Chemical aspects of Bohr's 1913 theory." *Journal of Chemical Education,* 54:208–10.

————. 1979. *On the History of Early Wave Mechanics, with special emphasis on the role of relativity.* Roskilde, Denmark: Roskilde Universitetscenter, IMFUFA, Tekst nr. 23.

KRAICHNAN, R. H. 1959. "Condensation of an Imperfect Bose Gas." *Physics of Fluids,* 2:463–65.

KRAMERS, H. A. 1936. "Zur Theorie des Ferromagnetismus." *Rapports et Communications, No. 29, 7ᵉ Congres International du Froid, la Haye-Amsterdam, juin* 1930 (La Haye: M. Nijhoff), pp. 279–300. Reprinted, *Communications from the Kamerlingh Onnes Laboratory of the University of Leiden,* supplement no. 83.

KRAMERS, H. A., & WANNIER, G. H. 1941. "Statistics of the two-dimensional ferromagnet" *Physica Review,* ser. 2, 60:252–62, 263–76.

KRAMERS, H. C. 1957. "Liquid Helium below 1°K." In *Progress in Low Temperature Physics,* C. J. Gorter, ed., vol. II, pp. 59–82. Amsterdam: North-Holland.

KRÖNIG, August Karl. 1856. "Grundzuge einer Theorie der Gase." *Annalen der Physik,* ser. 2, 99:315–22.

KUBLI, F. 1970. "Louis de Broglie und die Entdeckung der Materie-wellen." *Archive for History of Exact Sciences,* 7:26–68.

KUBO, Ryogo. 1979. "Statistical mechanics: A survey of its One Hundred Years." In *Scientific Culture in the Contemporary World,* V. Mathieu & P. Rossi, eds., pp. 131–57. Milan: Scientia.

KUHN, Thomas S. 1959. "Energy conservation as an example of simultaneous discovery." In *Critical Problems in the History of Science,* M. Clagett, ed., pp. 321–56. Madison: University of Wisconsin Press.

————. 1974. "Second Thoughts on Paradigms." In *The Structure of Scientific Theories,* Frederick Suppe, ed., pp. 459–82.

———. 1975. "The Quantum Theory of Specific Heats: A Problem in Professional Recognition." *Proceedings of the XIV International Congress of History of Science*, Tokyo, 1974, 1:170–82; 4:207.

———. 1978. *Black-body theory and the quantum discontinuity: 1894–1912*. New York: Oxford University Press.

LAGOWSKI, J. J. 1966. *The Chemical Bond*. Boston: Houghton Mifflin.

LAKATOS, Imre. 1971. "History of Science and its Rational Reconstructions." *Boston Studies in the Philosophy of Science*, 8:91–136 (PSA 1970).

LANDAU, L. D. 1930. "Diamagnetismus der Metalle." *Zeitschrift für Physik*, 65:629–37. English translation in Landau (1965).

———. 1932. "On the Theory of Stars." *Physikalische Zeitschrift der Sowjetunion*, 1:285–88.

———. 1937. "K teorii fazovnykh perekhodov." *Zhurnal eksperimentalnoi i teoreticheskoi fiziki*, 7:19–32. "On the theory of phase transitions," in Landau (1965).

———. 1938. "The intermediate state of supraconductors." *Nature*, 141:688.

———. 1941. "The Theory of Superfluidity of Helium II." *Journal of Physics, Academy of Sciences of the USSR*, 5:71–90.

———. 1941. "The Theory of Superfluidity of Helium II." *Journal of Physics* (USSR), 5:71–90. Reprinted in Galasiewicz (1971) and in Landau (1965).

———. 1943. "K teorii promeshutochnogo sostoyaniya sverkhprovodnikov." *Zhurnal eksperimentalnoi i teoreticheskoi fizike*, 13:377–87. "On the Theory of the Intermediate State of Superconductors," *Journal of Physics of the USSR*, 7:99–107.

———. 1947. "On the theory of Superfluidity of Helium II." *Journal of Physics* (USSR), 11:91–92. Reprinted in Landau (1965).

———. 1948. "K teorii sverkhtekuchesti," *Doklady Akademiii Nauk SSSR*, 61:253–256. "On the Theory of Superfluidity." *Physical Review*, ser. 2, 75:884–85 (1949). Reprinted in Landau (1965).

———. 1965. *Collected Papers of L. D. Landau*. Edited and with an Introduction by D. ter Haar. New York: Pergamon Press.

LANDÉ, Alfred. 1953. "Probability in Classical and Quantum Theory." In *Scientific Papers presented to Max Born*, pp. 59–64. New York.

LANDSBERG, P. T. 1954. "On Bose-Einstein Condensation." *Proceedings of the Cambridge Philosophical Society*, 50:65–76.

LANE, C. T. 1967. "Resource Letter LH-1 on Liquid Helium." *American Journal of Physics*, 35:367–75. Reprinted in AAPT, *Liquid Helium*.

LANGEVIN, Paul. 1913. "La Physique de Discontinu" [Conférence faite a la Société française de Physique le 27 novembre 1913]. In *Les Progrès de la Physique Moléculaire*, pp. 1–46. Paris.

LANGMUIR, Irving. 1919 "The structure of atoms and the octet theory of valence." *Proceedings of the National Academy of Sciences*, USA, 5:252–59. "The arrangement of electrons in atoms and molecules," *Journal of the Franklin Institute*, 187:359–64.

LAPLACE, P. S. de. 1816. "Sur la vitesse du son dans l'air et dans l'eau." *Annales de Chimie*, 3:238–41.

LARKIN, A. I., & KHMEL'NITSKII, D. E. 1969. "Phase transitions in uniaxial ferroelectrics." *Soviet Physics JETP*, 29:1123–28.

LEE, T. D. 1952. *See also* Yang (1952)

LEE, T. D., & MOHLING, F. 1959. "Possible Determination of the Helicity of Elementary Excitations in Liquid He II." *Physical Review Letters*, 2:284–85.

LEE, T. D., & YANG, C. N. 1952. "Statistical theory of equations of state and phase transitions. II. Lattice gas and Ising model." *Physical Review*, ser. 2, 87:410–19.

LE GUILLOU, J. C., & ZINN-JUSTIN, J. 1977. "Critical exponents for the n-vector model in three dimensions from field theory." *Physical Review Letters*, 39:95–98.

LEIBFRIED, Günther. 1947. "Über die Anwendbarkeit der Sattelpunkts-methode bei tiefen Temperaturen am Beispiel des idealen Bose-Gases." *Zeitschrift für Naturforschung*, 2A:305–10.

[LENNARD-] JONES, J. E. 1925a. "On the Atomic Fields of Helium and Neon." *Proceedings of the Royal Society of London*, A107:157–70.

LENNARD-JONES, J. E. 1925b. "On the Forces between Atoms and Ions." *Proceedings of the Royal Society of London*, A109:584–97.

———. 1929. "The electronic structure of some Diatomic Molecules." *Transactions of the Faraday Society*, 25:668–86.

———. 1931. "Cohesion." *Proceedings of the Physical Society of London*, 43:461–82.

———. 1936. "Interatomic Forces." Chapter X in *Statistical Mechanics*, 2d ed., by R. H. Fowler. Cambridge: At the University Press.

LENNARD-JONES, J. E., & DEVONSHIRE, A. F. 1937. "Critical Phenomena in Gases—I." *Proceedings of the Royal Society of London*, A163:53–70.

———. 1938. "Critical phenomena in gases. II. Vapour pressures and boiling points." *Proceedings of the Royal Society of London*, A165:1–11.

————. 1939. "Critical and cooperative phenomena." *Proceedings of the Royal Society of London*, A169:317–38; A170:464–84.

LENZ, W. 1920. "Beitrag zum Verständnis der magnetischen Erscheinungen in fester Körpern." *Physikalische Zeitschrift*, 21:613–15.

LEVELT SENGERS, J.M.H. 1975. "Liquidons and Gasons: Persistence of the Liquid State beyond the Critical Point." Colloquium at National Bureau of Standards, Gaithersburg, Maryland.

————. 1976. "Critical exponents at the turn of the century." *Physica* 82A:319–51.

————. 1979. "Liquidons and gasons: controversies about the continuity of states." *Physcia* 98A:363–402.

LEVELT SENGERS, J.M.H., & SENGERS, J. V. 1981. "How close is 'close to the critical point'?" In *Perspectives in Statistical Physics*, H. J. Raveche, ed., pp. 240–71. Amsterdam: North-Holland.

LEWIS, G. N. 1916. "The atom and the molecule." *Journal of the American Chemical Society*, 38:762–85.

————. 1923. *Valence and the Structure of Atoms and Molecules*. Reprint, New York: Dover, 1966.

LIEB, E. H. 1979. "Why matter is stable." *Chinese Journal of Physics*, 17:49–62.

LIFSHITZ, E. 1944. "Radiation of Sound in Helium II." *Journal of Physics* (USSR), 8:110–14. Translated from *Zhurnal eksperimentalnoi i teoreticheskoi fiziki*, 14:116–20.

————. 1969. "Lev Davidovich Landau (1908–1968)." *Uspekhi Fizicheskikh Nauk*, 97:169–86. English translation in *Soviet Physics Uspekhi*, 12: 135–45.

LINDSAY, Robert Bruce. 1973. *Julius Robert Mayer: Prophet of Energy*. New York: Pergamon Press.

LINUS, Franciscus [alias Francis Hall]. 1661. *Tractatus de corporum inseparabilitate, in quo experimenta de vacuo tam Torricelliana quam Magdeburgica et Boyliana examinantur*. London: Martin.

LIPPICH, F. 1870. "Ueber die Breite der Spectrallinien." *Annalen der Physik*, ser. 2, 139:465–79.

LIVANOVA, Anna, 1980. *Landau: A Great Physicist and Teacher*. Translated from Russian by J. B. Sykes. New York: Pergamon Press.

LONDON, F. 1930. "Über einige Eigenschaften und Anwendungen der Molekularkrafte." *Zeitschrift für Physikalische Chemie*, B11:222–25.

————. 1936. "On Condensed Helium at Absolute Zero." *Proceedings of the Royal Society of London*, A153:576–83.

————. 1938a. "The λ-Phenomenon of Liquid Helium and the Bose-Einstein Degeneracy." *Nature*, 141:643–44. Reprinted in AAPT, *Liquid Helium*.

————. 1938b. "On the Bose-Einstein Condensation." *Physical Review*, ser. 2, 54:947–54.

————. 1939. "The State of Liquid Helium near Absolute Zero." *Journal of Physical Chemistry*. 43:49–69.

————. 1945. "Planck's Constant and Low Temperature Transfer." *Reviews of Modern Physics*, 17:310–20.

————. 1950. *Superfluids*. Vol. 1. *Macroscopic Theory of Superconductivity*. New York: Wiley. 2d rev. ed. with essays by L. W. Nordheim, Edith London, and M. J. Buckingham. New York: Dover, 1961.

————. 1954. *Superfluids*. Vol. II. *Macroscopic Theory of Liquid Helium*. New York: Wiley.

LONDON, Fritz, & LONDON, Heinz. 1935. "Supraleitung und Diamagnestismus." *Physica*, 2:341–54.

LONDON, F., & RICE, O. K. 1948. "On Solutions of He^3 in He^4." *Physical Review*, ser. 2, 73:1188–93.

LONDON, H. 1960. "Superfluid Helium." *Yearbook of the Physical Society of London*, 34–48.

LONGUET-HIGGINS, H. Christopher, & FISHER, Michael E. 1978. "Lars Onsager 1903–1976." *Biographical Memoirs of Fellows of the Royal Society*, 24:443–71.

LORENTZ, H. A. 1905. "The motion of electrons in metallic bodies." *Proceedings of the Section of Sciences, Akademie van Wetenschappen, Amsterdam*, 7:438–53, 585–93, 684–91.

————. 1909. *The Theory of Electrons and its Applications to the Phenomena of Light and Radiant Heat*. Leipzig: Teubner. New York: Stechert. 2d ed., 1915. Reprinted, New York: Dover, 1952.

LORENTZEN, Hans Ludvig. 1953. "Studies of critical phenomena in carbon dioxide contained in vertical tubes." *Acta Chemica Scandinavica*, 7:1335–46. Reprinted in Stanley (1973).

————. 1965. "Preliminary study of carbon dioxide at critical conditions." In *Statistical Mechanics of Equilibrium and Non-Equilibrium*, J. Meixner, ed., pp. 262–74. Amsterdam: North-Holland.

LOSCHMIDT, Josef. 1865. "Zur Grösse der Luftmolecüle." *Sitzungsberichte, K. Akademie der Wissenschaften in Wien, Math.-Naturwiss. Kl.*, 52:395–413.

————. 1876. "Über den Zustand des Wärmegleichgewichtes eines Systemes von Körpern mit Rucksicht auf die Schwerkraft." *Sitzungsberichte, K. Akademie der Wissenschaften in Wien, Math.-Naturwiss. Kl.*, 73:128–42.

LUDWIG, G. 1968. *Wave Mechanics*. New York: Pergamon Press.

MacKINNON, Edward. 1976. "De Broglie's thesis: a critical retrospective." *American Journal of Physics*, 44:1047–55.

————. 1982. *Scientific Explanation and Atomic Physics*. Chicago: University of Chicago Press.

MAHY, Margaret. 1975. *Ultra-Violet Catastrophe! or the Unexpected Walk with Great-Uncle Magnus Pringle*. New York: Parents' Magazine Press.

MARCH, N. H., & GALASIEWICZ, Z. M. 1976. "Superfluidity, Ground-State Wave Function and Bose Condensation in Liquid Helium Four." *Physics and Chemistry of Liquids*, 5:103–8.

MARGENAU, Henry. 1950. *The Nature of Physical Reality*. New York: McGraw-Hill.

MARGENAU, Henry, & KESTNER, N. R. 1971. *Theory of Intermolecular Forces*, 2d ed. New York: Pergamon Press.

MARIOTTE, Edme. 1679. *Essay de la Nature de l'Air*. Paris.

MARIS, Humphrey J. 1974. "Hydrodynamics of Superfluid Helium below 0.6°K. III. Propagation of Temperature Waves." *Physical Review*, ser. 3, A9:1412–26.

MASON, E. A., & RICE, W. E. 1954. "The Intermolecular Potentials of Helium and Hydrogen." *Journal of Chemical Physics* 22:522–35.

MATSUBARA, Takeo, & MATSUDA, Hirotsugu. 1956, 1957. "A Lattice Model of Liquid Helium." *Progress of Theoretical Physics*, 16: 416–17, 569–82; 17:19–29; 18:357–66.

MATTHIAS, B. T.; GEBALLE, T. H.; & COMPTON, V. B. 1963. "Superconductivity." *Reviews of Modern Physics*, 35:1–22.

MAURER, R. D., & HERLIN, Melvin A. 1949. "Second Sound Velocity in Helium II." *Physical Review*, ser. 2, 76:948–50.

MAXWELL, Emanuel. 1950. "Isotope effect in the superconductivity of Mercury." *Physical Review*, ser. 2, 78:477.

MAXWELL, James Clerk. 1860a. "Illustrations of the Dynamical Theory of Gases." *Philosophical Magazine*, ser. 4, 19:19–32, 20:21–37.

————. 1860b. "On the Results of Bernoulli's Theory of Gases as applied to their Internal Friction, their Diffusion, and their Conductivity for Heat." *Report of the 30th Meeting of the British Association for the Advancement of Science*, pt. 2, 15–16.

————. 1867. "On the Dynamical Theory of Gases." *Philosophical Transactions of the Royal Society of London*, 157:49–88. Reprinted in Brush (1966).

————. 1871. *Theory of Heat*. London: Longmans, Green.

————. 1873. "Molecules." *Nature*, 8:437–41.

————. 1875a. "On the Dynamical Evidence of the Molecular Constitution of Bodies." *Nature*, 11:357–59, 374–77.

————. 1875b. "Atom." *Encyclopedia Britannica*, 8th ed., vol. 3, pp. 36–49.

————. 1877. "The Kinetic Theory of Gases." *Nature*, 16:242–46.
————. 1879. "On Boltzmann's theorem on the average distribution of energy in a system of material points." *Transactions of the Cambridge Philosophical Society*, 12:547–70.
————. 1965. *The Scientific Papers of James Clerk Maxwell*. Reprint of 1890 ed. New York: Dover Pubs.
MAYER, J. E. 1937. "The statistical mechanics of condensing systems. I." *Journal of Chemical Physics*, 5:67–73.
————. 1938. *See* Harrison, S. F.
————. 1951. "Note on the theory of critical phenomena." *Journal of Chemical Physics*, 19:1024–26.
MAYER, J. E., & HARRISON, S. F. 1938. "Statistical Mechanics of condensing systems. III." *Journal of Chemical Physics*, 5:87–100.
MAYER, J. E., & MAYER, M. G. 1940. *Statistical Mechanics*. New York: Wiley.
————. 1977. *Statistical Mechanics*, 2d ed. New York: Wiley.
MAYER, Maria Goeppert, & JENSEN, J. Hans D. 1955. *Elementary Theory of Nuclear Shell Structure*. New York: Wiley.
McCORMMACH, Russell. 1967. "Henri Poincaré and the Quantum Theory." *Isis* 58:37–55.
McCREA, W. H. 1929. "The Hydrogen Chromosphere." *Monthly Notices of the Royal Astronomical Society*, 89:483–97.
McGUCKEN, William. 1969. *Nineteenth-Century Spectroscopy*. Baltimore: Johns Hopkins University Press.
McGUINNESS, B., ed. 1974. *Ludwig Boltzmann, Theoretical Physics and Philosophical Problems*. Dordrecht & Boston: Riedel.
McLENNAN, J. C.; SMITH, H. D.; & WILHELM, J. O. 1932. "The Scattering of Light by Liquid Helium." *Philosophical Magazine*, ser. 7, 14:161–67.
McMILLAN, W. G., Jr., & MAYER, J. E. 1945. "The statistical thermodynamics of multicomponent systems." *Journal of Chemical Physics*, 13:276–305.
MEISSNER, W., & OCHSENFELD, R. 1933. "Ein neuer Effekt bei Eintritt der Supraleitfähigkeit." *Naturwissenschaften*, 21:787–88.
MENDELSSOHN, Kurt. 1977. *The Quest for absolute zero: The meaning of low Temperature Physics*, 2d ed. London: Taylor & Francis.
MESHKOVSKII, A., & SHALNIKOV, A. 1947. "Poverkhnostnie yavleniya u sverkhprovodnikov v promeshutochnom sostoyanii." *Zhurnal eksperimentalnoi i teoreticheskoi fizike*, 17:851–61. "The structure of superconductors in the intermediate state, II." *Journal of Physics of the USSR*, 11:1–15.

MEYER, L. 1969. "Phase transitions in van der Waal's [sic] lattices." *Advances in Chemical Physics*, 16:343–87.

MICHELS, A.; BIJL, A.; & DE BOER, J. 1937. "Effect of an excitation energy on the Specific Heat of Liquid Helium II and its Relation to the Exchange Effect in a Non-Ideal Bose-Einstein gas." *Nature*, 144:594–95.

MICHELS, A.; BLAISSE, B.; & MICHELS, C. 1937. "The isotherms of CO_2 in the neighborhood of the critical point and round the coexistence line." *Proceedings of the Royal Society of London*, 160:358–75.

MICHAELS, Stella; GREEN, Melville S.; & LARSEN, Sigurd Y. 1970. *Equilibrium Critical Phenomena in Fluids and Mixtures: A comprehensive Bibliography with Key-Word Descriptors*. Washington, D.C.: National Bureau of Standards, Spec. Pub. 327.

MILLER, Arthur I. 1976. "On Einstein, light quanta, radiation, and relativity in 1905." *American Journal of Physics*, 44:912–23.

———. 1978. "Visualization lost and regained: the genesis of the quantum theory in the period 1913–27." In *On Aesthetics in Science*, Judith Wechsler, ed., pp. 72–102. Cambridge, Mass.: MIT Press.

MILLIKAN, Robert A. 1916. "A direct photoelectric determination of Planck's 'h,'" *Physical Review*, ser. 2, 7:355–88.

———. 1924. "The electron and the light-quant from the experimental point of view." Nobel Lecture. In *Nobel Lectures, Physics, 1922—1941*, pp. 54–66. New York: Elsevier, 1965.

MILNE, E. A. 1921. "Ionization in stellar atmospheres." *Observatory*, 44:261–69.

———. 1924. "Recent work in stellar physics." *Proceedings of the Physical Society of London*, 36:94–113.

MISRA, B., & PRIGOGINE, I. 1982. "Time, probability, and dynamics." In *Long-Time Prediction in Dynamics*, C. W. Horton, L. E. Reichl, & V. G. Szebehely, eds. New York: Wiley.

MITROPOL'SKII, Y. A., & TYABLIKOV, S. V. "Nikolai Nikolaevich Bogolyubov (on the occasion of his fiftieth birthday). *Soviet Physics Uspekhi*, 2:765–70.

MOELWYN-HUGHES, E. A. 1961. *Physical Chemistry*, 2d rev. ed. Oxford: Pergamon Press.

MØLLER, Chr., & CHANDRASEKHAR, S. 1935. "Relativistic Degeneracy." *Monthly Notices of the Royal Astronomical Society*, 95:673–76.

MONTROLL, E. W. 1977. "Obituary. Lars Onsager." *Physics Today*, 30 (2):77.

MONTROLL, E. W.; POTTS, R. B.; & WARD, J. C. 1963. "Correlations and spontaneous magnetization of the two-dimensional Ising model." *Journal of Mathematical Physics*, 4:308–22.

MOORE, Ruth. 1966. *Niels Bohr*. New York: Knopf.

MOSER, J., ed. 1975. *Dynamic Systems, Theory and Applications*. New York: Springer.

MUELLER, Erwin W. 1956. "Resolution of the atomic structure of a metal surface by the field ion microscope." *Journal of Applied Physics*, 27:474–77.

MUELLER, K. H.; POBELL, F.; & AHLERS, G. 1975. "Thermal-expansion coefficient and universality near the superfluid transition of ^{4}He under pressure." *Physical Review Letters*, 34:513–16.

MULLIKEN, Robert S. 1928. "Assignment of Quantum Numbers for Electrons in Molecules. Part I." *Physical Review*, ser. 2, 32:186–222.

———. 1932. "Electronic Structures of Polyatomic Molecules and Valence. II. General Considerations." *Physical Review*, ser. 2, 41:49–71.

———. 1975. *Selected Papers*. Chicago: University of Chicago Press.

MUTO, Y. 1948. "On the order-disorder transition in solids." *Journal of Chemical Physics*, 16:519–23, 524–25, 1176.

NACHTRIEB, Norman H. 1975. "Interview with Robert S. Mulliken." *Journal of Chemical Education*, 52:560–64.

NAGEL, Ernest. 1955. "Principles of the Theory of Probability." In *International Encyclopedia of Unified Science*, O. Neurath et al., eds., vol. 1. Chicago: University of Chicago Press. Reprint of 1939 ed.

NAKAJIMA, S.; KURODA, Y.; & KURIHARA, Y. 1971. "On the Pair Theory of the Many Boson System." In *Proceedings of the 12th International Conference on Low Temperature Physics*, Kyoto, 1970, E. Kanda, ed., pp. 197–198. Tokyo: Academic Press of Japan.

NEEDHAM, Joseph. 1962. *Science and Civilisation in China*, vol. 4. New York: Cambridge University Press.

NERNST, Walther. 1906. "Ueber die Berechnung chemischer Gleich-gewichte aus thermischer Messungen." *Nachrichten von der Königl. Gesellschaft der Wissenschaften zu Göttingen, Mathematisch-Physikalische Klasse*, 1–39.

———. 1911. "Zur Theorie der spezifischen Wärme und über die Anwendung der Lehre von den Energiequanten auf physikalisch-chemische Fragen überhaupt." *Zeitschrift für Elektrochemie*, 17:265–75.

———. 1914. "Über die Anwendung des neuen Wärmesatzes auf Gase." *Zeitschrift für Elektrochemie*, 20:357–60.

———. 1916. "Über einen Versuch, von quantentheoretischen Betracht-ungen zur Annahme stetiger Energieänderungen zurückzukehren." *Verhandlungen der Deutschen Physikalischen Gesellschaft*, 18:83–116.

———. 1918. *Die Theoretische und Experimentellen Grundlagen des neuen Wärmesatzes*. Halle: Knapp.

———. 1919. "Einige Folgerungen aus der sogenannten Entartung-stheorie der Gase." In *Sitzungsberichte der preussische Akademie der Wissenschaften* (Berlin), pp. 118–27.

———. 1922. "Zum Gultigkeitsbereich der Naturgesetze." *Naturwissenschaften*, 10:489–95.

———. 1926. *The New Heat Theorem*. Translated from the 2d German ed. (1924). London: Methuen.

———. 1969. *The New Heat Theorem*. Translated from the 2d German ed. (1924). New York: Dover Pubs.

NERNST, W., & LINDEMANN, F. A. 1911. "Spezifische Wärme und Quantentheorie." *Zeitschrift für Elektrochemie*, 17:817–27.

NEWTON, Isaac. 1672. "A new theory about light and colours." *Philosophical Transactions of the Royal Society of London*, no. 80, pp. 3075–.

———. 1687. *Philosophiae Naturalis Principia Mathematica*, 1st ed. London. 2d ed., 1713; 3rd ed., 1726. *Sir Isaac Newton's Mathematical Principles of Natural Philosophy and His System of the World*, translated by A. Motte, rev. and annotated by F. Cajori. Berkeley: University of California Press, 1934.

———. 1701. "Scala graduum Caloris." *Philosophical Transactions of the Royal Society of London*, 22:824–29. English translation (1809) reprinted with original, in *Isaac Newton's Papers & Letters on Natural Philosophy*, I. B. Cohen, ed., pp. 259–68. Cambridge, Mass.: Harvard University Press, 1958.

———. 1730. *Opticks*, 4th ed. London: Innys. Reprinted, New York: Dover Pubs., 1952.

NISIO, Sigeko. 1967. "The role of the chemical considerations in the development of the Bohr atom model." *Japanese Studies in History of Science*, 6:26–40.

NYE, Mary Jo. 1972. *Molecular Reality: A Perspective on the Life of Jean Perrin*. New York: American Elsevier.

NYHOF, John. 1981. *Instrumentalism and Beyond*. Ph.D. Dissertation, University of Otago, Dunedin, New Zealand.

ÖPIK, E. J. 1951. "Stellar models with variable composition. II. Sequences of models with energy generation proportional to the fifteenth power of temperature." *Proceedings of the Royal Irish Academy*, 54:49–77.

OGG, Richard A., Jr. 1946. "Bose-Einstein condensation of trapped electron pairs, Phase separation and superconductivity of metal-ammonia solutions." *Physical Review*, ser. 2, 69:243–44, 544.

ONO, S. 1947. "Statistical thermodynamics of critical and surface phenomena." *Memoirs of the Faculty of Engineering, Kyushu University*, 10 (4):196–225.

ONSAGER, Lars. 1944. "Crystal Statistics. I. A Two-Dimensional Model with an Order-Disorder Transition." *Physical Review*, ser. 2, 65: 117–49.

———. 1949. Discussion Remark. *Nuovo Cimento*, supplement, 6:249.

———. 1958. Discussion remark. *Nuovo Cimento*, ser. 10, supplement, 9:261.

OSBORNE, Darrell W.; WEINSTOCK, Bernard B.; & ABRAHAM, Bernard M. 1949. "Comparison of the Flow of Isotopically Pure Liquid He^3 and He^4." *Physical Review*, ser. 3, 75:988.

OSTGAARD, E. 1971. "Liquid ^{4}He. I. Binding Energy and Excitation Energy spectrum from Two-Body Correlations" & "Liquid ^{4}He. II. Three-body correlations and their contribution to the binding energy and the depletion." *Journal of Low Temperature Physics*, 4:239–62, 585–604.

OSTRIKER, J. P., & BODENHEIMER, P. 1968. "Rapidly rotating stars. II. Massive white dwarfs." *Astrophysical Journal*, 151:1089–98.

OSTRIKER, J. P.; BODENHEIMER, P.; & LYNDEN-BELL, D. 1966. "Equilibrium models of differentially rotating zero-temperature stars." *Physical Review Letters*, 17:816–18.

PAIS, A. 1979. "Einstein and the quantum theory." *Reviews of Modern Physics*, 51:861–914.

PALEVSKY, H.; OTNES, K.; LARSSON, K. E.; PAULI, R.; & STEDMAN, R. 1957. "Excitation of rotons in helium II by cold neutrons." *Physical Review*, ser. 2, 108:1346–47.

PALMER, W. G. 1965. *A History of the Concept of Valency to 1930*. New York: Cambridge University Press.

PARADOWSKI, Robert J. 1972. *The Structural Chemistry of Linus Pauling*. Ph.D. Dissertation, University of Wisconsin.

PARRY, W. E., & TER HAAR, D. 1962. "On the Theory of Relaxation Processes in Liquid Helium." *Annals of Physics*, 19:496–539.

PARTINGTON, J. R. 1962. *A History of Chemistry*, vol. III. London: Macmillan.

PAULI, Wolfgang. 1925. "Über den Zusammenhang des Abschlusses der Elektronengruppen im Atom mit der Komplexstruktur der Spektren." *Zeitschrift für Physik*, 31:765–83. English translation in ter Haar (1967).

———. 1927. "Über Gasentartung und Paramagnetismus." *Zeitschrift für Physik*, 41:81–102.

———, ed. 1955. *Niels Bohr and the Development of Physics*. London: Pergamon Press.

PAULING, Linus. 1928a. "The Application of the Quantum Mechanics to the Structure of the Hydrogen Molecule and Hydrogen Molecule-Ion and to Related Problems." *Chemical Reviews*, 5:173–213.

————. 1928b. "The shared-electron chemical bond." *Proceedings of the National Academy of Sciences*, 14:359–62.

————. 1931. "The nature of the Chemical Bond. Application of Results obtained from the Quantum Mechanics and from a Theory of Paramagnetic Susceptibility to the structure of molecules." *Journal of the American Chemical Society*, 53:1367–1400.

————. 1939. *The Nature of the Chemical Bond*. Ithaca: Cornell University Press. 2d ed., 1940; 3d ed., 1960.

————. 1970. "Fifty Years of Progress in Structural Chemistry and Molecular Biology." *Daedalus*, 99:988–1014.

PAULING, Linus, & WHELAND, G. W. 1933. "The Nature of the Chemical Bond. V. The Quantum-Mechanical Calculation of the Resonance Energy of Benzene and Naphthalene and the Hydrocarbon Free Radicals." *Journal of Chemical Physics*, 1:362–74. Errata, *ibid.*, 2:482 (1934).

PAULING, Linus, & WILSON, E. B. 1935. *Introduction to Quantum Mechanics with Applications to Chemistry*. New York: McGraw-Hill.

PAYNE, Cecilia H. 1925a. "Astrophysical data bearing on the relative abundance of elements." *Proceedings of the National Academy of Sciences* (USA), 11:192–98.

————. 1925b. *Stellar Atmospheres*. Harvard Observatory Monographs, No. 1. Cambridge, Mass.: Harvard University Press.

PEIERLS, R. 1936. "On Ising's Model of Ferromagnetism." *Proceedings of the Cambridge Philosophical Society*, 32:477–81.

PELLAM, John R., & SCOTT, Russell B. 1949. "Second Sound Velocity in paramagnetically cooled Liquid Helium II." *Physical Review*, ser. 2, 76:869–70.

PENROSE, O. 1951. "On the quantum mechanics of Helium II." *Philosophical Magazine*, ser. 7, 42:1373-77.

PENROSE, O., & ONSAGER, L. 1956. "Bose-Einstein Condensation and Liquid Helium." *Physical Review*, ser. 2, 104:576–84.

PESHKOV, V. 1944. "'Second Sound' in Helium II." *Journal of Physics* (USSR), 8:381.

————. 1946. "Determination of the Velocity of Propagation of the Second Sound in Helium II." *Journal of Physics* (USSR), 10:389–98. Reprinted in Galasiewicz (1971).

————. 1948. "Skorost' vtorogo zvuka ot 1.3 do 1.pe°K." *Zhurnal eksperimentalnoi i teoreticheskoi fiziki*, 18:951–52.

————. 1960. "Second Sound in Helium II." *Soviet Physics JETP*, 11:580–84. Translated from *Zhurnal eksperimentalnoi i teoreticheskoi fiziki*, 38:799–805.

PLANCK, Max. 1898. "Über irreversible Strahlungsvorgänge. 4. Mitteilung." *Sitzungsberichte, Preuss. Akademie der Wissenschaften, Berlin, Phys.-Math. Kl.*, 449–476.

———. 1899, "Üeber irreversible Strahlungsvorgänge. 5. Mitteilung." *Sitzungsberichte der Preussischen Akademie der Wissenschaften, Physikalisch-mathematische Klass, Berlin*, 440–80.

———. 1900a. "Ueber eine Verbesserung der Wien'schen Spectralgleichung." *Verhandlungen der Deutschen physikalischen Gesellschaft*, 2:202–04. English translation in ter Haar (1967), Kangro (1972).

———. 1900b. "Zur Theorie des Gesetzes der Energieverteilung im Normalspectrum." *Verhandlungen der Deutschen physikalischen Gesellschaft*, 2:237–45. English translation in ter Haar (1967), Kangro (1972).

———. 1900c. "Ueber irreversible Strahlungsvorgänge." *Annalen der Physik*, ser. 4, 1:69–122.

———. 1909. "Die Einheit des physikalischen Weltbildes." *Physikalische Zeitschrift*, 10:62–75. English translation in Planck (1960).

———. 1910. "Zur Theorie der Wärmestrahlung." *Annalen der Physik*, ser. 4, 31:758–68.

———. 1911. "Eine neue Strahlungshypothese." *Verhandlungen der Deutsche Physikalische Gesellschaft*, 13:138–48.

———. 1915. "Die Quantenhypothese für Molekein mit mehreren Freiheitsgraden." *Verhandlungen der Deutsche Physikalische Gesellschaft*, 17:407–18, 438–51.

———. 1924. "Zur Quantenstatistik des Bohrschen Atommodells." *Annalen der Physik*, ser. 4, 75:673–84.

———. 1932. *Der Kausalbegriff in der Physik*. Leipzig: Barth. English translation in *The Philosophy of Physics*. New York: Norton, 1936.

———. 1950. *A Scientific Autobiography and other Papers*. New York: Philosophical Library.

———. 1960. *A Survey of Physical Theory*. New York: Dover. Reprint of 1923 ed.

POINCARÉ, Henri. 1889. "Sur le probleme des trois corps et les equations de dynamique." *Acta Mathematica*, 13:1–270 (published 1890). "On the three-body problem and the equations of dynamics," excerpt in S. G. Brush (1966).

———. 1900. "Relations entre la physique expérimentale et la physique mathématique." *Rapports présentés au Congrès International de Physique réuni à Paris en 1900*, vol. 1, pp. 1–29. Paris.

———. 1912. "Sur la théorie des quanta." *Journal de physique théorique et appliquée*, ser. 5, 2:5–34.

POPOV, V. N. 1973. "Quantum vortices and phase transitions in Bose systems." *Soviet Physics JETP*, 37:341–45. Translated from *Zhurnal eksperimentalnoi i teoreticheski fiziki*, 64:672–80.

POPPER, Karl. 1962. *Conjectures and Refutations*. New York: Basic Books.

PORTER, Theodore M. 1981. "A statistical survey of gases: Maxwell's social physics." *Historical Studies in the Physical Sciences*, 12:77–116.

POST, H. R. 1971. "Correspondence, invariance and heuristics: In praise of conservative induction." *Studies in History and Philosophy of Science*, 2:213–55.

POWER, Henry. 1663. *Experimental Philosophy*. London: Martin & Allestry. Reprinted with an introduction by Marie Boas Hall. New York: Johnson Reprint Corp., 1966.

QUAY, Paul M. 1978. "A philosophical explanation of the explanatory functions of ergodic theory." *Philosophy of Science*, 45:47–59.

RAMAN, V. V., & FORMAN, PAUL. 1969. "Why was it Schrödinger who developed de Broglie's ideas?" *Historical Studies in the Physical Sciences*, 1:291–314.

RAYFIELD, G. W., & REIF, F. 1963. "Evidence for the Creation and Motion of Quantized Vortex Rings in Superfluid Helium." *Physical Review Letters*, 11:305–8. Reprinted in AAPT, *Liquid Helium*.

RAYLEIGH, Lord [Third Baron]. 1900a. "Remarks upon the Law of Complete Radiation." *Philosophical Magazine*, ser. 5, 49:539–40.

———. 1900b. "On the Viscosity of Argon as affected by Temperature." *Proceedings of the Royal Society of London*, 66:68–74.

REICHENBACH, Hans. 1956. *The Direction of Time*. Berkeley: University of California Press.

REID, Robert. 1974. *Marie Curie*. New York: Saturday Review Press.

REYNOLDS, C. A.; SERIN, B.; WRIGHT, W. H.; & NESBITT, L. B. 1950. "Superconductivity of isotopes of Mercury." *Physical Review*, ser. 2, 78:487.

RICE, O. K. 1947. "On the behavior of pure substances near the critical point." *Journal of Chemical Physics*, 15:314–32.

———. 1949. "Critical phenomena in binary liquid systems." *Chemical Reviews*, 44:69–92.

RICH, Alexander, & DAVIDSON, Norman, eds. 1968. *Structural Chemistry and Molecular Biology, A Volume Dedicated to Linus Pauling by his Students, Colleagues, and Friends*. San Francisco: Freeman.

RIECKE, Eduard. 1898. "Zur Theorie des Galvanismus und der Wärme." *Annalen der Physik*, ser. 3, 66:353–89, 545–81, 1199–1200.

RITTER, A. 1878. "Untersuchungen über die Höhe der Atmosphäre und die Constitution gasförmiger Welkörper." *Annalen der Physik*, ser. 3, 5:405–25, 543–58.

RØMER, Ole. 1676. "Démonstration touchant le mouvement de la lumière trouvé par M. Roemer." *Journal des Scavans* (Dec. 7, 1676), pp. 233–36. English version (not an exact translation), "A Demonstration concerning the Motion of Light," in *Philosophical Transactions of the Royal Society of London* (June 1677), no. 136, pp. 893–94. Both reprinted in I. B. Cohen, "Roemer and the First Determination of the Velocity of Light (1676)," *Isis*, 31:327–79 (1940).

ROSS, Marvin, & ALDER, Berni J. 1967. "Shock compression of argon. II. Nonadditive repulsive potential." *Journal of Chemical Physics*, 46:4203–10.

ROSSELAND, S. 1925. "On the distribution of hydrogen in a star." *Monthly Notices of the Royal Astronomical Society*, 85:541–46.

ROTH, Laura M. 1976. Review of "Women in Science, Illustrated Interviews," produced by Dinah L. Moche. *American Journal of Physics*, 44:1020.

RUELLE, D. 1969. *Statistical Mechanics—Rigorous Results*. New York: Benjamin.

RUSHBROOKE, G. S. 1938. "A note on Guggenheim's theory of strictly regular binary liquid mixtures." *Proceedings of the Royal Society of London*, A166:296–315.

———. 1963. "On the thermodynamics of the critical region for the Ising problem." *Journal of Chemical Physics*, 39:842–43.

RUSSELL, C. A. 1971. *The History of Valency*. Leicester: Leicester University Press.

RUSSELL, H. N. 1922. "The theory of ionization and the sun-spot spectrum." *Astrophysical Journal*, 55:119–44.

———. 1929. "On the composition of the Sun's atmosphere." *Astrophysical Journal*, 70:11–82.

RUTHERFORD, Ernest, & SODDY, Frederick. 1902. "The Radioactivity of Thorium Compounds. II. The Cause and Nature of Radioactivity." *Journal of the Chemical Society, Transactions*, 81:837–60.

SACKUR, Otto. 1911. "Die Anwendung der kinetischen Theorie der Gase auf chemische Probleme." *Annalen der Physik*, ser. 4, 36:958–80.

———. 1912. "Die Bedeutung des elementaren Wirkungsquantums für die Gastheorie und die Berechnung der chemischen Konstanten." In *Festschrift W. Nernst*, pp. 405–23. Halle: Knapp.

SAHA, Megh Nad. 1920a. "Ionization in the Solar Chromosphere." *Philosophical Magazine*, ser. 6, 40:472–88.

———. 1920b. "Elements in the sun." *Philosophical Magazine*, ser. 6, 40:809–24.

———. 1921. "On a physical theory of stellar specta." *Proceedings of the Royal Society of London*, A99:135–53.

SALMON, Wesley C. 1971. "Determinism and Indeterminism in Modern Science." In *Reason and Responsibility*, J. Feinberg, ed., pp. 316–32. Encino, Ca.: Dickenson Pub. Co. 2d ed.

SALPETER, E. E. 1952. "Nuclear reactions in stars without hydrogen." *Astrophysical Journal*, 115:326–28.

———. 1961. "Energy and Pressure of a Zero-Temperature Plasma." *Astrophysical Journal*, 134:669–82.

SCHAFROTH, M. R. 1954. "Superconductivity of a charged Bose gas." *Physical Review*, ser. 2, 96:1149.

———. 1955. "Superconductivity of a charged ideal Bose gas." *Physical Review*, ser. 2, 100:463–75.

———. 1960. "Theoretical aspects of superconductivity." *Solid State Physics*, 10:293–498.

SCHLIER, Christoph. 1969. "Intermolecular Forces." *Annual Review of Physical Chemistry*, 20:191–218.

SCHMIDT, G. G. 1823. "Theoretische und experimentale Bemerkung über die Perkins'sche Dampfmaschine." *Annalen der Physik*, 75: 343–54.

SCHRIEFFER, J. Robert. 1973. "Macroscopic quantum phenomena from pairing in superconductors." *Physics Today*, 26 (7):23–28.

SCHRÖDINGER, Erwin. 1919. "Der Energieinhalt der Festkörper im Lichte der neueren Forschung." *Physikalische Zeitschrift*, 20:420–28, 450–55, 474–80, 497–503, 523–26.

———. 1922. "Was ist ein Naturgesetz." *Naturwissenschaften*, 17:9–11 (published 1929).

———. 1926. "Quantisierung als Eigenwertproblem." *Annalen der Physik*, ser. 4, 79:361–76, 489–527; 80:437–90; 81:109–39. "Quantization as an eigenvalue problem," in *Collected Papers on Wave Mechanics*, translated by J. F. Shearer and W. M. Deans. London: Glasgow, 1928. Partial translation in Ludwig (1968).

———. 1931. "Indeterminism in Physics." Translated from a paper read before the Congress of the Society for Philosophical Instruction, Berlin, 1931. In *Science, Theory and Man*, pp. 52–80. New York: Dover, 1957.

———. 1935. "Die gegenwärtige Situation in der Quantenmechanik." *Naturwissenschaften*, 23:807–12, 823–28, 844–49.

———. 1978. *Collected Papers on Wave Mechanics*, 2d English ed. New York: Chelsea.

SCHUBERT, G. 1946. "Zur Bose-Statistik." *Zeitschrfit für Naturforschung*, 1:113–120. Erratum, *ibid.* 2a, 250–51 (1947).

SCHWEIDLER, Egon von. 1905. "Über Schwankungen der radioaktiven Umwandlung." *Premier Congrès International pour l'Etude de la*

Radiologie et de l'Ionisation tenu à Liège du 12 au *14* septembre *1905, Comptes Rendus, Communications Présentées, Section de Physique, Langue Allemande,* pp. 1–3.

SCOTT, E. L. 1970. "Andrews, Thomas." *Dictionary of Scientific Biography,* 1:160–61.

SCOTT, Wilson. 1970. *The Conflict between Atomism and Conservation Theory, 1644 to 1860.* New York: Elsevier.

SEARS, Francis W., & ZEMANSKY, Mark W. 1970. *University Physics.* Pt. 1. *Mechanics, Heat, and Sound.* 4th ed. Reading, Mass.: Addison-Wesley.

SHALNIKOV, A. I. 1945. "The structure of the supraconductors in the intermediate state. I." *Journal of Physics of the USSR,* 9:202–10.

SHAPERE, Dudley. 1974. "Scientific Theories and their Domains." In *The Structure of Scientific Theories,* F. Suppe, ed., pp. 518–570. Urbana: University of Illinois Press.

SHARLIN, Harold I., with SHARLIN, Tiby. 1979. *Lord Kelvin: The Dynamic Victorian.* University Park, Pa.: Pennsylvania State University Press, 1979.

SHIMONY, Abner. 1977. "Comments on the papers of Brush and Tisza." In *PSA 1976, Proceedings of the 1976 Biennial Meeting of the Philosophy of Science Association,* F. Suppe & P. D. Asquith, eds., pp. 609–16. East Lansing, Mich.: Philosophy of Science Association.

SHOENBERG, David. 1978. "Forty odd years in the cold. Reminiscences of work in low temperature physics." *Physics Bulletin,* 29 (1):16–19.

SHOSTAK, Arnold. 1955. "Interaction energy among three helium atoms." *Journal of Chemical Physics,* 23:1808–13.

SITTERLY, Bancroft W. 1970. "Changing Interpretations of the Hertzsprung-Russell Diagram, 1910–1940: A Historical Note." *Vistas in Astronomy,* 12:357–66.

SIMON, F. 1927. "Zum Prinzip von der Unerreichbarkeit des absoluten Nullpunktes." *Zeitschrift für Physik,* 41:806–9.

———. 1934. "Behaviour of Condensed Helium near Absolute Zero." *Nature,* 133:529.

SINGH, Virendra; SUDARSHAN, E.C.G.; BHATTACHARJEE, Girijapati; & GHOSH, Parimalkanti. 1974. "The Eighty Years of Satyen Bose." *Science Today* (Jan.), 28–45.

SKLAR, L. 1973. "Statistical Explanation and Ergodic Theory." *Philosophy of Science,* 40:194–212.

SLATER, John C. 1928. "The normal state of helium." *Physical Review,* ser. 2, 32:349–60.

———. 1929. "The theory of complex spectra." *Physical Review,* ser. 2, 34:1293–1322.

———. 1967. "The current state of solid-state and molecular theory." *International Journal of Quantum Chemistry*, 1:37–102.

———. 1975. *Solid-State and Molecular Theory: A Scientific Biography.* New York: Wiley-Interscience.

SMITH, Crosbie. 1976. "Natural philosophy and thermodynamics: William Thomson and 'the Dynamical Theory of Heat.'" *British Journal for the History of Science*, 9:293–319.

———. 1977. "William Thomson and the creation of thermodynamics, 1840–1855." *Archive for History of Exact Sciences*, 16:231–88.

———. 1978. "A new chart for British natural philosophy: the development of energy physics in the nineteenth century." *History of Science*, 16:231–79.

SOMMERFELD, A. 1911. "Das Plancksche Wirkungsquantum und seine allgemeine Bedeutung für die Molekularphysik." *Physikalische Zeitschrift*, 12:1057–69.

———. 1927. "Zur Elektronentheorie der Metalle." *Naturwissenschaften*, 15:825–32; 16:374–81.

———. 1928. "Zur Elektronentheorie der Metalle auf Grund der Fermischen Statistik." *Zeitschrift für Physik*, 47:1–32, 43–60.

STANLEY, H. Eugene. 1971. *Introduction to phase transitions and critical phenomena.* New York: Oxford University Press.

———, ed. 1973. *Cooperative phenomena near phase transitions: A Bibliography with Selected Readings.* Cambridge, Mass.: MIT Press.

STEFAN, Josef. 1879. "Über die Beziehung zwischen der Wärmestrahlung und der Temperatur." *Sitzungsberichte, K. Akademie der Wissenschaften, Wien, Mathematisch-Naturwissenschaftliche Klasse*, 79: 391–428.

STONER, Edmund C. 1930. "The equilibrium of dense stars." *Philosophical Magazine*, ser. 7, 9:944–63.

STRÖMGREN, B. 1932. "The capacity of stellar matter and the hydrogen content of the stars." *Zeitschrift für Astorophysik*, 4:118–52.

———. 1933. "On the interpretation of the Hertzsprung-Russell Diagram." *Zeitschrift für Astrophysik*, 7:222–48.

———. 1938. "On the Helium and Hydrogen content of the interior of the stars." *Astrophysical Journal*, 87:520–34.

STRUVE, Otto, & ZEBERGS, Velta. 1962. *Astronomy of the 20th Century.* New York: Macmillan.

STUEWER, Roger H. 1975. *The Compton Effect: Turning Point in Physics.* New York: Science History Pubs.

SUCHER, J. 1977. "Long range forces in Quantum Theory." In *Cargese Lectures in Physics*, vol. 7, pp. 45–110. New York: Gordon & Breach.

SUPPES, Patrick. 1961. "A Comparison of the Meaning and Uses of Models in Mathematics and the Empirical Sciences." In *The Concept and the Role of the Model in Mathematics and Natural and Social Sciences*, pp. 163–76. New York: Gordon & Breach.

SUTHERLAND, William. 1893. "The Laws of Molecular force." "The viscosity of gases and molecular force." *Philosophical Magazine*, ser. 5, 35:211–95; 36:150–51, 507–31.

SYDORIAK, S. G.; GRILLY, E. R.; & HAMMEL, E. F. 1949. "Condensation of pure He^3 and its vapor pressures between 1.2° and its critical point." *Physical Review*, ser. 2, 75:303–5.

SYKES, M. F.; MARTIN, J. L.; & HUNTER, D. L. 1967. "Specific heat of a three-dimensional Ising ferromagnet above the Curie temperature." *Proceedings of the Physical Society of London*, 91:671–77.

TAMMANN, G. 1897. "Ueber die Grenzen des festen Zustandes." *Annalen der Physik*, ser. 3, 62:280–99.

———. 1903. *Kristallisieren und Schmelzen*. Leipzig: Barth.

TANAKA, T.; KATSUMORI, H.; & TOSHIMA, S. 1951. "On the theory of cooperative phenomena." *Progress of Theoretical Physics*, 6: 17–26.

TER HAAR. *See* HAAR

TETRODE, H. 1912. "Die chemische Konstante der Gase und das elementare Wirkungsquantum." *Annalen der Physik*, ser. 4, 38:434–42; 39:255–56.

———. 1913. "Bemerkungen über den Energieinhalt einatomiger Gase und über die Quantentheorie für Flüssigkeiten." *Physikalische Zeitschrift*, 14:212–15.

THOMSON, James. 1849. "Theoretical considerations on the effect of pressure in lowering the freezing point of water." *Transactions of the Royal Society of Edinburgh*, 16:575–80.

———. 1871. "Considerations on the abrupt change at boiling or condensation in reference to the continuity of the fluid state of matter." *Proceedings of the Royal Society of London*, 20:1–8.

THOMSON, J. J. 1897. "Cathode Rays." *Philosophical Magazine*, ser. 5, 44:293–316.

———. 1907. *The Corpuscular Theory of Matter*. London: Constable.

THOMSON, William. See Kelvin

THOULESS, D. J. 1970. "The BCS model Hamiltonian as an exactly soluble problem in statistical mechanics." In *Methods and Problems of Theoretical Physics, In Honour of R. E. Peierls*, J. E. Bowcock, ed., pp. 29–35. New York: American Elsevier.

TISZA, Laszlo. 1938a. "Transport phenomena in Helium II." *Nature*, 141:913.

———. 1938b. "Sur la supraconductibilité thermique de l'hélium II liquide et la statistique de Bose-Einstein." *Comptes Rendus hebdomadaires des Séances de l'Académie des Sciences, Paris,* 207: 1035–37.

———. 1940. "Sur la Théorie des Liquides Quantiques. Application à l'Hélium Liquide." *Journal de Physique et Radium,* ser. 8, 1:164–72, 350–58.

———. 1947. "The Theory of Liquid Helium." *Physical Review,* ser. 2, 72:838–54.

———. 1948. "Helium, The Unruly Liquid." *Physics Today,* 1 (4):4–6, 26.

———. 1949. "On the Theory of Superfluidity." *Physical Review,* ser. 2, 75:885–86.

———. 1951. "On the General Theory of Phase Transitions." In *Phase Transformations in Solids* (Symposium at Cornell, August 1948), R. Smoluchowski et al., eds., pp. 1–37. New York: Wiley.

———. 1958. "Generalized Phase Rule and the Symmetry of the Superfluid State." *Physical Review,* ser. 2, 110:587–89.

———. 1966. Discussion Remark. In *Critical Phenomena,* M. S. Green & J. V. Sengers, eds., pp. 101–2. Washington, D.C.: National Bureau of Standards, Misc. Pub. 273.

TISZA, Laszlo, & QUAY, P. M. 1963. "The Statistical Thermodynamics of Equilibrium." *Annals of Physics,* 25:48–90.

TOULMIN, Stephen, ed. 1970. *Physical Reality.* New York: Harper & Row.

TRIGG, George L. 1975. *Landmark Experiments in Twentieth Century Physics.* New York: Crane, Russak & Co.

TRUESDELL, C. 1956. "I. The first three sections of Euler's Treatise on Fluid Mechanics, 1766; II. The Theory of Aerial Sound, 1687–1788; III. Rational Fluid Mechanics, 1765–1788." Editor's introduction to *Eulerii Opera Omnia,* series II, 13:ix–cxviii. Zürich: Orell Füssli.

———. 1968. *Essays in the History of Mechanics.* New York: Springer-Verlag.

———. 1980. *The Tragicomical History of Thermodynamics, 1822–1854.* New York: Springer-Verlag.

UHLENBECK, George E. 1927. *Over statistische methoden in de theorie der quanta.* 'sGravenhage: M. Nijhoff.

———. 1966. "The classical theories of the critical phenomena." In *Critical Phenomena,* M. S. Green & J. V. Sengers, eds., pp. 3–6. Washington, D.C.: National Bureau of Standards, Misc. Pub. 273.

———. 1978. "Some historical and critical remarks about the theory of phase transitions." In *The Ta-You Wu Festschrift, Science of Matter*, S. Fujita, ed., pp. 99–107. New York: Gordon & Breach.

———. 1980. "Some reminiscences about Einstein's visits to Leiden." In *Some Strangeness in the Proportion, A Centennial Symposium to Celebrate the Achievements of Albert Einstein*, H. Woolf, ed., pp. 524–25. Reading, Mass.: Addison-Wesley.

UHLENBECK, G. E., & GOUDSMIT, S. 1925. "Ersetzung der Hypothese vom unmechanischen Zwang durch eine Forderung bezüglich des inneren Verhaltens jedes einzelnen Electrons." *Naturwissenschaften*, 13:953–54. "Spinning Electrons and the Structure of Spectra." *Nature*, 117:264 (1925).

UNSÖLD, Albrecht. 1928. "Über die Struktur der Fraunhoferschen Linien und die quantitative Spektralanalyse der Sonnenatmosphäre." *Zeitschrift für Physik*, 45:765–81.

UREY, H. C. 1924. "The distribution of electrons in the various orbits of the hydrogen atoms." *Astrophysical Journal*, 59:1–10.

VAKS, V. G., & LARKIN, A. I. 1966. "On phase transitions of second order." *Soviet Physics JETP*, 22:678–87.

VAN DER HANDEL, J. 1973. "Kamerlingh Onnes, Heike." *Dictionary of Scientific Biography*, 7:220–22.

VAN DER WAALS, Johannes Diderik. 1873. *Over de continuiteit van den gas- en vloeisoftoestand*. Leiden: Sijthoff. "The Continuity of the Liquid and Gaseous States of Matter," translated from the revised 1881 German version by R. Threlfall & J. F. Adair, in *Physical Memoirs*, vol. 1, pt. 3. London: Taylor & Francis, for the Physical Society, 1890.

———. 1893. "Thermodynamische theorie der capillariteit in de onderstelling van continue dichtheidsverandering." *Verhandelingen der Koninklijke Akademie van Wetenschappen, Amsterdam*, 1; no. 8. English translation by J. Rowlinson in *Journal of Statistical Physics*, 20:197–244 (1979).

VAN DER WAERDEN, B. L. 1967. *Sources of Quantum Mechanics*. Amsterdam: North-Holland Pub. Co. Reprinted, New York: Dover Pubs.

———. 1968. "Mendel's Experiments." *Centaurus*, 12:275–88.

VAN LEEUWEN, Hendrika Johanna. 1919. *Vraagstukken uit de electronen-theorie van het magnetisme*. Proefschrift, Leiden.

———. 1921. "Problemes de la Theorie Electronique du Magnetisme. *Journal de Physique*, ser. 5, 2:361–77.

VAN VLECK, J. H. 1932. *The Theory of Electric and Magnetic Susceptibilities*. Oxford: University Press.

———. 1970. "Spin, The Great Indicator of Valence Behavior." *Pure & Applied Chemistry*, 24:235–55.

VERWEY, E.J.W. 1967. Discussion Remark. In *Study Week on Molecular Forces. April 18–23, 1966* (Pontificiae Academie Scientiarum Scripta Varia 31), p. 171. New York: Wiley.

VINEN, W. F. 1958. "Detection of single quanta of circulation in rotating helium II." *Nature*, 181:1524–25.

———. 1961. "The Detection of Single Quanta of Circulation in Liquid Helium II." *Proceedings of the Royal Society of London*, A260: 218–36.

VORONEL, A. V.; CHASHKIN, Y. R.; POPOV, V. A.; & SIMKIN, V. G. 1963. "Measurement of the specific heat c_v of oxygen near the critical point." *Soviet Physics JETP*, 18:568–69. Translated from *Zhurnal eksperimentalnoi i teoreticheski fiziki*, 45:828–30.

VORONEL, A. V.; SNIGIREV, V. G.; & CHASHKIN, Yu. R. 1965. "Behavior of the specific heat of pure substances near the critical point." *Soviet Physics JETP*, 21:653–55. Translated from *Zhurnal eksperimentalnoi i teoreticheski fiziki*, 48:981–84.

WALI, Kameshwar C. 1982. "Chandrasekhar vs. Eddington—an unanticipated confrontation." *Physics Today*, 35 (10):33–40.

WANG, Shou Chin. 1927. "Die gegenseitige Einwirkung zwier Wasserstoffatome." *Physikalische Zeitschrift*, 28:663–67.

WATERSTON, J. J. 1846. "On the physics of media that are composed of free and perfectly elastic molecules in a state of motion." Abstract only, in *Proceedings of the Royal Society of London*, 5:604. Published in full in *Philosophical Transactions of the Royal Society of London*, 183*A*:5–79 (1893).

———. 1928. *The Collected Scientific Papers of John James Waterston*, ed., with a biography, by J. S. Haldane. Edinburgh: Oliver and Boyd.

WEBSTER, C. 1965. "The Discovery of Boyle's Law, and the Concept of the Elasticity of Air in the Seventeenth Century." *Archive for History of Exact Sciences*, 2:441–502.

WEINBERGER, M. A., & SCHNEIDER, W. G. 1952. "On the liquid-vapor coexistence curve of xenon in the region of the critical temperature." *Canadian Journal of Chemistry*, 30:422–37.

WEISS, Pierre, & KAMERLINGH ONNES, H. 1910. "Recherches sur l'aimantation aux très basses températures." *Journal de Physique*, ser. 4, 9:555–84.

WESSELS, Linda. 1979. "Schrödinger's route to wave mechanics." *Studies in History and Philosophy of Science*, 10:311–40.

WESTFALL, Richard S. 1973. "Newton and the fudge factor." *Science*, 179:751–58.

WHEELER, Lynde Phelps. 1951. *Josiah Willard Gibbs: The History of A Great Mind*. New Haven: Yale University Press.

WHELAND, G. W. 1934. "The Quantum Mechanics of Unsaturated and Aromatic Molecules: A Comparison of Two Methods of Treatment." *Journal of Chemical Physics*, 2:474–81.

WHITMORE, S. C., & ZIMMERMANN, W., Jr. 1965. "Observation of stable superfluid circulation in liquid-helium II at the level of one, two, and three quantum units." *Physical Review Letters*, 15:389–92. Reprinted in AAPT, *Liquid Helium.*

WICHELHAUS, H. 1869. "Die Hypothesen über die Constitution des Benzols." *Berichte der Deutsche Chemische Gesellschaft*, 2:197–99.

WIDOM, B. 1965. "Surface tension and molecular correlations near the critical point." "Equation of state in the neighborhood of the critical point." *Journal of Chemical Physics*, 43:3892–97, 3898–3905.

———. 1979. "In Memoriam: Oscar Knefler Rice, 1903–1978." *Journal of Statistical Physics*, 21:341–44.

WIEGEL, F. W. 1975. "Path integral methods in statistical mechanics." *Physics Reports (Physics Letters C)*, 16:57–114.

———. 1978. "The interacting Bose fluid: path integral representations and renormalization group approach." In *Path Integrals and their applications in Quantum, Statistical, and Solid State Physics*, G. J. Papadopoulos and J. T. Devreese, eds., pp. 419–54. New York: Plenum Press.

WIEN, W. 1893. "Eine neue Beziehung der Strahlung schwarzer Körper zum zweiten Hauptsatz der Wärmetheorie." *Sitzungsberichte der K. Preussischen Akademie der Wissenschaften, Physikalisch-mathematische Klass, Berlin*, pp. 55–62.

———. 1896. "Ueber die Energievertheilung im Emissionsspectrum eines schwarzes Körpers." *Annalen der Physik*, ser. 3, 58:662–69. English translation in *Philosophical Magazine*, ser. 5, 43:214–20 (1897).

WIGHTMAN, Arthur S. 1979. "Introduction." In *Convexity in the Theory of Lattice Gases*, by R. B. Israel, pp. ix–lxxxv. Princeton: Princeton University Press.

WILKS, J. 1957. "The Theory of Liquid ^{4}He." *Reports on Progress in Physics*, 20:38–85.

———. 1967. *The Properties of Liquid and Solid Helium.* London: Oxford University Press.

WILSON, Kenneth G. 1971. "Renormalization group and critical phenomena." *Physical Review*, ser. 3, B4:3174–83, 3184–3205.

———. 1979. "Problems in physics with many scales of length." *Scientific American*, 241 (2):158–79.

WILSON, Kenneth G., & FISHER, Michael E. 1972. "Critical exponents in 3.99 dimensions." *Physical Review Letters*, 28:240–32.

WILSON, Kenneth G., & KOGUT, J. 1974. "The renormalization group and the ϵ expansion." *Physics Reports, Physics Letters*, 12C (2): 75–200.

WONG, K. W. 1970. "Possible existence of pair-condensate in superfluid helium." *Physics Letters*, 32A:195–96.

———. 1971. "A New Theory of Liquid Helium II." In *Proceedings of the 12th International Conference on Low Temperature Physics, Kyoto, 1970*, E. Kanda, ed., pp. 195–96. Tokyo: Academic Press of Japan.

WONG, K. W., & HUANG, Y. H. 1969. "A Numerical Calculation of the Excitation of Helium II based on the Hard-Sphere Model." *Physics Letters*, 30A:292–94.

YANG, C. N. 1962. "Concept of off-diagonal long range order and the quantum phases of liquid He and of superconductors." *Reviews of Modern Physics*, 34:694–704.

YANG, C. N., & LEE, T. D. 1952. "Statistical theory of equations of state and phase transitions. I. Theory of condensation." *Physical Review*, ser. 2, 87:404–9.

ZERMELO, Ernst. 1896a. "Ueber einen Satz der Dynamik und die mechanische Wärmetheorie." *Annalen der Physik*, ser. 3, 57:485–94. "On a Theorem of Dynamics and the Mechanical Theory of Heat," in Brush (1966).

———. 1896b. "Über mechanische Erklärungen irreversibler Vorgänge." *Annalen der Physik*, ser. 3, 59:793–801. "On the Mechanical Explanation of Irreversible Processes," in Brush (1966).

ZIFF, R. M.; UHLENBECK, G. E.; & KAC, M. 1977. "The ideal Bose-Einstein gas, revisited." *Physics Reports*, 32C:169–248.

ZIMM, Bruno H. 1950. "Opalescence of a two-component liquid system near the critical mixing point." *Journal of Physical & Colloid Chemistry*, 54:1306–17.

———. 1951. "Contribution to the theory of critical phenomena." *Journal of Chemical Physics*, 19:1019–23.

ZINOV'EVA, K. N. 1956. "The Coefficient of Volume Absorption of Second Sound and the Viscosity of the Normal Component of Helium II down to 0.83°K." *Soviet Physics JETP*, 4:36–40. Translated from *Zhurnal eksperimentalnoi i teoreticheskoi fiziki*, 31:31–36.

ZWANZIG, Robert. 1981. "Recent developments in non equilibrium statistical mechanics." *Kinam: Revista de Fisica*, ser. 3, A:5–13.

INDEX

Cherwell, Lord [Lindemann, Frederick
Alexander] (1886–1957, German-British
physicist, adviser to Winston Churchill,
villain of C. P. Snow's Science and
Government), 151, 176
Chester, G. V., 194, 196, 200
Chomsky, Noam (b. 1928, American
theoretical linguist), 8
Christiansen, C., 119
Clairaut, Alexis Claude (1713–1765, French
mathematician), 8
Clapeyron, Emilo (1799–1864, French
engineer who applied Carnot's theory
of steam engines), 46n
Clark, Peter, 72, 262
Claus, 229
Clausius, Rudolf (1822–1888, German
physicist), 46–55, 64, 66, 84–86, 88,
104, 234, 250, 266–67
Clausius-Clapeyron equation, 234–35
Clifford, Tony, 286
clockwork universe, 5, 6, 79
Clusius, Klaus (German physicist), 175–76
cluster integrals, 247–48, 255–56
Cohen, E.G.D., 285
Cohen, Michael, 195
Colding, Ludvig August (1815–1888,
Danish engineer and physicist), 39, 264
collective excitations, 182–85, 187
collisions, 7, 8–9, 24, 36, 52, 54, 55, 63,
68, 87, 115, 125, 127, 251
combinatorial analysis, 110, 158, 160,
193–94, 196, 243, 255–56
combining volumes, law of, 33–34
complementarity principle, 143
compressibility, 26, 41, 211
Compton, Arthur Holly (1892–1962,
American physicist), 103, 127
Compton, Karl Taylor (1887–1954,
American physicist), 155
Compton effect, 126–27
computer (electronic), 250–51, 253n
Conant, J. B., (1893–1978, American
chemist, president of Harvard, promoted
case studies in science education), 12, 15
condensation, 69–71, 166, 247, 250, 251
Coniglio, A., 200
conservation laws, 7–9, 20, 24, 27, 127,
164, 264
Cook, W. R., 210

Cooper, Leon (b. 1930, American
physicist, Nobel Laureate, co-author of
BCS theory of superconductivity),
170, 172
Copeland, D. A., 217
Copenhagen interpretation, 118, 144, 230
Copernicus, Nicholas, (1473–1543,
Polish astronomer), 4, 68, 79
Cornu, Alfred (1841–1902, French
physicist), 116
correlation functions, 281–82
correspondence principle, 123, 193
corresponding states, law of, 173
Coulomb's law, 221
Coulson, Charles Alfred (1910–1974,
British theoretical chemist), 229
crassitude, 26
Crawford, Adair (1748–1795, British
physician and chemist), 36
critical point (gas-liquid), 69–71, 233,
246–50, 256–57; exponents, 201–3,
233–34, 249–50, 253–55, 257–58,
273, 276
critical point (liquid-solid), 235–56
Croune [Croone], William (1633–1684,
English physician, first registrar of the
Royal Society), 18
Croxton, C. A., 284
Culverwell, Edward Parnell (1855–1931,
Irish scientist), 92
Cummings, F. W., 201
Curie, Marie Sklodowska (1867–1934,
Polish-French physicist), 237, 239n
Curie, Pierre (1859–1906, French
physicist, husband of Marie Curie),
119, 237–39, 241, 252
Curie point/temperature, 175, 237–39, 272
Curtiss, C. F., 214
Cuthiell, D., 203

Dahler, J. S., 281
Dalton, John (1766–1844, English
chemist), 31–35, 204
Dana, Leo, American physicist, 174
Darwin, Charles Galton (1887–1962,
British physicist), 153
Darwin, Charles Robert (1809–1882,
English biologist, proposed theory of
evolution by natural selection), 79

Hobbes, Thomas (1588–1679, English philosopher), 16
Hohenberg, Pierre Claude (b. 1934, French-American physicist), 200, 202n
Holm, E., 78
Hooke, Robert (1635–1703, English scientist), 14
Hoover, William, 286
Hooykaas, R., 27
Hopkinson, John (1849–1898, English engineer), 236–37
Hoyer, Ulrich, 114, 126
Hoyle, Fred (b. 1915, English astronomer), 165
Huang, H., 200
Huang, Y. H., 200
Hückel, Erich, (d. 1980, German physicist and chemist), 228
Hudson, R. P., 191
Huggins, William (1824–1910, English astronomer who used Doppler shift of spectra of stars to estimate their motion), 117
Hund, Friedrich (b. 1896), 226, 228
Hunter, Douglas Lyle (b. 1940, Canadian physicist), 253n, 254, 257
Hunter, John, 82
Hutchison, K., 46n
Hutton, James (1726–1797, Scottish geologist), 81
Huygens, Christian (1629–1695, Dutch physicist), 8, 9, 19, 21, 24, 36
hydrodynamics, 27; quantized, 182, 184
hydrogen, 116–18, 125–26, 135–37, 147–48, 155, 157, 210, 214, 224–26, 278
Hyland, G. J., 201
hypothetico-deductive method, 183, 267

ideal gases, 52, 68, 154, 166
ideal gas law (equation of state), 68, 206, 259–60, 263
idealism, 141
identity, 131
Ihde, A., 31
Imry, Y., 200
indeterminacy principle, 100, 103, 140–41, 143–44, 157, 161
indeterminism, 79–80, 86, 94, 98–104, 270
indistinguishability, 131, 180, 229
induction (electrical), 209–10, 215

Ingold, Christopher Kelk (1893–1970, British organic chemist), 229
instrumentalism, 143
intelligence, 142
ionization, 150–55; by pressure, 162–63, 166
ionization potential, 124, 152
iron, 235, 237–38
irreversibility, 4, 79–81, 83–85, 88–93, 102, 106, 261, 269
Isihara, A., 285
Ising, Ernst (b. 1900, German-American physicist), 240–42, 271–72
Ising model. See Lenz–Ising model
Israel, W., 284

Jackson, H. W., 200
Jacobi, Karl Gustav Jacog (1804–1851, German mathematician), 133
Jammer, Max, 108n, 132, 141
Jancovici, B., 284
Jauch, J. M., 101
Jeans, James Hopwood (1877–1946, English physicist and astronomer, author of popular books on science), 107, 120, 151, 166
Jeffreys, Harold (b. 1891, British geophysicist), 155
Jensen, J. Hans D. (1907–1973, German physicist, Nobel Laureate, proposed nuclear shell model), 162
Jevons, William Stanley (1835–1882, English economist and philosopher), 86–87
Johnson, M. C., 175
Jones, H., 177
Jones, J. E. See Lennard–Jones
Joule, James Prescott (1818–1889, English physicist), 38, 39, 49, 50, 53, 207n, 264
Joule–Thomson experiment, 207, 264

Kac, Mark (b. 1914, Polish-American mathematician), 177n, 258
Kadanoff, Leo P. (b. 1937, American physicist), 257, 284
Kamerlingh Onnes, Heine (1853–1926, Dutch physicist, Nobel Laureate, discovered superconductivity and investigated low-temperature properties or helium), 168, 173–74, 233, 237n

LIBRARY OF CONGRESS CATALOGING IN PUBLICATION DATA

Brush, Stephen G.
  Statistical physics and the atomic theory of matter.

  (Princeton series in physics)
  Bibliography: p.
  Includes index.
  1. Matter—Properties.   2. Quantum theory.
3. Statistical physics.   4. Atomic theory.
I. Title.   II. Series.
QC173.3.B78   1983   530.1   82—61357
ISBN 0-691-08325-8
ISBN 0-691-08320-7 (pbk.)